Mastering Audio

the art and the science

Bob Katz

Focal Press

OXFORD AMSTERDAM BOSTON LONDON NEW YORK PARIS
SAN DIEGO SAN FRANCISCO SINGAPORE SYDNEY TOKYO

Focal Press is an imprint of Elsevier Science.
Mastering Audio: The Art and the Science
Copyright © 2002 by Robert A. Katz. All rights reserved.

Recognizing the importance of preserving what has been written, Elsevier Science prints its books on acid-free paper whenever possible.

Library of Congress Cataloging-in-Publication Data

ISBN: 0-240-80545-3

British Library Cataloguing-in-Publication Data
A catalogue record for this book is available from the British Library.

The publisher offers special discounts on bulk orders of this book.
For information, please contact:

Manager of Special Sales
Elsevier Science
200 Wheeler Road
Burlington, MA 01803
Tel: 781-313-4700
Fax: 781-313-4882

For information on all Focal Press publications available, contact our World Wide Web home page at: http://www.focalpress.com

10 9 8 7 6 5 4 3 2 1

Printed in Canada

DEDICATION

This book is dedicated to **Mary Kent**, my best friend (and wife) for over 17 years, without whose love and support I would be absolutely nowhere today!

I also dedicate this book to those mastering engineers whose work I especially admire, **Bernie Grundman, Ted Jensen, Bob Ludwig, Glenn Meadows, Bob Olhsson**, and **Doug Sax**. Although I have not yet met all of you in person, but I want to say that your fine work has brought great pleasure into my life through the many LPs and CDs you have mastered.

And finally, a dedication to those who will come after... For they will be working with tools that we have barely dreamed of. If you're young, remember to protect your hearing, because that precious gift can never be returned once it is taken away.

— B. K.

CREDITS AND THANKS TO...

International thanks to Britain's (but I knew him in Hong Kong) **Dr. Eric James**, (*www.urmaudio.com*) whose zealous and empathic editing helped make this book a lot shorter and more cogent than it would otherwise have been!

Thanks to Orlando's **Mary Kent, Deborah Dunkle, David Hudzik, Mark Corbin** and **Dale Drumheller** at **Digital Domain**, whose valuable help kept book production in motion, ultimately contributing to its smooth look and feel. Thanks to Orlando's **Gail Kent** for coming up with the punny title! And to Orlando Jazz Producer **Charlie Bertini**, for finding and preserving a perfect print of the Carnegie Hall chart found inside the front cover of this book.

As for the other images, I thank **Mary Kent** for the rack and Mytek photos of Chapter 1, the lovely color studio photo seen at the end of the color section, plus thanks for making me look years younger! I took the Dorrough and scope photos of Chapter 2, car radio photo of Chapter 6, and the manufacturers supplied the balance of the equipment photos. All the flowcharts and diagrams I generated using Macdraft.

Thanks to New Jersey's **Jim Johnston**,[*] one of the smartest audio experts I know, whose fact-checks on critical chapters and shared knowledge kept me on the straight and narrow. He also clarified some of the more esoteric inner workings of equipment and the human ear. Please blame me, not Jim, for any inaccuracies that may have seeped through. Thanks to **Konrad Strauss**[†] of Illinois, who reviewed the work and made many helpful suggestions. Also to **Bob Ludwig** of Maine's Gateway Mastering, one of the busiest and nicest guys in audio, who read the manuscript and provided valued input.

I thank Orlando's graphic designers **Toni Gonzalez** and **Thuan Nguyen** of **Art Tested Graphics** (*arttested@cfl.rr.com*) for beautifully and colorfully realizing a cozy visual dream of a book. And muchas gracias to Miami's **José Pacheco** (*xilval1@bellsouth.net*) for the evocative, mysterious and deep cover design and layout.

Thanks to **Roger Nichols** for writing the foreword, for making some of the best-sounding mixes and masters on this planet, and for just being Roger! The great Florida weather has attracted Roger to Miami where he runs a mastering studio and can be contacted at *Roger24bit@aol.com*.

Thanks to New Zealand's **Richard Hulse** for refining the parallel compression technique in the analog domain, and bringing it to my attention, where I converted it to the digital domain, described in Chapter 11. Thanks to Britain's **Mike Collins**, writer, for stressing the problems with QC and Electronic Delivery through articles in **One to One** magazine, which inspired my own investigation of such problems in Chapter 1. Thanks to the state of Washington's mastering engineer **Jim Rusby** for telling us about CD Text (Chapter 20). Thanks, Dan Stout, of Chicago's Collossal Mastering, for some alerts on the current state of low-class CD duplicators. Thanks to Germany's **Ralph Kessler** of Pinguin Audio, *http://www.masterpinguin.de*, for helping to rigidly specify the attack and decay constants of the K-System meters, which Ralph has implemented in his system.

Thanks to San Leandro, California's **Bob Orban**, *http://www.orban.com*, and **Frank Foti** *http://www.omniaaudio.com*, radio processing gurus, for providing the text of their excellent article, in Appendix 1. Thanks to San Francisco's **Tardon Feathered** and **Marvin Humphrey** of Mr. Toads, *http://www.mrtoads.com*, in San Francisco, for producing the **What is Hot** CD, collaborating and organizing the **What is Hot** competition on the mastering webboard, and for locating Bob Orban to answer that essential question.

Thanks to **BJ Buchalter**, **Stu Buchalter**, and **Dan Metivier** of Metric Halo Labs, Castle Point, New York, *http://www.mhlabs.com*, for devising and supporting one of the most powerful and economical audio analysis tools in the universe, **SpectraFoo**. B.J., one of the unsung audio geniuses, taught me enough about FFT to wade through it, but I could never reach his amazing level of Einsteinian gestalt.

Thanks to Tennessee's **Seva**, formerly of **Waves**, for helping to develop their products to true usefulness and inviting serious user feedback. The Transfer Function images in the chapter on dynamics are adapted from screen shots of the Waves C1 Comp and Waves Gate.

Thanks to Nashville's Producer/Engineer/Designer **George Massenburg**, *http://www.gmlinc.com*, for his inspiration, consistent search for excellence, and endless knowledge of audio history, music and technology. George has produced some of the best-sounding albums in history. I'm proud that we share many of the same feelings about what makes good sound. Thanks to New York's **Al Grundy**, founder of the Institute of Audio Research;

* Researcher in audio, former research scientist at Bell Labs/AT&T, and expert on the science and art of perceptual coding, both in audio and video.

† Acting Director of Recording Arts, Visiting Associate Professor of Music, Indiana University School of Music.

Boulder, Colorado's **Ray Rayburn**; Hartford, Connecticut's **Steve Washburn;** and New York City's **Noel Smith**, whose friendship and guidance over the years led me through many difficult audio topics. Similarly, Britain's **Michael Gerzon** and California's **Deane Jensen** were short-lived geniuses who contributed so much to this audio world, and liberally shared their thoughts with me over the years.

Thank you, Britain's **Julian Dunn**, for your selfless help explaining the more esoteric aspects of jitter and other topics! I thank Denmark's **Thomas Lund** of TC Electronic, a good friend and one of the best technical support engineers on this planet, for helping to make the System 6000 one of the easiest to use and powerful audio processors. He also edited earlier versions of the manuscripts that form the cornerstones of a couple of these chapters. Thank you, **Glenn Zelniker** of Florida's Z-Systems, and Switzerland's **Daniel Weiss** and **Andor Bariska** of Weiss Audio, for making audio excellence and perfection your first priority, and letting me nudge in almost all the features a mastering engineer could desire!

Thanks to **Diane Wurzel**, and the unsung heros at Focal Press, who exemplify just as much care and attention to detail in publishing as we do in audio!

Finally, I must thank all my Internet comrades, too numerous to mention, who participate in the Mastering Webboard and liberally share ideas. You will see your names mentioned now and again in our footnotes. And once again, Nashville's own **Glenn Meadows**, who founded the Mastering and Sadie Webboards (*http://webbd.nls.net:8080/~mastering*) and unselfishly participates in the worldwide spirit of audio education and enlightenment.

—B. K.

contents

CONTENTS CONTINUED

foreword

By Roger Nichols

When a recording artist I produced heard a great song on the radio he would turn to me and say, "I was going to write that song!" After reading this book my reaction was, "I was going to write this book!" Well, I am glad Bob beat me to it because it looks like he did a much better job than I could have.

What places this book head and shoulders above the rest is the attention to useful detail. Instead of some hyperbole, the reader can actually put these methods to good use. The descriptions of how to perform a task are augmented with the reason that you should perform the task. Not just how downward compressors work, but when and why you would want to use them. Science is meaningless without art.

How do I tell if the digital signal is 16 bit or 24 bit? What does noise shaping do? Should I mix at 96 kHz? How do you make something 3 dB louder when it is already lighting up the over lights? Should I mix to analog or digital? How do I set up my speakers for mixing surround? Which weighs more, a pound of gold or a pound of feathers? These are some of the questions that Bob answers in a clear and concise style.

Bob enters each mastering session with his eyes wide open. Each project is unique, and each mastering session will require a unique approach to bring out the very best results. Bob's musical background helps him select the proper tools for the job. Knowing that a string quartet record does not require the same approach as the Back Street Boys record is half the battle.

Every day clients ask for louder and louder CDs when they come to a mastering session. It is very hard to find Hi-Fidelity CDs these days. Now that you can do your own recording to a digital workstation, buy your own multi-band compressors and burn your own CDs, who needs mastering? My answer is that if you record your own projects at home, you need mastering more than the producer who works with the top engineers in the top studios. The key is outside reference. No, I don't mean that your neighbor came over and said, "Hey, that sounds really great!" I mean reference to other projects, and reference to other engineers who have worked on great sounding CDs.

Bob does an excellent job of dispelling the myth that the louder you make your CD, the louder it will be on the radio. Read this part more than once. Once the reality sinks in, then maybe we will have more viable candidates for a *Best Engineered CD* Grammy, instead of having to choose a CD for the *Least Offensive Engineering* award.

The professional mastering engineer works on material from all corners of the music business. This is the last stop before the CD hits the radio and the record stores. The smartest thing any mixing engineer can do is leave the final loudness tools to the loudness professional.

Limiters and compressors should be treated just like firearms. There should be guides for the proper use and classes you must take before you can own one. That class is here in this book. After you read this "audio firearms manual" you will have a much better understanding of the mastering process. You will know when and how to use these tools yourself and when to leave it to the professional. Treat every compressor/limiter as a loaded weapon, and don't point it at anyone unless you intend to use it. It's the **LAW**!

I get e-mail quite often from independent artists who are recording their music at home and want to know what gear to buy to help them mix before they send it to me for mastering. I tell them that the first piece of equipment they should buy is Bob Katz's **Mastering Audio, The Art and the Science**.

Roger Nichols
Miami, August 2002

INTRODUCTION

What Is Mastering?

Mastering is the last creative step in the audio production process, the bridge between mixing and replication—your last chance to enhance sound or repair problems in an acoustically-designed room—an audio microscope. Mastering engineers lend an objective, experienced ear to your work; we are familiar with what can go wrong technically and esthetically. Sometimes all we may do is—nothing! The simple act of approval means the mix is ready for pressing. Other times we may help you work on that problem song you just couldn't get right in the mix, or add the final touch that makes a record sound finished and playable on a wide variety of systems.

The Approach of this Book

The mastering studio is the place where experience in the musical art is combined with the science of audio, but the dividing line between art and science is nebulous, and so my book constantly tries to integrates the art and the science.

Technology changes so fast in today's world; for example, no one predicted that a rapid proliferation of digital cameras would threaten the once-mighty Polaroid Corp. Five years after this book is published, one-third of its technical information will be outdated. Ten years from today, one-third or more of its technical information will be obsolete. But old-fashioned craftsmanship and attention to detail will always be in demand. I hope that even fifty or one hundred years from today, mastering engineers will still be considered crafts persons. I hope that the artistic and procedural information provided herein will always be precious to students of the art of audio mastering.

Attention Gearheads

This book is designed to help you learn to make informed decisions on your own; how audio equipment works, and what happens when you turn the knobs. Just about every day I get a letter like this one from engineers asking me to approve or bless their particular list of equipment:

> Dear Bob, I always master with a Sis-boom-bah brand compressor and equalizer, then I follow it off with a touch of a Franifras enhancer. On the next pass I use a Caramba tool to maximize the sound and then Whosizats dither before going to CD. Please tell me what you think of my choices?   Sincerely, Gearhead.

On Language

Sex is good! And being sexy can be fun! I feel that language should be sexy, too, and our centuries-old male-centric language must be rather wearying to the women in our society. It's time to put some vitality back into our syntax. Thus, you will find that in one chapter of this book, the Mastering Engineer may be a female, and in another, male! Vive la différence!

I usually reply, politely,

> Dear Gearhead, your equipment list sounds pretty extensive, but much more important is how you use it. For example, some of the gear you describe would be entirely inappropriate for some kinds of music....

If there is one essential piece of information you can get from this book it is this aphorism written by master engineer Glenn Meadows.

{ *"There is no magic silver bullet. There is no one magic anything that will be 'best' in all situations. The ability of the operator to determine what it is that needs to be done and pick the best combination of tools is more important than what tools are used."*
— GLENN MEADOWS }

Glenn's statement also applies to the amount or setting of each knob or control within your equipment. There is no magic threshold, or EQ setting, or ratio, or preset that will turn ordinary sound into magic. Sonic magic comes from the hard work you put into using your tools (musical magic can only come from the music itself). The truth is that in a typical mastering session, each tool makes only an incremental improvement, and the final result comes from the synergistic totality of the tools working together. In these days of mass-gear-marketing by competitive manufacturers there is too much emphasis on the glitz, fashion and style of the gear rather than on its sound quality and principles of operation. While this book is definitely for gearheads (in the sense that it has lots of glitzy pictures and description of gear designed to produce good sound), serious engineers who want to improve their techniques will also find out how their devices function. Audio principles never go out of style, but models of gear will always fade away.

The theories and background covered here are what I consider to be the minimum necessary to become a competent audio engineer in this digital age. I do not include any heavy mathematical formulas in the main text (you will find more thorough explanations in the footnotes). There are plenty of good foundational basics for beginners, and the most experienced digital design engineer will find useful detail. I include practical examples at every stage; but if the going gets difficult at any point, simply move to the next section. As you grow in experience, when you revisit those sections you may have skipped everything will seem less abstract. I try to define any special terms the first time you meet them; terms can also be found in the glossary (Appendix 15) and in the index. Just like a well-sequenced record album, the chapters in this book were designed to be read sequentially.

A Taste of This Book: Chapter by Chapter

Part I of the book is called **Preparation**. The mastering engineer has tremendous power, and with that power comes great responsibility. Although it is possible to turn an ordinary mix into a glorious-sounding production, sadly it is also possible to ruin a piece of delicate music by applying the wrong approach.

Chapter 1: No Mastering Engineer is an Island, outlines the steps taken in producing a record album, our mastering philosophy, workflow and procedures.

Chapter 2: Connecting It All Together, presents the block diagram of a mastering studio and a general equipment description.

Chapter 3: An Earientation Session, shows how we develop listening skills.

Chapter 4: Word lengths and Dither, is a simplified explanation of one of digital audio's technical mysteries.

Chapter 5: Decibels for Dummies, describes how level meters work, the *myths of normalization*, and how to effectively interface analog and digital equipment.

Chapter 6: Monitoring, demonstrates the need for accurate monitoring and proper room acoustics.

Part II is called **Mastering Techniques**, the important techniques and processes we use in a mastering session.

Chapter 7 shows that **Putting The Album Together** is a critical art and science.

Chapter 8: Equalization, differentiates EQ practice for mastering from that used in tracking/mixing.

Next comes our *dynamics trilogy*: **How To Manipulate Dynamic Range For Fun And Profit**, in three parts, **Chapters 9-11**, covering dynamics processing, theory and philosophy from A-Z.

Chapter 12: Noise Reduction, includes both manual and automatic noise reduction techniques.

Chapter 13: Other Processing, includes such tricks of the trade as M/S processing, classic and not-so-classic specialized analog and digital processors.

Part III: Advanced Theory and Practice, begins with a two-part series: **How To Make Better Recordings in the 21st Century, Chapters 14-15.**

Chapter 14: Monitor Level Calibration, shows how to set up and calibrate a stereo or 5.1 monitor system, and how to use the simple tool of the monitor knob's *position* to help judge

MYTH:
Digital Audio requires less technical skill to use than analog.

program loudness and quality.

Chapter 15: The K-System, is my proposal for a 21st century approach to metering and monitoring to help us produce more consistent and better-sounding recordings.

Chapter 16: Analog And Digital Signal Processing, describes some of the analytical tools we use to *look at sound* and investigates the non-linear relationship between equipment measurements and auditory perception.

Chapter 17: How To Achieve Depth and Dimension in Recording, Mixing and Mastering, studies the powerful classic techniques for obtaining space and depth in 2-channel stereo so as to make an effective move on to surround.

Chapter 18: High Sample Rates, Is This Where It's At? tells us why it's still important to use a high-bandwidth system even though our ears are only good to 20 kHz (on a good day!).

Chapter 19: Jitter: Separating the Myths From the Mysteries, is a direct and definitive layman's explanation of the topic.

Chapter 20: Tips and Tricks, digs into the practical aspects of making AES/EBU and S/PDIF work for you and provides other little-known tips to ease your audio life.

Part IV: Out of the Jungle, presents some of my personal conclusions.

In **Chapter 21: Education, Education, Education,** we get to *preach what we practice!*

Chapter 22: At Last, is a contemplative poem, my hopes and dreams of our musical and audio future.

Part V: Appendices contains some very useful information, including:
- **How to prepare** tapes and files for mastering
- **Radio Ready, The Truth,** largely written by guest authors Robert Orban and Frank Foti, with contribution by Tardon Feathered, shows how radio processing severely affects our mixes and debunks for all time the myth that super-hot recordings sound better over the radio
- **I Feel The Need For Speed,** a comparison of transfer speeds
- **Recommended Reading**
- **Glossary**
- **Audio File Formats**

Plus, **visit the digido.com** website for an online companion to this book:
- An **Honor Roll** of great-sounding Pop CDs, newly compiled for this book
- **URLs** and websites with mastering resources

Now that you've had a taste, let's begin Mastering Audio…

PART I: PREPARATION

"GETTING READY IS HALF THE JOB."

—Anon

CHAPTER 1

No Mastering Engineer Is An Island

I. In The Beginning

This chapter is about the philosophy of mastering and the mastering engineer's approach to audio. We begin by reviewing the place of mastering in the overall scheme of producing and manufacturing a record.

The Record Album:
from Conception to Finished Product

In the beginning was the word (and the music). And that shall never change. But consumer formats do change, and I'm going to miss the Compact Disc when it becomes obsolete—it is probably the first and last professional audio medium that can be created, nurtured and mastered by a single individual. The CD is much easier to produce than the LP, because computer technology has removed forever the words *rewind* and *razor blade* from our working vocabulary. But now even the simplest DVD-A demands a team effort, specialists in audio, video (menus or stills), and interactivity. And quality control for multichannel requires great time and attention to detail.

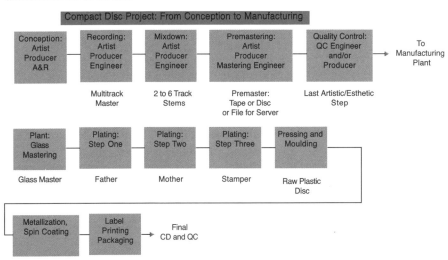

The preceding figure outlines the major artistic and technical steps in Compact Disc or SACD production, from the conceptual beginning, through to the finished technical product.

The song composition and overall conception of the album takes shape in a gestation period that can last for years, with contributions from the artist, the producer, the record company A&R or all three. Then arrangements are written, musicians are hired, and the artists go into the recording studio or on location for the recording to multitrack. This may seem terribly antiquated to those who can record an entire "virtual orchestra" in their project studio, but my personal hope is that the rich art of musical collaboration, with musicians actually playing together "live," never goes away.

Tracking...in the not-so-distant future

The accepted medium for the multitrack recording is rapidly becoming the computer hard disc as a replacement for tape-based formats. In the not-so-distant future universal storage will be so large, and Internet communication so speedy, that the need for a local physical multitrack "machine" may eventually entirely disappear. A single central server will provide all our computing and audio needs. The artist will be able to fly from Seattle to San Francisco without carrying anything, plug into the Internet, and continue overdubbing! However, before this can happen, Internet bandwidth to the home and studio will have to increase by a few orders of magnitude. This also means that the mastering process will involve the mastering engineer simply accessing the relevant tracks from the central server instead of being sent tapes by FedEx.

Mixing

After the tracking is complete the producer, artist and mixing engineer produce the mix of each song or section of the work. If mixing to stereo, the mix goes to two tracks, but even then it may be divided into several 2-track *stems* so that the mastering engineer can tweak the interrelationship between leads and rhythm if it proves necessary after mastering processing, or in the light of the reference monitoring at the mastering house. If mixing for surround, the mix may go to six or more tracks; and if divided in stems, the vocal, rhythm and lead stems may take up 18 or more tracks!

Editing and Premastering

The next step, *editing*, may be carried out at either the recording studio or at the mastering house. It is followed by *premastering*, which is the official name of our profession, to distinguish it from the technical mastering that takes place at the plant (though everyone calls us *mastering engineers* for short). Premastering can include the artistic and technical tasks of **sequencing** (putting the album in song order), **dynamics processing**, **leveling**, **equalization**, **noise reduction**, even some mixing, described in detail in later chapters. Naturally, the output medium of *premastering* is officially called the *premaster*, but we usually label it *master*.

At the Plant

At the plant, the premaster is used to create the *glass master* — an ephemeral product that actually gets destroyed during the production process! At many plants, glass mastering is performed in a class 10 clean room (or better) by engineers wearing white "space suits" (affectionately known as monkey suits). But an alternative is that some plants house their LBRs (laser beam recorders) in a self-contained clean room that can be loaded up in the morning by one suited individual and run all day without intervention, just observation through a Plexiglas window. The LBR is a multi-million dollar machine that takes the digital information for the master, encodes it[*] to the proper format and then sends an encoded laser beam onto a light sensitive emulsion applied to the surface of a 9.5" glass disc. The on-off laser pattern generates a series of pits and lands after the emulsion is developed. The coated glass disc is then moved to another clean room, where the emulsion is sputtered with a fine nickel alloy in a process called *metallization*. Next, the glass plate is put in a vat where an electrical charge is applied, allowing the surface to be plated, in a process called *electroforming*. After plating, the metal plate is peeled off and the glass surface can be cleaned and reused for a new master.

This first metal plate is called the *father* and is the inverse of the final CD (pits are lands and vice versa). For small runs, the father can be used directly as a stamper. But for any significant quantity, the father is electroformed to create a *mother* (which is the inverse of the father) from which many stampers can be produced. Each stamper goes into a press, where a clear polycarbonate disc is inserted and molded. Afterwards, the disc is metallized with an aluminum reflective layer (gold can be used in specialty pressings) and coated with a protective lacquer. Finally, a silk-screened or offset label is applied to the top of the disc, which is then packaged with booklets into the CD boxes by automated machinery. Every element must be carefully inspected for defects—booklets must be properly trimmed, cardboard seams must not tear, CD surfaces must not be stained, labeling should look clean, and the CD itself must meet the proper tests for pit depth and spacing (e.g. jitter and RF output tests). It's an exacting process but....

DVDs are even more complex

Although producing a DVD or DVD-A is very similar to producing a CD, it requires a much greater magnitude of precision. This is because a one-sided DVD contains about 7 times the information density of CD, and thus costs more, in the creative, technical and manufacturing stages. The creative department has to generate the graphics and menu copy and the plan for interactivity well in advance of the authoring stage; furthermore, all of these elements might be in constant flux until the reference audio track has been firmly edited and mastered. Finally, at the plant, DVDs require much more stringent QC standards than CDs, especially because of the delicate bonding process for a multi-layer DVD.

II. Mastering Philosophy and Procedures

For every good mastering engineer, meticulousness and attention to detail is the norm, not the exception. We've always been called upon to keep careful track of a project from the time it arrives until it becomes the final product. Days, weeks, or perhaps years later, if revisions are called, the

---

[*] The encoding includes EFM modulation and error correction information. The exact nature of compact disc and DVD encoding is beyond the scope of this book. Further references can be found in Appendix 10.

client has a reasonable chance of ascertaining which processes were used by consulting with the mastering engineer. At RCA Records through the 80's, analog tape box labels included "dash numbers" (e.g. –1, -2, -3), for each copy generation, and a card catalog carefully logged the tape's status and which one was the correct *master* to use for LP or Cassette duplication. When masters were sent for disc cutting, the cutting engineer inserted a written log indicating the Pultec or other equalizer settings they used, left/right channel gains, and so on.

{ *Attention to detail. The last 10% of the job takes 90% of the time.* }

Today, the situation is far more complicated than simply looking in a tape box for cutting information and marking the box with the generation number. Audio-only projects may arrive in multiple forms, from DATs to Pro Tools Hard discs to CD ROMs to analog tapes. Projects may be two channel or multichannel surround; they may arrive as full mixdowns, partial mixdowns (stems) or combinations. The definition of what is the **Master** becomes even more vague, since multimedia projects may be finished at the audio mastering studio, or authoring added at some studio down the road. Metadata (see Chapter 15) including watermarking may be added during a later authoring stage, further complicating the situation.

But one thing has not changed: it is the responsibility of the mastering engineer to ensure that the audio quality which leaves the mastering studio is the same quality that will be represented on the final medium. We must be familiar with what may happen to the project when it leaves our office, and we must familiarize the producer with what is necessary to preserve the audio quality. I believe in the concept of the Mastering Studio as the **Mothership**, the coordinator of audio quality, and perhaps more, if we've also taken over the authoring duties.[1] In these days of Multimedia, DVDs and SACDs, it is possible that the sound we mastered may be further manipulated by a video engineer, or by some individual who is not skilled in audio production—which is truly counterintuitive.[2] All the more reason for the mastering studio to take on the **Mothership role**.

Now, let's examine the steps, tools and processes involved in mastering a project.

Load-In	Load-Out	Revisions
Date: 09/07/2001	Mon = -7 dB Ref. RP 200	Rev. 2
	Sound comments:	
48K/24 bit loadin from tracks 7/8 of an ADAT 48 kHz tape.	This recording is ADAT grainy. Fun music, good stereo image. Sometimes too much reverb. Needs some fullness.	He sent a dry version of Margaret's Waltz. Tk. 5. I added my "better" verb in the session, revised the settings for the TC 6000 only, created a new sequence, and captured just track 5 to 48K. Then SRC and insert into the EDL 44.1K ver. 2. Then increased the space between tunes by 1/2 second and out to DDP with POW-R 3 dither.
1/Sure Beats Me. Western Swing. Was truncated (somewhere) to 16 bits. Lead violin is a little too loud compared to the rhythm. I could make it swing a bit more. Add a hair more bottom.	For 48/24 capture, used capt w/CDR with no filters and no session. Session: Router Patch: Z Sys #9D, with K-Stereo, TC and Weiss.	
2/Lucky Old Sun. "I notice some of the lead vocals on the word "old" sound abit held back. We tried to compensate for their peakiness". BK: vocals seem too laid back. Try to give it a bit more bounce. "Noise in the tail out". Bs is all there but a hair washy in some frequencies. Violin is a bit grainy? Try CS.	Route: M3/M4 out to -K-Stereo, switched memories by sequence to -TC, doing EQ and sometimes low level comp. There's a reverb module in the chain but it's not used (all dry).	
3/Hustle And bustle. Hoedown instrumental. Bounce it more and make it bigger!	Cranesong to Tape 5 is inserted by automated routing in the TC To	
4/Don't It Just Make You Wonder. Mix #1. Country rock. A bit thin. Fatten it. (If we use this mix). Should I add some reverb?	-Weiss, varied with snapshots (L2 is bypassed) Studio Vision sequence changes the parameters.	

This Excerpt from a Preparation Log Contains details of Load-in, Load-out, and any Revisions made.

III. Logging

Preparation Logs

As we have seen, a multimedia project may have video elements, graphics, menus, etc. CD Audio projects are usually a lot simpler. Here is a sample preparation log of a CD project, containing information about load-in, load-out, and any revisions.

Every mastering engineer has a different approach, but the object of all logging is to be able to reconstruct what was done during the mastering session so as to make revisions or changes easier. In column 1 I put my notes on the original sources (with client's comments in quotation marks to distinguish from my own), column 2 is used for loadout notes, and column 3 for revision notes. Of particular interest is the monitor gain which is logged, and the settings of the processors. Note that most of the digital processor settings are digitally stored in the processor's memory and then saved on floppy disc or Sysex dump or other medium. If analog processors are used, we make verbal descriptions or pictures of the positions of the controls (e.g., "band four boosted 2 clicks at 4.7 kHz, Q = 0.7"). In this revision, since settings for the TC 6000 were changed for tune #5, the floppy disc for the TC contains two files, one labeled **revision 2**, for a complete historical record. During loadout, I use a fully-automated technique controlled by a MIDI sequencer; the only processor with a manual setting in the above master is the Cranesong HEDD, whose "tape" control has been set to position #5.

At our studio, an automatic computer network tape backs up audio logs and sequences as well as all the mundane items such as word processing and accounting. Since computer systems and processors are evolving at Roadrunner pace, we also keep a high-resolution capture of the master just in case processors, applications or operating systems won't recover the old settings. Some clients are insisting on analog tape safeties, since this seems to be the only medium exempt from the technical obsolescence ironically known as progress.

```
LABEL:   Boa Music              DATE    November 24, 2001
TITLE:   Alma De Buxo           SOURCE  DIG    ANALOG  X
ARTIST:  Susana Seivane         FORMAT  DDP, v.1.0, PQ @ Head
CD NO.   10002028               MASTER  X    SAFETY
NO EMPHASIS   X    EMPHASIS     SAMPLING FREQ.   44.1 KHz
DIGITAL HEADROOM  0/0  dB       MASTERING ENG:  BK
This master was created on Sonic Solutions V5. All levels, fades & PQ times are client
approved. Please do not alter in any way. Please refer all technical questions to
Digital Domain at (407) 831-0233.
UPC/EAN CODE  : 0804071020727
T-X  TITLE/ISRC   COPY EMPH    NO OFFSET      OFFSET        OFFSET        CD
                               TIME           TIME          DURATION      TIME
                               hh:mm:ss:ff    hh:mm:ss:ff   hh:mm:ss:ff     mm:ss:ff
--------------------------------------------------------------------------------------
1    ES6080132801 OFF   OFF   A
0    Pause                     -00:00:00:19  -00:00:00:29   00:00:02:00    00:00:00
1    1/Vai De Polcas            00:00:01:11   00:00:01:01   00:03:48:04    00:02:00
                                             TOTAL:         00:03:50:04
--------------------------------------------------------------------------------------
2    ES6080132802 OFF   OFF   A
1    2/A Farándula              00:03:49:10   00:03:49:05   00:02:53:16    03:50:10
                                             TOTAL:         00:02:53:16
--------------------------------------------------------------------------------------
3    ES6080132803 OFF   OFF   A
1    3/Sainza-Riofrio           00:06:42:26   00:06:42:21   00:04:04:02    06:43:50
                                             TOTAL:         00:04:04:02
--------------------------------------------------------------------------------------
4    ES6080132804 OFF   OFF   A
0    Pause                      00:10:46:21   00:10:46:23   00:00:03:14    10:47:55
1    4/Roseiras De Abril        00:10:50:12   00:10:50:07   00:03:59:23    10:51:15
                                             TOTAL:         00:04:03:07
--------------------------------------------------------------------------------------
5    ES6080132805 OFF   OFF   A
0    Pause                      00:14:49:28   00:14:50:00   00:00:02:02    14:50:72
1    5/Xoaniña                  00:14:52:07   00:14:52:02   00:02:58:00    14:53:02
                                             TOTAL:         00:03:00:02
--------------------------------------------------------------------------------------
6    ES6080132806 OFF   OFF   A
0    Pause                      00:17:50:00   00:17:50:02   00:00:02:21    17:51:02
1    6/Rumba Para Susi          00:17:52:28   00:17:52:23   00:04:28:08    17:53:55
                                             TOTAL:         00:04:30:29
--------------------------------------------------------------------------------------
7    ES6080132807 OFF   OFF   A
1    7/Vals Bretón-Muiñeira Pica00:22:21:06   00:22:21:01   00:04:51:06    22:22:00
                                             TOTAL:         00:04:51:06
--------------------------------------------------------------------------------------
8    ES6080132808 OFF   OFF   A
0    Pause                      00:27:12:05   00:27:12:07   00:00:02:10    27:13:15
1    8/Na Terra De Trasancos    00:27:14:22   00:27:14:17   00:03:22:13    27:15:40
                                             TOTAL:         00:03:24:23
--------------------------------------------------------------------------------------
9    ES6080132809 OFF   OFF   A
1    9/Muiñera De Alén          00:30:37:05   00:30:37:00   00:02:28:23    30:37:72
                                             TOTAL:         00:02:28:23
--------------------------------------------------------------------------------------
10   ES6080132810 OFF   OFF   A
0    Pause                      00:33:05:21   00:33:05:23   00:00:02:03    33:06:55
1    10/Ti E Máis Eu            00:33:08:01   00:33:07:26   00:03:11:20    33:08:62
                                             TOTAL:         00:03:13:23
--------------------------------------------------------------------------------------
11   ES6080132811 OFF   OFF   A
0    Pause                      00:36:19:14   00:36:19:16   00:00:02:00    36:20:37
1    11/Chao/Xose Seivane       00:36:21:21   00:36:21:16   00:03:00:28    36:22:37
                                             TOTAL:         00:03:02:28
--------------------------------------------------------------------------------------
12   ES6080132812 OFF   OFF   A
1    12/Chao-Curuxeiras         00:39:22:17   00:39:22:14   00:03:25:21    39:23:32
                                             TOTAL:         00:03:25:21
--------------------------------------------------------------------------------------
13   ES6080132813 OFF   OFF   A
0    Pause                      00:42:48:03   00:42:48:05   00:00:01:26    42:49:10
1    13/Marcha Procesional Dos C00:42:50:06   00:42:50:01   00:04:35:16    42:51:00
                                             TOTAL:         00:04:37:12
--------------------------------------------------------------------------------------
     LeadOut                    00:47:25:15   00:47:25:17                  47:26:40
--------------------------------------------------------------------------------------
Total                                                       00:47:26:16
```

PQ Listing showing engineer's comments, track times, ISRC codes and other information.

PQ Lists

The name **PQ** comes from the letter-code abbreviations for the information contained in the subcode of the Compact Disc. The **P** flag is the most primitive flag; it changes state to indicate the beginning of a new track. The Q subcode contains information such as timing and program length, copy prohibit or permit, emphasis condition, and ISRC codes (see Chapter 20), most of which will be stored in the final disc's TOC (table of contents). The written PQ log is actually a redundant log, since nowadays the master medium contains all the tracks and an electronic version of the PQ codes. In the old days,[*] the replication plant would take the written information from the PQ log and enter it electronically into a PQ editor, since most mastering houses did not have a PQ editor. Today, while most mastering houses generate their own PQ codes, all responsible replication plants still require a written PQ list. This is the only place they can see the names of the titles, and engineer's comments. Mastering engineers appreciate good quality control procedures after the master has left their possession. A reliable plant will cross-check all the information in the written PQ log against the electronic version on the master medium, and call the engineer if there are any discrepancies. An exceptional plant will even note noises they hear or over levels, and ask for engineer's approval before pressing. Sadly, this has become very rare, so the burden for quality control has fallen heavily upon the mastering house.

---

[*] Not so long ago, but computer years are like dog years, and in this fast-paced world, ten years feels like seventy!

IV. Mastering Output Formats

While we can accept recordings in nearly any format, only four media are suitable to be used by the replication plant for CD-Audio (CD-A): DDP (Disc Description Protocol, on Exabyte 8 mm tape), PCM-1630 (on 3/4" video cassette), CDR (Orange Book, write-once media), or Sony PCM-9000 Optical disc. As of this publication, the PCM-1630 format is rapidly becoming obsolete, and the PCM-9000 never took off and is also considered obsolete. Of all the above formats, the PCM-9000 was probably the most reliable. Almost as reliable and most popular is the DDP, which can be duplicated at 4X and greater speeds (not necessarily producing better sound quality, see Chapter 19). The least reliable is the audio CDR, first because its error rate is not as good as the DDP, and also because it is easily susceptible to fingerprints and mishandling. A DAT is generally not considered a suitable medium for glass mastering, though one or two plants have adapted their systems to work with timecode DAT. The master must be recorded in one continuous pass, without stopping, under the control of a computer. Some recording engineers attempt to deliver "masters" on CDRs recorded on a stand-alone CD recorder, but this is usually unsatisfactory because of the inaccuracy of the track points, inability to put separate track end marks (which creates extra-long track times), and E32 errors introduced every time the recorder stops its laser (breaking the "one continuous pass" rule). CDRs make reasonable sources for premastering (I like them better than DATs), but not good masters for glass mastering.

The master medium which may take over is the DLT (Digital Linear Tape). It has much higher capacity, typically 40 to 80 gigabytes. In theory, the DLT can carry the DDP protocol, and could take over from Exabyte, but no one has implemented it.[*] DLT is the specified medium for DVD and DVD-A masters. Another up-and-coming medium is a CD-ROM or DVD-ROM with DDP image files, since the CD-ROM has excellent error correction. More masters are now being sent to the factory via high-speed Internet lines, which brings up legal questions of just what medium is the physical master.

> *"Audio CDRs make good sources for mastering, but not good output masters."*

V. Picking the Right DAW

By the mid-80's Sonic Solutions Digital Audio Workstations (DAW) had taken over the mastering field. As soon as engineers discovered the virtues of non-linear editing, and a workstation that could integrate PQ coding with audio, they quickly abandoned their slow Sony DAE-3000 editors. Sonic workstations use a powerful Source-to-Destination editing model that many editors prefer, have extremely high data integrity (producing clones of the source when not processing), and can make those "impossible" audio edits through the use of a very flexible crossfade editor. The crossfade

---

[*] There are actually two versions of the DDP protocol, version 1.0 and 2.0. In addition, PQ code may be put at the head of the tape or at the tail. If put at the tail, PQ codes and ISRC can be changed without rewriting the master. However, some plants do not accept PQ codes at the end of the tape. Check with the plant in advance if making anything other than version 1.0, PQ at head.

editor is the main reason why Sonic and its brethren are very popular editors in the classical music field. We'll see the editor in action in other chapters. To this day, only a few other workstations or software programs have been qualified or dedicated to mastering: Audiocube, Pyramix (Merging Technologies), SADiE, Sequoia and Wavelab. SADiE has recently caught up to Sonic with converts, and Sequoia has garnered a good number of dedicated users. SADiE is now the only workstation to incorporate a dedicated SCSI (hard disk) bus, which makes it very stable and free from operating system interference; you can even purposely crash Windows and SADiE will keep on cutting a CDR. The race is not yet won, since not each workstation has the ability to do multichannel and high sample rates with equal facility, and not all manufacturers offer an upgrade to DVD-A and SACD authoring.

Other criteria appropriate to picking a DAW include software and hardware reliability and economic stability of the company. Consider the number of man-years that have gone into software and DSP development and make sure that development is ongoing. Five man-years is the minimum time I would consider required to make a powerful, dependable mastering program. Be wary of marketing promises: *if the product does not have the features you want today, don't buy it on the basis of "real soon now."* Find out if the company has fast and efficient technical support. Ask if there is an upgrade policy. Another valuable approach before buying is to get feedback from users, especially those doing similar work. Is there an established user base and support group?

All these criteria raise the short-term purchase price of a good workstation, but greatly lower the long-term cost of ownership.

Don't Be a Complete Bithead

Far less successful are engineers who attempt to perform mastering on software platforms not specifically dedicated to mastering—largely because of lack of integrated PQ editing, low data integrity, low sound quality, inflexible editing, and so on. If you are going to dedicate yourself to mastering, you must get a dedicated workstation. These workstations have other attributes besides data integrity: high calculation accuracy, which translates to low distortion. They all implement proper dithering (see Chapter 4), and high precision, with the highest precision award going to the Audiocube (64-bit floating point) or Sonic Solutions HD (48-bit fixed point), which yield excellent-sounding equalization and noise reduction algorithms. But don't be a bithead, because all things are never equal; the skill of the programmer can turn everything around—one programmer's 32-bit float can sound better than another's 64-bit (see Chapter 16).

VI. Mastering Procedures

Mastering With or Without a Producer Present?

Mastering engineers are independent beasts and can master quite comfortably without a

{ *"The Customer is Always Right."*
— DALE CARNEGIE }

producer or artist present. Once there was a certain type of mastering engineer who had a specific sound—if you went to that engineer, you would send your tape, and get her sound. But there are very few (if any) of those kinds of mastering engineers, and the reason is quite plain: every piece of music is unique, and requires a special approach that is sympathetic to the needs of that music and the needs of the producer and artist.

A good mastering engineer is familiar with and comfortable with many styles of music. She knows how acoustic and electric instruments and vocals sound, plus she's familiar with the different styles of music recording and mixing that have evolved. In addition, a good mastering engineer knows how to take a raw tape destined for duplication and make it sound like a polished record. Upon listening to a tape, a good mastering engineer should be able to tell what she likes and doesn't like about a recording, and what she can do to make the recording sound better. Then, by sympathetically listening to, and working with, the producer, the engineer can produce a product that is a good combination of her ideas and the producer's intentions, a better-sounding product than if the engineer had simply mastered on her own.

The best masters are produced when both the producer and the engineer solicit feedback, use empathy, courtesy, and understanding, and are willing to experiment and listen to new ideas. My approach is to welcome and encourage the producers' input for they are the ones most familiar with the music and what they want it to say.

If the producer cannot attend the mastering session, then we'll have discussions prior to and during the session of how they perceive their music, and how I think it sounds. Sometimes it helps if the producer sends in existing CDs as examples of their tastes. Then I'll send a reference or evaluation CD prior to the final mastering. Usually by that time we are enough in sync so there is no need to produce a second reference, or just some minor changes.

Weeks or even months prior to the mastering session, an exceptional producer will send a preliminary mix to solicit the mastering engineer's feedback, because there are things which are better fixed in the mix or not possible to fix in mastering. We don't hesitate to suggest a remix if there is a severe problem. The better the mix, the better we look! How much can the sonics of a mix be improved in the mastering? I like to answer: about a letter grade, which can turn a B plus mix into an A plus master!

The Mastering Workflow

The mastering engineer's workflow comprises editing, cleanup, leveling, processing and output to the final medium. Every engineer has a unique approach, using analog or digital processing or a hybrid. Currently, most engineers work with DAWs in very much the same way we worked before there were any DAWs[3]: First, we take the source for each tune (e.g., DAT, CDR, Masterlink, AIFF or WAV file), and process one song at a time. If that source is digital and if analog processing is to be used, we send it to a high-quality D/A converter, pass it through one or more analog audio processors and

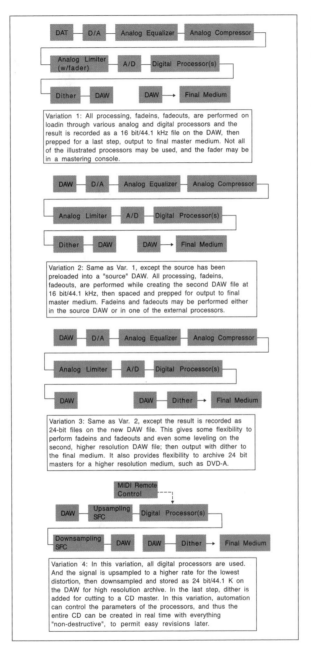

Variation 1: All processing, fadeins, fadeouts, are performed on loadin through various analog and digital processors and the result is recorded as a 16 bit/44.1 kHz file on the DAW, then prepped for a last step, output to final master medium. Not all of the illustrated processors may be used, and the fader may be in a mastering console.

Variation 2: Same as Var. 1, except the source has been preloaded into a "source" DAW. All processing, fadeins, fadeouts, are performed while creating the second DAW file at 16 bit/44.1 kHz, then spaced and prepped for output to final master medium. Fadeins and fadeouts may be performed either in the source DAW or in one of the external processors.

Variation 3: Same as Var. 2, except the result is recorded as 24-bit files on the new DAW file. This gives some flexibility to perform fadeins and fadeouts and even some leveling on the second, higher resolution DAW file; then output with dither to the final medium. It also provides flexibility to archive 24 bit masters for a higher resolution medium, such as DVD-A.

Variation 4: In this variation, all digital processors are used. And the signal is upsampled to a higher rate for the lowest distortion, then downsampled and stored as 24 bit/44.1 K on the DAW for high resolution archive. In the last step, dither is added for cutting to a CD master. In this variation, automation can control the parameters of the processors, and thus the entire CD can be created in real time with everything "non-destructive", to permit easy revisions later.

At left: Infinite Variations on a Mastering Theme. Four examples of approaches to audio mastering.

possibly control the level, EQ, or fade via a customized analog mastering console. The signal is then passed to a high quality A/D converter, optionally through various digital processors, dithered to 16 bits, and then recorded into the DAW. We then move on to the next song, resetting processors until the best sound is achieved for that song. And so on as illustrated in **Variation 1 of the figure at left.**

In this variation, all leveling, fading, processing and equalization has already been accomplished, and the DAW is only used for assembling and spacing, which is a very efficient approach. When we reach the end of the tune, if it requires a fadeout and we missed it, instead of reloading the entire song we may back up before the fadeout, do a simple punch-in on the workstation, perform the fade, and then a matched edit. Chapter 4 tells why this 16-bit file should not be further processed. What this means is that, in Variation 1, if the client orders any revisions, the engineer must repatch the entire chain, reset the processors, make any processing changes, and re-record/replace the old destination file with a new one.

Often there is no real time "load in," since sources may arrive as high resolution or high sample rate computer files on CD ROMs or other media, and can be loaded at high speed directly into the workstation (Var. 2). The mastering engineer then has to listen to each tune to get the feel of the whole album and check for noises or other problems that may need fixing. She may begin by putting the material in order, cleaning up heads and tails, perform fadeouts and spacing, and then proceed as in Var. 1, except she uses the workstation as the new "source" as well as destination. In **Variation 3**, the mastering engineer waits until the final output to dither, which gives some flexibility to perform fade-ins and fadeouts on the final DAW file and perhaps some leveling. (Although most of the leveling should have been performed beforehand to avoid cumulative loss of resolution). After digital limiting, levels cannot be raised, only lowered, and equalization should not be performed on a previously-limited signal, as the peak protection will be undone. Digital filtering of any type can cause overloads on a digitally limited signal, because it creates higher-level intersample peaks. Thus it is best to return to the source and reprocess in order to change levels between tunes.

With the increasing number of high sample rate projects, another variation is to use two workstations, one to play back the high sample rate material, the other to record a sample-rate-converted and dithered output for CD prep. Yet another variation is to use **upsampling** followed by **downsampling (Var. 4)**. Even if the source material is ready for CD at 44.1 kHz, it is well-known that digital audio processing and conversion at a higher rate sounds better (see chapters 16 and 18). The engineer may reproduce the source material at the lower rate, feed an upsampling **sample rate converter** (abbreviated SFC, SRC), then perhaps D/A convert using a high-resolution, high sample rate D/A for analog processing, then record the material into a high sample rate A/D converter for optional further digital processing, then finally downsample and dither (if the result must be 16-bit). If the source material is at 44.1 kHz, a CD can be cut in real time using this chain. But two steps (and two DAWs) may be necessary if the source material is not recorded at the target rate, since most DAWs can only work at a single rate. First, the material is stored at 24 bits/44.1 K on the new DAW file, then it is dithered in the last step to the 16-bit master medium.[*]

Material that arrives at multiple sample rates (different songs at different rates) is particularly problematic, often necessitating sample rate conversion to a common rate before the mastering can get started.

Tune by Tune or Fully-Automated?

All of the above descriptions have one thing in common: they follow a tune by tune approach to mastering, i.e. master one tune, reset the processors, then move on to the next one. Although engineers have been making excellent albums using this method for years, an increasing number of digital audio processors are remote-controllable via MIDI (Var. 4), which permits them to be automated and thus completely integrated with the workflow. Most engineers already use some sort of automation in their work, since advanced workstations provide automated equalization, leveling, fades, dynamics, and even automated plug-ins. If a revision is requested, the mastering engineer can save the previous EDL (edit decision list) and instantly make changes in the amounts or timing of the

---

[*] One unique workstation (Sequoia) permits working at two rates simultaneously, so only one workstation is needed!

workstation's internal equalization. **The MIDI technique extends this ability to the outboard equipment.** For me this is a revolution—finally I can work with the album in the making in a comfortable, fluid, non-linear manner. I work with a song until it is cooked, save the parameters in the memories of the processors, and then move on to the next song without having to capture to a DAW file. I save those parameters in another processor memory, then return to near the end of the previous song and play the two together with the MIDI automation following along, nondestructively. This makes it easy to integrate two dissimilar songs, e.g. if one ends big and the other begins soft and easy (more details on this technique in Chapters 7 and 10). It's also non-linear—having the context of the whole album in development makes it possible to revisit and reprocess any portion of the album. For example, we may make a great climax, then recheck the first song in context and reprocess it if necessary without having to reload or recapture. Full automation also permits special effects—for example, as we approached the climax on one tune, upon the entrance of a big vocal chorus, I created MIDI-automated changes in the K-Stereo Processor that increased step by step the spaciousness and depth, producing a gigantic sound in the final chords. After we're satisfied that the album sounds good, we then go back to the beginning and cut a CDR reference in real time with full automation.

The biggest advantage of full-automation is the ease of revision, especially if you have a critical clientele. Processing is always applied in a non-destructive, non-cumulative manner; anything can

be undone without going down another generation or forcing a reload. Another advantage of this method is that the raw, unaltered sources can be immediately compared with the master and demonstrated to the client. We try to ensure the master is better than the source in every possible aspect; it's a sobering moment if we discover that the source is *better* than the processed master, in which case—back to the drawing board!

One sonic advantage of this method is that the highest resolution processor can be used to change gain. Thus, the MIDI automation accomplishes the changes in levels from song to song. I often use the output gain of the mastering compressor, since the 40-bit float Weiss has a more transparent-sounding gain change than the DAW or any other device in the chain, and this also avoids additional DSP. The biggest disadvantage of this method is the amount of technical know-how and concentration required to run a MIDI sequencer and control the parameters of external equipment.

Here's how the MIDI-automated chain is hooked together:

The audio resides in the mastering DAW on a PC, for example, SADiE, which feeds a series of external rack processors, and returns back into SADiE. With SADiE or Sonic Solutions, the CD master can be cut in real time using this routing if the source audio is at 44.1 kHz SR.

The timecode master is SADiE, and this timecode feeds another computer, in this case, a Macintosh running a sequencer called **Digital Performer**.

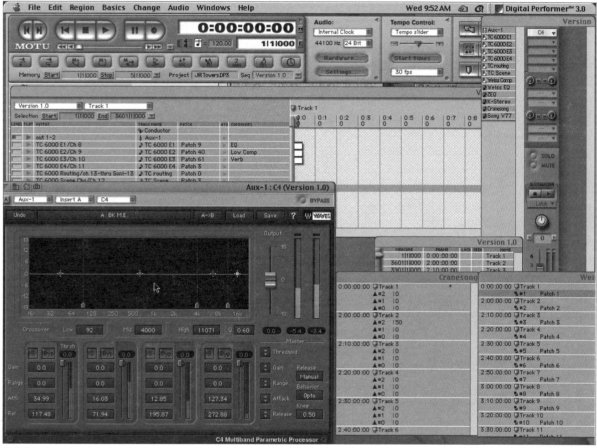

Digital Performer in action, slaved to the mastering DAW. Performer automates both external devices via MIDI and plug-ins acting as outboard processors to the main workstation, for example, *SADiE*.

The MIDI instructions are fed from Performer to the external rack processors, and in a cute trick, automate a native plug-in, the **Waves C4**, implemented directly in Performer, illustrated in the above figure. We treat the C4 functionally as another rack device external to SADiE. Native processors are not always used in mastering, but I've created this illustration to show how it can be done even when the mastering DAW does not support "live" plug-ins.

VII. Media Verification, Archiving/Backups

Listening Quality Control

At the end of the project, the **art** of mastering has to turn back into a **science**. In larger mastering studios, this is performed by a separate QC department. The QC engineer must have musical/artistic ears, technical prowess, but also a lot of common sense: the project has already been auditioned by the mastering engineer and producer and all the noises presumably were accepted, perhaps even welcomed as "part of the music." If a single unacceptable tic or noise is discovered anywhere in a master, the entire full-length master has to be remade and listened to/evaluated. There is no shortcut. During the QC pass, we have to utilize as many objective criteria as possible. For example, a critical listener using headphones is bound to hear more noises than someone using loudspeakers. Does this mean that we have to use headphones to verify a project? If the monitoring acoustic is less than ideal, then QC must be performed with headphones, but the loudspeakers in a critically-designed mastering room are more than adequate. Mastering engineer Bob Ludwig has reported that headphone listening becomes essential when the number of channels multiplies. Potentially embarrassing noises or glitches hidden in the surround channel when auditioned on loudspeakers become quite audible when that channel is isolated in a pair of headphones. To complicate the situation even further, one consumer may be auditioning all channels using surround headphones while others will be hearing stereo reductions (folddowns). Clearly, we have to give much greater attention to detail, and costly time to evaluate a final master in surround, even to the point of requiring 3-4 hours to QC an hour program, including any extra passes necessary to check a folddown! We than have to decide how to deal with each noise that is encountered during QC. We follow the practice of noting the timecode of each offending noise, then checking with the mastering engineer and/or producer to see if the noise had already been accepted.

> *"If a single unacceptable tic or noise is discovered anywhere in a master, the entire full-length master has to be remade and listened to/evaluated. There is no shortcut."*

QC also includes verification that the proper songs are in the proper place, based on client-supplied lists of the song lengths, lyric sheets, etc. We must ensure that the correct master goes out for duplication, and must be especially wary of misidentifying individual CDs of a multiple-CD set. With the advent of authoring and DVDs, more than one QC may be needed, including the final watermarked and MLP'ed* master. And with electronic delivery comes the legal issues of which "physical master" has been officially evaluated.

Objective Media Verification/Error check

Digital media are susceptible to data dropouts which cause errors, which is why all the digital audio storage formats, DAT, Exabyte, PCM-1630, and DLT

---

* MLP is Meridian Lossless Packing, see Chapter 15.

tapes, and optical discs, CDR and DVD-R, utilize error correction algorithms.[*] Uncorrected errors result in glitches, clicks, and other noises. Normally, when playing a digital tape or disc, we do not know the amount of error correction which is occurring. It can sound great, but the tape or disc could be near dying! If the error correction system is working very hard, the next time that tape is played, a speck of dust or head alignment problem, or simply wear and tear, will cause a signal dropout during playback. Our job is to look behind the scenes using specialized measurement tools. Listening alone is like having a doctor look at the patient without taking his temperature. So media verification is a thorough internal examination.

There is also the issue of *error concealment*, which is the last defense mechanism in digital playback. If the error correction does not work, that is, if there is an uncorrectable error, then the playback machine uses an interpolator. The interpolator looks at the audio level before and after a dropout and supplies an intermediate replacement. If performed well, error concealment can sound very good, but professionals never use a master medium that is so degraded. On the PCM-1630, error concealment can be turned off, and the result is an audible mute that purposely lasts a second or more, to call attention to itself.

The PCM-1630 system uses an evaluator known as the DTA-2000, and each plant and mastering facility decides on objective criteria for acceptability. For example, some houses reject tapes with CRC (**Cyclical Redundancy Check**, correctable errors, aka *soft errors*) counts over 50 in any minute, or over 200 total on any tape. Other houses accept up to 300 or even 400 CRCs in an hour, though this is considered exceptional or rare, and an indication of poor master tape quality. Of course, any uncorrectable error is cause for rejection of a master.

MYTH:
An audio loadback/null test shows the integrity of a CD Master.[†]

---

* Hard discs, however, generally do not require error-correction, since their error rates are extremely small.

† On the contrary, all the null test proves is that there were no uncorrectable errors, but it is not a measure of media reliability or error-count. The null test is post the error correction. You could be one bit away from failure and not know it. The next time an error-prone disc plays, there could be an interpolation or a mute if the error count is high. Thanks to Glenn Meadows for pointing out these facts.

Title			TEST2, 00, , 00, PREMAST,						
Test Start	C-Time		0 01 53	A-Time	0 00		Samples		4851
Test Stop			AA 00 15		79 42		De-tracks		0
Parameter			**Value**	**C-time**	**A-time**	**Avg**	**Thr**	**Cnt**	
BLER	1 Sec	Max	29	1 00 18	0 20	2.5	220	0	
	10 Sec	Max	17	1 52 30	52 32		220	0	
E-11	1 Sec	Max	17	1 75 03	75 05	0.9	200	0	
	10 Sec	Max	3	1 01 07	1 09		200	0	
E-21	1 Sec	Max	9	1 01 43	1 45	0.4	200	0	
	10 Sec	Max	3	1 05 07	5 09		200	0	
E-31	1 Sec	Max	20	1 52 27	52 29	1.2	200	0	
	10 Sec	Max	13	1 52 29	52 31		200	0	
BRST	1 Sec	Max	3	1 00 19	0 21	0.5	7	0	
	10 Sec	Max	3	1 52 18	52 20		7	0	
E-12	1 Sec	Max	237	1 52 27	52 29	8.6	300	0	
	10 Sec	Max	155	1 52 30	52 32		300	0	
E-22	1 Sec	Max	12	1 03 45	3 47	0.0	15	0	
	10 Sec	Max	1	1 03 45	3 47		15	0	
E-32	1 Sec	Max	0			0.0	1	0	
	10 Sec	Max	0				1	0	
I11 / Itop		Min	0.580 **	1 00 13	0 15	0.596	0.600		
I3 / Itop		Min	0.277 **	0 02 10	0 00	0.329	0.300		
		Max	0.354	1 79 14	79 16		0.700		
SYM*		Min	-12	1 64 53	64 55	-8.5	-20.0		
		Max	-5	0 02 51	0 00		20.0		
Rad Noise		Max	18	1 51 55	51 57	12.1	30.0		
PP Mag		Min	0.086	1 00 00	0 02	0.086	0.040		
		Max	0.086 **	1 00 00	0 02		0.070		
Cross talk		Max	50	1 00 00	0 02	49.5	50.0		

CD-A Report from Clover Brand Analyser. Note the BLER value of 29 in any 1 second maximum at :20 Abs time.

```
      Exabyte drive reports: retry
level of 0 percent, and an error
level of 0.0443386 percent

      This error rate is within the
factory standard for a new
drive.

      Delivery Job Complete
```

Exabyte Error Report from Sonic Solutions showing the total error rate for the duration of the tape.

```
                CD201032 - 20011028.txt


Audio Length consistent between subcode and mapstream data = 33837
8 CD Frames

SD (PQ descr) file - 960 bytes read. This is 15 blocks of size 64
bytes
Expected no subcode packets = 15, Actual no packets = 15
Comparing PQ List read in from DDP medium with list in memory.
Verification successful, lists are identical


AUDIO FILE
==========

Verifying audio integrity...

        block    block   Elapsed time    kilobytes     soft error
        size     count   (mm:ss.msec)    per second    count

        9408     56      00:00.109       4720          0
        9408     56      00:00.285       1805          0
        9408     56      00:00.534       963           0
        9408     56      00:00.588       875           0
        9408     56      00:00.543       947           0
        9408     56      00:00.589       873           0
        9408     56      00:00.570       902           0
        9408     56      00:00.546       942           0
        9408     56      00:00.586       877           1
        9408     56      00:00.544       945           0
        9408     56      00:00.589       873           0
        9408     56      00:00.563       913           0
        9408     56      00:00.546       942           0
        9408     56      00:00.585       879           1
        9408     56      00:00.571       901           0
        9408     56      00:00.564       912           0
        9408     56      00:00.569       904           0
        9408     56      00:00.544       945           0
        9408     56      00:00.588       875           0
        9408     56      00:00.545       944           0
        9408     56      00:00.589       873           0
        9408     56      00:00.569       904           0
```

Comprehensive Exabyte Error Report from SADiE System, showing errors at each block, which goes on for 30 more pages! The plant is satisfied with a one page graphic summary showing the total count of errors and that no large error amounts occur in any short period.

The Clover system is a popular CDR media evaluator. The most critical criterion for CD-A and CDR quality is called BLER (Block Error Rate). A very good CD can have a BLER as low as 10, yet CDs will still play with BLERs of 1000 or even above—which illustrates just how robust the error correction system is for CD-A. CD ROMs use an additional layer of error correction. One conservative mastering house's standard rejects any CDR with BLER over 100, or any CDR with an E32 (uncorrectable) error.

For Exabyte tapes, reports can get as complex as a multipage document showing error count in each block, or simply a one paragraph total report indicating error percentage (see figures at left). Many mastering houses will reject Exabytes with error percentages over 0.1%, though 0.2% or even 0.3% error is quite acceptable, as long as there were no read-after-write retries in the error report.

Other QC Issues

The responsibility for QC must be accepted by someone, but the movements of technology and economics are making it difficult to guarantee standards. The PCM-1630 has obtained legendary status for its sonic quality, and it also forces glass mastering to be at 1X speed, where the master may be auditioned, thus gaining one critical stage of Quality Control. However, the 1630 technology is now old enough to be causing concern about its reliability and many plants copy from 1630 to Exabyte to avoid problems during expensive glass mastering.

There is usually no press proof except when very large quantities are pressed. There used to be a listening room at each pressing plant where masters were auditioned prior to glass mastering. But now when the master arrives at the replication plant, whether in physical or electronic form, it will likely be copied high speed to an Exabyte tape or to the factory's central server, and there is no auditioning during glass mastering. The day has come when the home consumer is the first person to audition the product! Every project needs a **Mothership** to get through this mess.[*]

Since human QC at the plant seems to be decreasing, especially for electronic delivery, I propose that the approved electronic delivery have an error-detecting format built-in, as used by programs like ZIP for the PC and Stuffit for the Mac. On opening, an error will be generated if a stuffed file does not contain the identical data that was used to create it. Using such a *coded master* can confirm that the file remains intact through all transfers up to the point of glass mastering. The Meridian Lossless Packing format (**MLP**), used for the DVD-A, is a self-correcting medium, but its cost and encoding time make it overkill for simple stereo work.

Backups/Archives

After a project is finished, we wait until the client has approved the master (usually by listening to a copy of the master). We then may wipe the material from our hard discs, but not before saving

the logs on hard disc with all the material, and making an in-house audio backup on some form of computer tape. The in-house backup is mostly in case a revision is requested within a reasonable time since as we mentioned, digital technology is constantly changing. Some record labels require full backups of the masters, often on Sonic Solutions Exabyte tapes, or some other acceptable archive format.

The critical difference between a **backup** and an **archive** is that an archive is made to a medium

{ *Backups? We don't need no ba&*9 u.* }

which is supposed to last a long time (30 years or more). However, I wish good luck to those who have to decipher those multi-formatted ones and zeros; will the equipment still be around to read them even ten years from now? Computer manufacturers seem bent on obsolescence and equipment turnover, which makes the idea of full data-recovery frightening. Technological evolution is a serious issue.

---

[*] Thanks to Mike Collins, *One To One* Magazine, November 2001, and to various discussions on the Mastering Webboard, for inspiring this section.

1 I am reminded of an analogous situation in the film world. Prior to 1977, the role of **Sound Designer** was unheard of, but Ben Burtt received the honor of the title on the first Star Wars film. The Sound Designer is the Mothership for the entire film, coordinating the film from first recording, through transfers, editing, and the final mix. As a result, the film takes on a flowing, gestalt feel.

2 One mastering engineer reported a situation where another house added the CD ROM portion to an extended CD, and somehow in the process, changed the audio quality of the audio portion. Never assume that everything will be fine when the master goes out the door, even to the extent of (on critical projects) approving and testing the final product. It is possible to do **null tests** or **bit for bit comparisons** which compare the original audio master against the final pressing, assuring that the audio data had not been altered after it left the mastering house.

In another situation, a less than reputable plant copied **all** incoming masters using a consumer-based program which automatically shortens tracks to the end marks, and then puts 2-second silent gaps between all the tracks. Thus, the final pressing of a beautifully-engineered live concert sounded like it was edited with an axe! These are real horror stories from the trenches, so be sure to mind your Q's and C's!

3 Well, this is true for CD mastering. But if you go way back to the ages of LP cutting, the cutting engineer was forced to cut an entire record in one continuous pass. If you stop, you create a locked groove, which you could say was yesterday's E32 error. A sophisticated LP cutting engineer would note settings for each tune and manually change her processors during the banding between each track. Equalizers were developed with A and B settings, allowing her to press one switch during the intertrack gap, and then leisurely preset the opposite equalizer for the next track. Primitive, but roughly equivalent to the fully-automated process which I use today.

CHAPTER 2

Connecting It All Together

The Principle of Consistent Monitoring

The following page shows a block diagram of the audio connections in the *ideal* digital audio mastering studio, The heart of this studio is an integrated A/D/A system ①, typically 6 to 8 channels. Since our clients expect us to make consistent quality judgments, we audition all digital sources and pressed media through this single converter. Unfortunately, this principle of consistent monitoring has been subverted by the advent of new copy-protected media such as DVD-A and SACD, whose players do not have digital outputs; thus it is not always possible to proof the final product through the same D/A converters that were used for the mastering.

All channels of the A/Ds and D/As are housed in the same chassis, with internal clock connections designed for minimum jitter and immunity from external jitter. In Chapter 19 we will learn why this is the best architecture for minimum jitter. With a jitter-immune system, the mastering engineer avoids chasing ghosts and non-problem problems.

Routing It All

The router ②, switches all digital sources and destinations in any combination. A 16 x 16 router can be used in a smaller studio or one dedicated to stereo production, but at least 32 x 32 is required for surround work. The Z-Systems brand of routers can switch virtually any type of signal and support multiple sample rates and different synchronizations in the same chassis, can be configured for different voltage and impedance standards, and thus can be used for AES/EBU or S/PDIF (2 channels per

Block Diagram of A State of the Art, Jitter-Immune Digital Audio Mastering Studio.

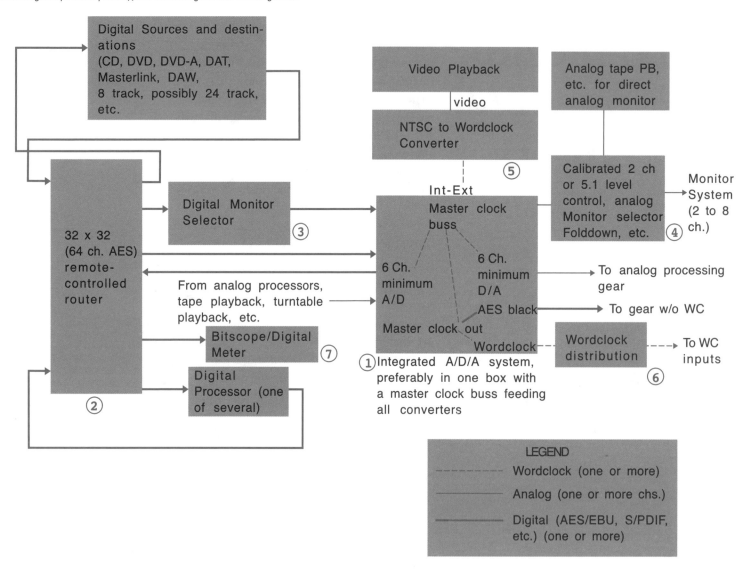

LEGEND
- - - - - - Wordclock (one or more)
———— Analog (one or more chs.)
———— Digital (AES/EBU, S/PDIF, etc.) (one or more)

connection for a total of 64 in and out at any standard sample rate) or Dolby E (8 channels per connector) or Dolby Digital (6 channels), MADI (multiple channels) or encoded formats such as MP3, even to distribute wordclock. Possible sources and destinations include DAW(s), tape, CD(R), digital compressors, equalizers, A/D and D/A converters, and so on. One digital source can be routed to multiple destinations, but any digital input can only accept a single source.

Complex chains with analog or digital components can be created at the push of a button, since the analog processors are connected to the converters, and the converters are connected to the router. For example, this figure shows the Macintosh computer-based remote control for a Z-Systems 16x16 router.

Individual setups can be saved and named for each project. For example, in this project, a stereo loop begins at the DAW and returns to DAW: Sonic Solutions M3/M4 feeds the Z Systems digital equalizer, which then feeds TC System 6000 inputs 1/2 for further processing, then to POW-R dither, and back to Sonic inputs L3/L4, where they are routed to the SCSI CD recorder or master tape machine. This router setup also handles 2-channel monitoring, and provides an auxiliary loop path to and from the DAW and a reverb unit (the Sony V77).

In my mastering studio, the TC System 6000 functions as the central A/D/A converter, calibrated digital monitor level control, Folddown[1] control, master clock, and insertion between digital points and analog processors. In other studios, some of these functions are relegated to the analog monitor/line-stage preamp ④, which follows the monitor DAC, but my line-stage preamp just serves to check the direct sound of analog-only sources (such as turntable or tape deck).

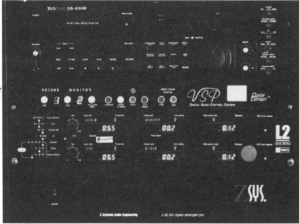

The top component in this rack is a Tascam DA-45, 24-bit DAT machine, below which is the Digital Domain model VSP, which selects from 6 digital sources for recording or dubbing, and 6 sources for monitoring. An A/B monitor selector allows for comparisons. Below the VSP is a Waves L2 digital limiter, below which is the front panel of a remote-controlled Z-Sys 16x16 router.

The digital monitor selector ③, is a smaller router (8x8 recommended) which takes any subset of the 32X32 and routes it to the monitoring DAC.

Mac-based remote control for a Z-Systems 16x16 router.

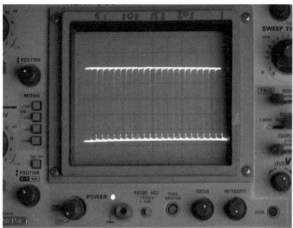

24 Bits active on the bitscope.

This allows A/B monitoring or comparison of any two digital sources, such as "before" and "after" mastering. Digital Domain manufactures a Digital monitor selector called the VSP (see photo on page 37), that allows instant A/B selection of any two stereo sources, and can preselect from 6 choices.

Normally, the converters perform best on internal sync, but when doing video, the converters must slave to the wordclock which comes from the NTSC to wordclock converter ⑤, and we have to depend on the quality of the converter's PLL to reduce jitter, explained in Chapter 19. A wordclock distribution amplifier ⑥, feeds multiple wordclock lines to the DAW, DAT machine, CD transport, and some processors which support wordclock input. Otherwise, we must depend on AES black or signal-carrying AES to synchronize the ancillary digital gear.

Other important equipment includes ⑦, a bitscope (see photos this page) and digital meter,

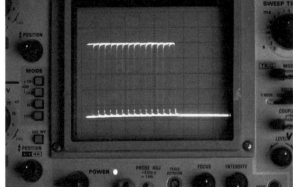

16 Bits active on the bitscope, truncated after the LSB.

which can be routed from any digital source. The bitscope serves to double-check the bit-integrity of the source, confirm that dither appears to be functional, and that there are no extra bits due to hardware or software bugs.

I usually connect the meter and bitscope to the same router output. Pictured are two examples of common meters used in mastering.

I. Block Diagram and Wire Numbers

When constructing a mastering studio, begin with a detailed block diagram, inserting wire numbers from a separate wire number list. On the opposite page is an example block diagram, with wire numbers in parentheses.

Proper grounding and wire layout techniques are critical.[2] A modern-day digital mastering

Mytek digital Meter DDD-603 with 96 kHz upgrade. Meter only responds to top 16 bits of signal, but indicates and counts overloads with a clever counter.

Dorrough Loudness Meter. It's extremely useful due to the dual-scales, but I quibble with calling it a "loudness monitor," since it does not correlate with loudness any better than a standard VU meter.

studio may contain only a few analog processors, so it is easy to put all the analog gear physically together in its own rack, at a distance from clock interference. Analog gear used for mastering can be customized for minimalist signal path, removing transformers and superfluous active stages, something which is not advisable in a large analog studio where ground loops are more difficult to chase down. I avoid analog patch bays, as they only deteriorate over time and their small contact area contributes to contact-resistance-distortion, preferring to use instead individual short interconnect cables.

Some mastering studios have constructed custom mastering consoles, which insert analog elements at will. My approach avoids a mastering console, since all the analog gear is patched manually, and the monitoring functions are absorbed by a custom-built analog monitor selector and level control. Every mastering engineer has his own variation on these themes. The digital equivalent of a mastering console is accomplished by a combination of the Z-Sys routing, the digital monitor selector, plus the TC System 6000, which has internal stereo and 5.1 processing including fold down, some internal mixing capability, analog-digital insert points and a remote control with a tiny acoustic footprint. Some mastering studios use digital mixing consoles for mastering. The DAW also contains some routing and can be used as part of the console concept.

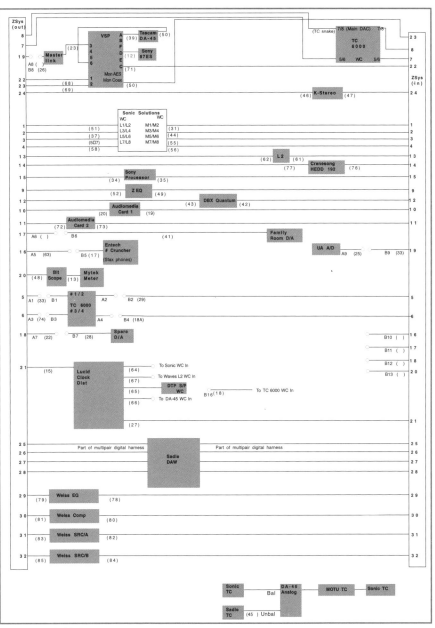

Mastering Studio block diagram with wire numbers in parentheses.

Tools that we're missing: Customized and special-purpose gear

One tool that I am missing is a more ergonomic method of routing. Instead of a crossword-puzzle routing matrix, I'd like to see specialized software to control routers that illustrates the audio chain the way we think, from source to output in a straight-forward linear fashion. A company called Crookwood has created modular control systems for this purpose.

---

1 Fold down is the ability to take a multichannel or stereo source and monitor a reduction to 2 channels or one (mono). We use this to help confirm compatibility of a 5.1 recording to stereo and/or stereo to mono.

2 See Appendix 10 for recommended reading.

CHAPTER 3

An Earientation Session

I. Introduction

Ear training is really mind training, because the appreciation of sound is a learned experience. Stereo imaging is an illusion that some people still don't get! The first listeners to Edison's acoustic phonograph felt that its reproduction was indistinguishable from real life. It is only with each advance in sound reproduction that most people become aware of the shortfalls of the previous technology. For example, whenever I work at a very high sample rate, and then return to the "standard" (44.1 kHz) version, the lower rate sounds much worse, although after a brief settling-in period, it doesn't sound that bad after all. [See Chapter 18]

As we become more sophisticated in our approach to listening, we develop a greater awareness of the subtleties of sonic and musical reproduction. We can also grow to like a particular sound, and each of us has slightly different preferences, which vary over the years. When I was much younger, I liked a little brighter sound, but from about the age of 20, I've tended to prefer a well-balanced sound and immediately recognize when any area of the spectrum is weak or over present. It's also important to recognize that a frequency emphasis that's too strong for one musical genre or song may be just right for another, as we explain in Chapter 8.

A mastering engineer requires the same ear training as a recording and mixing engineer, except that the mastering engineer becomes expert in the techniques for improving completed mixes, while the mixing engineer specializes in methods for improving the mix by altering the sound of individual instruments within it. As we move into the era of mastering from stems (sub mixes, or splits of a larger mix, e.g., vocals, bass, rhythm), there will be more overlap between mixing and mastering, since the mastering engineer will also then have some control over individual instruments or groups.

{ *"Make passive ear training a lifelong activity."* }

Ear training can either be a **passive** or a **hands-on** activity. Passive ear training goes on all the time ("what a tinny speaker in that P.A. system"), while active ear-training occurs while your hands are on the controls. **Make passive ear training a lifelong activity**—exercising your ear/brain connection regularly will increase your ability to discriminate fine sonic differences. Practice being consciously aware of the sounds around you and identifying their characteristics. Acousticians can't help judging the reverberation time of every hall they enter. Too much ear-training practice can ruin the enjoyment of a musical program or a good relationship, so rule number one is not to tell your spouse every time you notice the surrounds in the movie theatre are set too high or the left tweeter is blown! However, when the program material is sufficiently boring, work on ear-training. For me it's a curse that hits subconsciously at the strangest moments ("what a boxy-sounding reverb chamber they're using").

Hands-on ear training is the process of learning how to manipulate the controls of an audio system to arrive at the sound you have in your head; this is also known as developing *hand to ear coordination*. With practice, you can learn to get there quickly and efficiently. Before you work on a piece of music, try to visualize (audiolize?) the sound you are looking for; you should have a definite sonic goal in mind. I received a mix from a musician who is a fine jazz bass player. It was obvious to me that he had not listened to the mix over a variety of playback systems, for the bass sounded muddy, indistinct, and uneven, the last thing a bass player would want to hear, and the instrument was also much too loud. Fortunately the bass player agreed with me on all my judgments. I diagnosed this as a case of *small-speaker near-field-itis* and it wasn't long before I found the cure with equalization and dual-band upward expansion (explained in Chapter 11). Sometimes we don't know how we're going to solve a problem, but having a clear goal keeps us from fumbling.

Speaking the Language

The classic chart folded into the front cover was hand-drawn in 1941 by E.J. Quinby of room 801 within the depths of Carnegie Hall.* We've reproduced it for the benefit of musicians who want to know the *frequency language* of the engineer, and

---

* I've never visited that room, but it would be an interesting archeological voyage to find out who E.J. Quinby was.

for engineers who want to speak in a musical language. Sometimes we'll say to a client, "I'm boosting the frequencies around middle C," instead of "...around 250 Hz". Learn a few of the key equivalents, e.g., 262 Hz represents middle C, 440 is A above middle C, and then remember that an octave is a 2X or 1/2X relationship. For example, 220 Hz is the frequency of A below middle C in the equal-tempered scale. The ranges of the various musical instruments will also clue you to the characteristics of sound equalization—next time you boost at around 225 Hz, think of the low end of the English horn or viola.

Although it helps an engineer to have played an instrument and to be able to read music, many successful engineers can do neither. Nonetheless, they are not handicapped because they have good pitch perception, can count beats and understand the musical structure (verse, chorus, bridge...) very well.

This next chart is a graphic representation of the subjective terms we use to describe excesses or deficiencies of various frequency ranges.

Excess of energy is shown above the bar and a deficit below. The bar is also divided into eight approximate regions. There are no standard terms for these divisions: what some people call the **upper bass**, others call the **lower midrange**; some call the **upper midrange**

what others call **lower treble**. Notice that we have far more descriptive terms for areas that are boosted as opposed to those which are recessed. This is because the ear focuses much more on boosts or resonances than on dips or absences.[*]

A few subjective examples

With an equalizer, the sound can be made **warmer** in two ways: by boosting the range roughly between 200 and 600 Hz; or by dipping the range roughly between 3 and 7 kHz. These two ranges form a yin and yang, which we'll discuss in Chapter 8. Another way to make sound warmer (or its converse, edgier) is to add selective harmonics, as described in Chapter 16. Too much energy, and/or distortion, in the 4 to 7 kHz region can be judged as **edgy**, especially with high brass instruments. Equalizing in this region can exaggerate or de-emphasize the harmonic distortion of a preamplifier or converter. The term **presence** is associated with any sound that is strong and clear, which often means a strong upper midrange, but too much **presence** can be

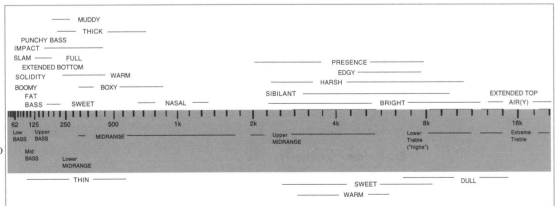

Subjective Terms we use to describe Excess or Deficiency of the various Frequency Ranges.

[*] Jim Johnston (in correspondence) points out that peaks change the partial loudness of a signal more than dips. It's all psychoacoustics!

fatiguing or harsh. If the sound is edgy, it can often be made **sweet(er)** by reducing energy in the 2.5 to 8 kHz range. Too much energy in the 300-800 range gives a **boxy** sound; go up another third octave and that excess is often termed **nasal**. A deficiency in the range from roughly 75 to 600 Hz creates a thin sound.

Ear Training Exercise #1:
Learn to Recognize the Frequency Ranges

Learning to recognize frequencies is an exercise in the perfection of pitch perception. To have *perfect pitch* means that you can identify each note blindfolded. At concerts it's a neat trick if you can identify the frequency of feedback before the mix engineer. But this ability is not just a trick: if you learn to identify the ranges by ear, this will greatly speed up your performance at the equalizer's controls. There was a time when I practiced until I could automatically identify each 1/3 octave range blindfolded, but now my absolute pitch perception is between 1/3 and 1/2 octave, which is about what you need to be a fast and efficient equalizer. Start ear training with pink noise and then move to music, boosting each range of a 1/3 octave graphic equalizer until you can recognize the approximate range. Get a friend to boost the EQ faders randomly and give a blindfold test. Don't be dismayed if you're only accurate to about an octave. This will get you close enough to the range of interest to be able to "focus" the equalizer the rest of the way.

Ear Training Exercise #2:
Learn the Effects of Bandwidth limiting

Less-expensive loudspeakers usually have a narrower bandwidth, as do lower-quality media and low sample rates (e.g. the 22.05 kHz SR audio files often used in computers). Train your ears to recognize when a program is naturally extended, and when it has been bandwidth-limited. It's surprising to discover how much high end filtering you can get away with, as can be heard when old films with optical sound tracks are shown on TV. Most musical information is safely tucked away in the midrange, the only frequencies that remain in an analog telephone connection. My career in television began when telephone landlines were still the primary means of network transmission, and I soon learned that a 5 kHz bandwidth takes away the life and clarity of the sound, even if all the informational content is there. Those were not pleasant days before satellite transmission and ISDN opened up network television sound to high fidelity. Practice learning to identify these effects using high and low pass filters on various musical examples. As for the bottom end, the human ear tends to supply the missing fundamentals. This can be observed when watching an old TV show that's been dubbed and filtered too many times; you may not notice the voice is very thin-sounding until it's been pointed out to you. Another way to study the contribution of the low bass range is to turn your subwoofers on and off, or listen to historic acoustic recordings.

Ear Training Exercise #3:
Learn to Identify Comb Filtering

About the only advantage of the English system of measurement is that the speed of sound is a nice round number, about 1000 feet per second, or even more approximately, one foot per millisecond. When a single sound source is picked up by two spaced microphones, and those microphones are

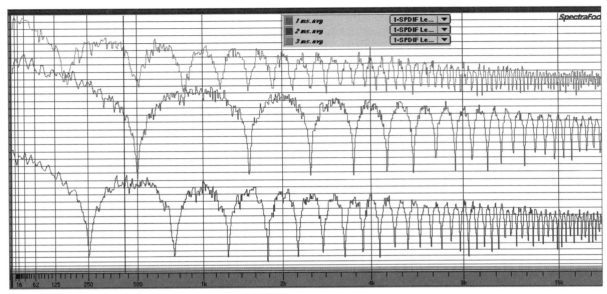

Severe Comb Filtering

combined into a single channel, audible comb filtering will result if

- the gain of each microphone is about the same and the microphones are identical or similar models. When one mike's gain is reduced at least 10 dB, the comb filtering becomes audibly insignificant.
- the relative mike distance from the source is in the critical area from about 1/2 foot (~150 mm) through about 5 feet (~1.5 M). At 5 feet, the attenuation of the more distant mike's signal also reduces the combing effect.

Comb-filtering can occur anytime a source and its delayed replica are mixed to a single channel. The above figure shows the frequency response resulting when the source and the delay are at equal gain. The vertical divisions are 3 dB. From top to bottom—a delay of 3 ms (approximately equivalent to a 3

feet/1M path difference), 1 ms, and 2 ms. In real life, the reflection (delay) will be diffused and somewhat attenuated, so the comb-filtering effect will be less severe.

It's amazing how many engineers think they can fix the reflections from a singer's music stand by adding a piece of carpet. But carpet has no meaningful effect in the range below about 5 kHz, and as you can see from the figure, that's where the major problems are. Another example of comb filtering is when the sound from an instrument reaches the microphone both directly and also via reflections from the floor. Nearfield monitoring is inherently inaccurate because the sound from the speakers reaches the ear directly and also via a bounce off the console top, yielding very uneven frequency response.

> *"Did you know that wearing a hat with a brim puts a notch in your hearing at around 2 kHz?"*

Television and film soundtracks provide excellent laboratory exercises in learning how comb-filtering can mutilate sound, since the proper operation of a lavalier microphone depends on indirect sound, including reflections from nearby surfaces. Listen to the weather report blindfolded and create a play-by-play based on your ear's perception of where the weatherperson must be: "Now she's crossed her hands on her chest, about 3" below the lavalier microphone. Now she's turned around to face the blue screen, about 2 feet away. Now she's uncrossed her hands and is walking away from the screen. She's sitting down at the anchor desk for the discussion and you can hear from the hollow dip at 500 Hz that her mike is about a foot above the desk. Uh-oh, the mix engineer has opened a second microphone and the anchorman's voice is leaking into her mike from a couple of feet away."[1] The ear really begins to notice comb filtering when the delay is changing, for example, the classic *flanging* effect when an artist sways to and fro in front of a reflecting music stand. That's why the best music stand is none at all; open-wire stands are second-best and careful placement does the rest.

What does comb-filtering have to do with audio mastering? The answer is that learning to identify its effects is an excellent *earientation* exercise. The figure shows that comb filtering is extremely difficult to remove with an equalizer. And a corrective equalizer would be especially problematic in mastering since the equalization affects the entire mix, not just the instrument that needs fixing. Ideally comb-filtering should be prevented before the mix gets to mastering by using acoustic know-how. Unfortunately, comb-filtering problems are more common than you'd believe. By the way, did you know that wearing a hat with a brim puts a notch in your hearing at around 2 kHz? Comb-filtering is all around us. To hear comb-filtering right now, talk into your cupped hands, then take them away while still talking. Learn to recognize the effect blindfolded. Or walk into an announce booth with your eyes closed, talk into the window and see how close you have to get to it before you notice the coloration.

Ear Training Exercise #4: The Sound of Great Recordings well-reproduced; Perception of Dynamics, Space and Depth

Many mastering engineers are privileged to work on a wide variety of music throughout the week; there's never a dull moment. Train your ears to recognize good recorded sound in each genre. Start by becoming familiar with the sound of great recordings made with purist mike techniques, little or no equalization or compression. Learn what wide dynamic range and clear transients sound like captured and reproduced, which will help you recognize limited dynamic range material when it is played. The percussive impact of real life is the standard that can never be bettered. It's an exhilarating, incomparable live experience to stand directly in front of a live big band. Next, compare the depth which can be captured with simple miking techniques and which is lost when multiple miking is used.

Ear Training Exercise #5: The Proximity Effect Game

Take the opportunity to experience and reference the sound of live, unamplified music. I'll never forget the wonderful artist who broke into song in my mastering room. There's no greater privilege than to receive a private, live unamplified concert given just for you by a world-class vocalist. Seek out those rare opportunities. Listen to your singer rehearsing without a microphone; check out the natural tonality, clarity and incredible dynamics of a voice that's singing and projecting.

Now compare that natural sound with engineers' use of proximity effect, which is the increase in bass response when a directional microphone is moved closer to the source. Most recorded pop vocals have greater lower midrange and presence than real life. The trick is to use just enough to make it sound "super-natural" but not muddy, thick, sibilant, bright or edgy.

Ear Training Exercise #6: The Sound of Overload

Many amplifiers have their own unique sound, probably attributable to subtle differences in harmonic structure. When solid-state amplifiers are driven into overload, they **clip**, the round part of their output waveform starts to square off. Clipping is a form of severe overload; some amps (particularly tube amps) overload gracefully, and can be used as a form of compressor, making sounds fatter when you push them past their linear region. Others clip drastically, producing lots of high, odd harmonic distortion. Learn to identify the sound of overload in all its forms: analog tape reaching saturation, analog tape in severe saturation, overdriven power amplifiers producing intermodulation distortion, optical film distortion (as in classic 1930's talkies), and so on. As a first training exercise, study the saturation on peaks of a classical or pop recording made from analog tape versus a modern all-digital recording. You may prefer one type of overload to the other. As a benefit of this ear-training, you will begin to learn the characteristics of each piece of gear you encounter; become a master of the gear instead of it mastering you. Soon you'll discover some rare digital gear that overloads more gently than others.

Ear Training Exercise #7: Identify the Sound Quality of Different Reverb Chambers

Artificial reverb chambers have progressed tremendously over the years. Become familiar with the artifacts of different models of reverbs. Some models exhibit extreme *flutter echo*, some sound very flat, while others produce an excellent simulation of depth. We'll learn a bit how they accomplish this in Chapter 17.

Non-Exercise: Recognize Bad Edits, Wow and Flutter, Polarity Problems

Bad Edits: I'm so paranoid I sometimes think I can hear edits at concerts! But seriously, an experienced mastering engineer should be able to recognize a bad edit in a tape, where the ambience or the sound is partially cut off, or the sound partially drops out. I don't have any specific exercises to recommend except to apprentice/ practice with an experienced editing engineer who will listen to your edits and point out their faults.

Wow and Flutter: Wow and flutter are caused by speed variations in recordings, and are no longer a problem with digital recording. But mastering engineers are often called upon to restore older analog recordings. So to enhance your perceptual acuity, make a cassette recording of a solo piano, and compare it side by side with a digital recording of the same instrument.

Polarity problems: Learn to recognize when the left channel of a recording is out of polarity with the right. Reverse the polarity of the wires to one loudspeaker and become familiar with the sound of the error, which is characterised by thin sound and a hole in the middle of the image. This will also help you to recognize when some instruments in a mix are out and others are in polarity.

In Summary

Earientation should be a lifelong activity and no one can become an expert in one fell swoop. These exercises will help get you up to speed.

---

1 There is a specialized television engineer's mixing technique to deal with mike leakage to avoid acoustic phase cancellation (comb filtering). Most women's voices require a bit more gain, so for this discussion we made the weatherperson a woman and the anchorperson a man. Ride the level of one mike only, drop it about 5 dB when the person is not talking; this should be the mike requiring the most gain (the quietest talker)—because her voice will hardly leak into the anchorman's mike, but his will leak into hers. Watch her lips closely so as not to up cut her words.

CHAPTER 4

Wordlengths and Dither

I. Introduction

This chapter is about (pick one):
a) the smallest, most subtle, insignificant problem in digital audio
b) the biggest, most important problem in digital audio

If you picked both a) and b), then you are correct. Audio engineers must learn how to deal with and take advantage of wordlengths and proper dithering, but we must also keep our problems in perspective. If everything else in a project is right, then proper dithering is very important. But if the mix isn't good, or the music isn't swinging, then dither probably doesn't matter very much. If we want to get **everything** right, and maintain the sound quality of the audio, we need to pay particular attention to the topics of this chapter.

II. Dither in the Analog Domain

In an analog system, the signal is *continuous*, but in a PCM digital system, the amplitude of the signal out of the digital system is limited to one of a set of fixed values or numbers. This process is called **quantization**. Each coded value is a discrete step. For example, there are exactly 65,536 discrete steps, or *values* available in 16-bit audio, and 16,777,216 discrete steps available in 24-bit audio. To calculate the approximate codable range of any PCM system, multiply the wordlength by 6; e.g. multiply 8 by 6 to get 48 db for an 8-bit system. So the lowest value that can be encoded in 16-bit is 96 dB down from the top; in 24-bit it's 144 dB. In a moment we will introduce the concept of dither, but if a signal is quantized without using dither, there will be

quantization distortion related to the original input signal. This can introduce harmonics, subharmonics, aliased harmonics, intermodulation, or any of a set of highly undesirable kinds of distortion. In order to prevent this, the signal is *dithered*, a process that mathematically removes the harmonics or other highly undesirable distortions entirely, and that replaces it with a constant, fixed noise level.

Here's a simple *thought experiment* that explains why dither is necessary and how it works.[*] Let's create a basic A/D converter. We'll make it sensitive to DC, and bipolar, so it responds to both positive and negative analog inputs, and we'll give it a very big LSB threshold of 1 volt to make the numbers easy. We'll construct our ADC so that an analog source over the range between -.5 volts and +.5 volts produces a digital output word of 0, and an analog source over the range between +.5 volts and 1.5 volts produces an output of 1, and so on. If, without applying any dither, we present a 0.25 volt DC (continuous) signal to the input of the ADC, the output of the ADC will be a string of zeros. In fact, any signal between -0.5 and 0.5 volt will result in an ADC output of zero. Any information below the LSB threshold is completely lost, as illustrated above.

Remove the 0.25 volt signal and apply dither to the input of the ADC in the form of a completely

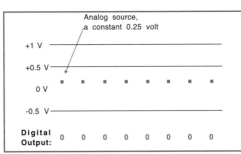

Graph of a hypothetical ADC whose LSB threshold is 1 volt (+ or − 0.5 volts). Each sampled analog input is represented by a small orange square; in this example, the analog source is held at a continuous 0.25 volt. Note that any input between -.5 volt and +.5 volt will be lost, because it is below the threshold of the LSB, producing a string of zeros. Because it is below threshold, a DC signal held continuously at 0.25 volts will not be detected.

random signal (i.e., noise) centered around 0 volts. Its peak amplitude randomly toggles the LSB of the ADC. The output of the ADC will be a stream of very small random values. However, the **average** of all these values will be zero.

Now let's apply our 0.25 volt signal again (with the dither on). The two analog voltages **sum together**, the dither and our signal. **At each sample**

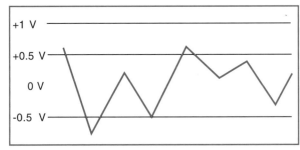

Random dither applied to the ADC whose highest peak-to-peak value is slightly greater than the LSB and whose average value is zero volts.

point (in time), the 0.25 value of our analog source is added to the random dither value. The output stream will again look like a stream of very small random numbers, but guess what? The AVERAGE of all those numbers will now be...you guessed it, 0.25. We have thus retained the information that was previously lost (even though it's buried in "noise"). In other words, our resolution has improved. The conversion is still essentially random, but the presence of the 0.25 volt signal *biases* the randomness. Put another way, the characterization of the system with dither on is transformed from completely deterministic to one

* Courtesy of Mithat Konar, director of engineering, biró technology. Also, many thanks to Jim Johnston for helping me to diagram this visually.

of statistical probability. The periodic alternation of the LSB between the states of 0 and 1 results in encoding a source value that is **smaller** than the LSB. In other words, on the average, the LSB puts out a few more ones than zeros because of our +0.25 volt signal. We say that dither *exercises* or *toggles* or *modulates* the LSB.[1]

With the dither on, we can now change the input signal over a continuous range and the average of the ADC output will track it perfectly. An input signal of 0.373476 volts will have an average ADC output of (the binary equivalent of) 0.373476. The same will hold true of inputs going over the LSB threshold: an input of 3.22278 will have an average ADC output of 3.22278. So not only has the dither enhanced the resolution of the system to many decimal places, but it has also eliminated "stepping," quantization effects!

Dither actually extends the resolution of a digital system, and in addition to being able to record and reproduce **all** the analog values at high and medium levels, dither lets us encode low level signals **below the -96 dB limit!**[2] These results — resolution enhancement and the elimination of quantization distortion — cannot be achieved by adding noise **after** the A/D conversion. So dither must be added at the proper point in the circuit and adding noise is not the same as dithering.

Dither's resolution enhancement is truly physical/mathematical in nature, not merely a trick which fools the ear. Dither is **not** simply a means "to mask the low level digital breakup." The psychoacoustic explanation is that it is because human beings are able to hear signals in the presence of noise of greater energy than the signal, i.e., with negative signal-to-noise ratios. In practice, we can hear signals about as far as 15 to 20 dB below the LSB, so a properly-dithered 16-bit recording can have a **perceived dynamic range** about as great as 115 dB. But its signal to noise (signal to dither) ratio will only measure about 91 dB, since the addition of the dither raises the noise floor about 5 dB.[3] Regardless, we can hear signals **below the noise**, which explains why the perceived dynamic range of the dithered system is greater than its *codability*.

Every well-made 16-bit A/D incorporates dither to linearize the signal. If you were lucky enough to have a 20-bit or 24-bit A/D and 24-bit storage to begin with, then dither is probably not necessary during the original analog encoding. Although the inherent thermal noise on their inputs is not shaped to perfectly dither the source, current 20-bit A/Ds self-dither **to some degree** around the 18-19 bit level because of this basic physical limitation. Similarly, a transfer from typical analog tape probably has enough hiss to self-dither any transfer to 16-bits, as long as there is no digital processing before storage. But I believe there is a slight advantage to encoding any transfer at 20 bits or above because the ear can hear signal below the noise; it certainly doesn't hurt to encode at 24 bits, except for taking up more storage space.

The dynamic range of an A/D converter at any frequency can be measured without an FFT analyzer. All that you need is an accurate test tone generator and a low-noise headphone amplifier with

MYTH:
Adding noise is the same as dithering.

sufficient gain. To conduct the test simply listen to the analog output and see when it disappears (use a real good D/A for this test). Another important test is to attenuate music in a workstation (about 40 dB) and listen to the output of the system with headphones. Listen for ambience and reverberation; a good system will still reveal ambience, even at that low level. Also listen to the character of the noise—it's a very educational experience.

III. The Need for (re)Dither in the Digital Domain

The First Secret of Digital Audio: How Wordlengths Expand

Even once the signal has been turned into numbers, under many circumstances we are still not exempt from the need for further dithering. Unfortunately, many processor and DAW manufacturers still have not recognized this fact,[4] and this partly explains why some digital devices sound pure and sweet, while others are cold and harsh. The reason: as soon as you transform audio by changing its level, equalizing, compressing, or nearly any other sort of calculation, **you have also increased its wordlength!** Which means that the sound quality of your music will be deteriorated if you simply truncate the output to 16 bits or any shorter wordlength. Let's see how that happens, and how we can prevent the problem.

Here's a simplified lesson in DSP (Digital Signal Processors). Digital audio is all arithmetic, but the accuracy of that arithmetic, and how the engineer (or the workstation) deal with the arithmetic product, can make all the difference between pure-sounding digital audio or digital sandpaper. All DSPs deal with digital audio on a sample by sample basis. At 44.1 kHz, there are 44,100 samples in a second (88,200 stereo samples). When changing gain, the DSP looks at the first sample, performs a multiplication, spits out a new number, and then moves on to the next sample. It's that simple.

To avoid unnecessarily complicated esoterica like 2's complement notation, fixed versus floating point, and other digital details, I'm going to invent the term *digital dollars*. Suppose that the value of your first digital audio sample is expressed in dollars instead of volts, for example, a dollar 51 cents—$1.51. And suppose you want to take it down (attenuate it) by 6 dB. If you do this wrong, you'll lose more than money, by the way. 6 dB is half the original value.[5] So, to attenuate our $1.51 sample, we divide it by 2.

Oops! $1.51 divided by 2 equals 75-1/2 cents, or $0.755. So, we've just gained an extra decimal place. What should we do with it? It turns out that dealing with extra places is what good digital audio is all about. If we just drop the extra five, we've theoretically only lost half a penny—but back in the audio world that 'half a penny' contains a great deal of the natural ambience, reverberation, decay, warmth, and stereo separation that was present in the original $1.51 sample! Lose the half penny, and there goes your sound. The dilemma of digital audio is that most calculations result in a longer wordlength than you started with. Getting more decimal places in our digital dollars is analogous to having more bits in our digital words. When a multiplication or division is performed, the wordlength can increase infinitely, depending on the precision we use in the calculation. A 1 dB gain boost involves multiplying by 1.122018454 (to 9 place accuracy). Multiply $1.51 by 1.122018454, and you get $1.694247866 (try it on your calculator). Each individual decimal place may seem insignificant, but DSPs require **repeated** precision calculations to perform filtering, equalization, and compression and the end number may not resemble the right product at all, *unless adequate precision is maintained*. Remember, the more precision, the cleaner your digital audio will sound in the end (up to a reasonable limit).

So this is the first critical secret of digital audio: *word lengths expand*. But if this concept is so simple, why is it ignored by too many manufacturers? The answer is simply cost. While DSPs are capable of performing double and triple precision arithmetic (all you have to do is store intermediate products in temporary storage registers), it slows them down, and complicates the whole process. It's a hard choice, entirely up to the DSP programmer/ processor designer, who has probably been put under the gun by management to fit more program features into less space, for less money. Questions of sound quality and quantization distortion can become moot compared to the selling price. In Chapter 16 we'll try to learn whether processors which measure better also sound better. It's a safe bet to say that high horsepower is both costly and better-sounding.

Inside a digital mixing console (or workstation), the mix bus must be much longer than 16 bits, because adding two (or more) 16-bit samples together and multiplying by a coefficient (the level of the master fader is one such coefficient) can result in a 32-bit (or larger) sample, with every little bit significant.[6] Since the AES/EBU standard can carry up to 24-bits, it is practical to take the internal long word, bring it down to 24 bits, then send the result to the outside world, which could be a 24-bit storage device (or another processor). The next processor in line may have an internal wordlength of 48 or more bits, but before output it too must reduce the precision back to 24 bits. The result is a slowly cumulating error in the least significant bit(s) from process to process. Fortunately, the least significant bit of a 24-bit word is 144 dB down, and most sane people recognize that degree of error to be inaudible,[7] but only as long as the processors reduce their respective long word lengths properly to 24 bits on the way out.

Something For Nothing?

But suppose we want to record the digital console's output to a CD Recorder, which only stores 16 bits. Frankly, it's a meaningful compromise to take a console's 24-bit output word and truncate it to 16 bits. Even if the source (multitrack) is 16-bit, there is an advantage to using the 24-bit output of the console or DAW. Similarly, there's only one right way to use a digital compressor or equalizer or reverb or other processor: Record its 24-bit output onto a 24-bit medium. And processors or consoles that purportedly produce a 16-bit output from a 16-bit input are throwing away bits! The same is true for those inexpensive programs built into computers which take in audio CDs and allow you to manipulate the sound and write a new CD. Critical listeners immediately realize you don't get something for nothing. Greater resolution and better audio quality can be achieved by mixing with an analog console to a 30 IPS, 1/2" analog tape than by passing the signal through a digital console that truncates its internal wordlength to 16 bits. If the console dithers its output to 16 bits instead of truncating (check with the manufacturer), the situation is a little better but even dithering has its compromises, too, as we shall see.

How Dither Works in the Digital Domain

Since truncation[8] is so bad, what about rounding? In our digital dollar example, we ended up with an extra 1/2 cent. In grammar school, they taught us to round the numbers up or down according to a rule (we learned "even numbers...round up, odd...round down"). But rounding produces little better results than truncation, perhaps adding half a bit additional precision, but with lots of correlated quantization distortion. So, when we're dealing with more numerical precision and small numbers that are significant, we still have to use dither noise to bring the information from the LSBs into the bits we intend to use.

The logic is the same as we described in the analog domain, except the processor must generate the dither digitally, as a series of random numbers, simulating the randomness of analog dither. This is often called *redithering*, because the signal may have been already dithered during the encoding (recording) process. But the advantage of the original dither becomes moot once we have reprocessed the audio, and we must dither all over again to preserve resolution before truncation. In the analog example, we learned that the encoded signal plus dither noise contains all the low level information below the LSB, because we **added** the analog dither to the low level analog signal. Similarly, in the digital domain, we can add two digital numbers together, one of which is a random number, representing random noise.

To do this, we calculate random numbers and add a different random number to every sample. Then, cut it off at 16 bits (or whatever shorter wordlength we desire). The random numbers must also be different for left and right samples, or else stereo separation will be compromised.

For example:

Starting with a 24-bit word (each bit is either a 1 or a 0 in binary notation):

The result of the addition of the Z's with the Y's gets carried over into the new least significant bit of the

```
                ---Upper 16 bits--- -Lower 8-
Original 24-bit   MXXX XXXX XXXX XXXW YYYY YYYY
Add random number                     ZZZZ ZZZZ
```

16-bit word (LSB, letter W above), and possibly higher bits if you have to carry. Just as in the analog example, the random number sequence combines with the original lower bit information, *modulating* the LSB. The result is that much of the sound quality of the long word is carried up into the shorter word.

Random numbers such as these translate to random noise (hiss) when converted to analog and this hiss is audible if listening carefully with headphones.

Some Tests for Linearity

Whether a digital audio workstation truncates digital words or does other nasty things, can be verified without any measurement instruments except your ears. Track 42 of *Best of Chesky Classics and Jazz and Audiophile Test Disc, Vol. III*[9] is a fade to noise without dither, demonstrating quantization distortion and loss of resolution. Track 43 is a fade to noise with white noise dither, and track 44 uses noise-shaped dither (to be explained). Using Track 43 as the test source; it is possible to hear smooth and distortion-free signal down to about -115 dB. Track 44 shows how much better it can sound. If we then process track 43 with digital equalization or level changes (both gain and attenuation, with and without dither) we can hear what they do to the sound. If the workstation is not up to par, the result can be quite shocking. Alternatively we can send the output of the test from the workstation to a CD recorder, load the CD back in, and raise the gain of the result 24 to 40 dB to help reveal the low level problems. The quantization distortion of the 40 dB boost will not mask the problems you are trying to hear, although it's theoretically better if dither can be added for the big boost.

So Little Noise—So Much Effect

-91 dB seems like so little noise. But strangely, astute listeners have been able to hear the effect of the dither noise, even at normal listening levels. Dither noise helps us recover ambience, but conversely it also obscures the same ambience we've been trying to recover! Dither at the 16 bit level adds a slight veil to the sound. That's why I say, *dither, you can't live with it, and you can't live without it.*

Improved Dithering Techniques

However, where there's a will, there's a way. Although the required amplitude of 16-bit dither is about -91 dB, it's possible to shape (equalize) the dither to minimize its audibility. Noise-shaping techniques re-equalize the spectrum of the dither while retaining its average power, effectively moving the noise away from the areas where the ear is most sensitive (circa 3 KHz), and into the high frequency region (10-22 KHz).

On the next page is a graph of the amplitude versus frequency of one of the most successful noise-shaping curves, POW-R dither, type 3.

This is clearly a very high-order filter, requiring considerable calculation, with several dips where human hearing is most sensitive. It is the inverse of the "F" weighting curve, which defines the low-level limit of human hearing. The sonic result is an incredibly silent background, even on a 16-bit CD. Chapter 16 studies these effects in more detail.

There are numerous noise-shaping redithering devices on the market. Very high precision (56 to 72 bit) arithmetic is required to calculate these random numbers with justice. One box uses the resources of an entire DSP chip just to calculate dither, with 72-bit precision arithmetic. The sonic results of these noise-shaping techniques range from very good to marvelous. The best techniques are virtually inaudible to the ear, all the dither noise has been

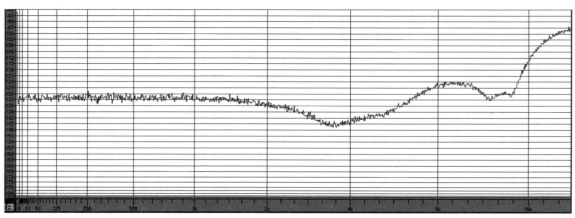

The noise-shaping curve of POW-R Dither, type 3

pushed into the high frequency region, which at -60 or -70 dB is still inaudible. Critical listeners were complaining that the high frequency rise of the early noise-shaping curves changed the tonality of the sound, adding a bit of brightness. But it turns out that psychoacoustically, it is the shape of the curve in the midband that affects the tonality, due to masking. A couple of the latest and best of these noise-shaping dithers are virtually tonally neutral, to my ears. It took a long time to get there (about 10 years of development), but I feel that the best of these processors yield 19-20 bit performance on a 16-bit CD, with virtually no tonal alteration or loss of ambience from the 24-bit source.

Noise-shapers on the market include: Lavry Engineering model 3000 Digital Optimizer, Meridian Model 618, Sony Super Bit Mapping (SBM), Waves L1 and L2 Ultramaximizers, Prism, POW-R, and several others.

Apogee Electronics produced the **UV-22** system, in response to complaints about the sound of earlier noise-shaping systems, and declaring that 16-bit performance is just fine. They do not use the word "dither" (because their noise is periodic, they prefer to call it a "signal"), but it smells like dither to me. Instead of noise-shaping, UV-22 adds a carefully calculated noise at around 22 KHz, without altering the noise in the midband.

Pacific Microsonics has produced the **HDCD (High Definition Compatible Disc)** system, which incorporates one of the best A/D converters with an *encode-decode* system. Special codes are buried in the 16th bit (LSB) along with standard dither; these codes inform HDCD-equipped D/A converters how to alter their gain structure so as to produce 20-bit or better quality, but only on the proper D/A converter. When an HDCD DAC is not used, the sound quality is reduced to that of a standard CD. However, if the mastering engineer manipulates some extra features of the HDCD system, known as *peak extension* and *low level*, then the music sounds compressed on a standard CD player and can only be properly reproduced (without compression) on an

HDCD player/DAC. Despite its name, HDCD, if manipulated aggressively, is not compatible with regular playback. The sound quality of the Pacific A/D is very nice; it's regretful that the license requires all CDs made from that converter to be HDCD-encoded; so we cannot legally choose to use another manufacturer's dither with the Pacific A/D.

We can effectively compare the sound and resolution of these redithering techniques, by performing a low level test with music. We simply feed low level 24-bit music (around -40 dB) into the processor, and listen to the output at high gain in a pair of headphones with a good quality 16-bit D/A converter, or a higher resolution D/A auditioned through a truncation device.[10] The sonic differences between the systems can be shocking: Some will be grainy, some noisy, and some distorted, indicating improper dithering or poor calculation. Though the winner of this test will probably be the best choice of dithering processor, also audition the music at normal monitor levels, because the psychoacoustic effect of the dither will be different and the high frequency noise less bothersome.

The Cost of Cumulative Dithering at 16 bits

As we have already seen, the measured amplitude of 16-bit dither is extremely low, approximately -91 dBFS. But a skilled listener does not have to listen at a very high level to hear the degradation of improper dithering. When feeding processors, DAWs or digital mixers to a shorter wordlength medium, dither should always be applied to the output of the processor because dithering always sounds better than truncation without dither.[11] But since dithering to 16 bits adds a slight veil to the sound[12]— cumulative dithering to 16 bit, multiple generations of 16-bit dither should be avoided: **redithering to 16-bit should be the one-time, final process in the project**. Mix to a long wordlength medium and send that file to the mastering house, which will apply 16-bit dither once, at the tail end of the project.

The Sound Effects of Defective Digital Processors

Since digital processors are computers programmed by human beings, we have to be sure to *Question Authority*, never taking a digital processor, or any DAW or computer that processes audio, for granted. For example, when software is changed or updated, we should never assume that the manufacturers have found all the bugs and we should assume that they may have created new ones. We even need to ensure that BYPASS mode, which seems seductively simple, actually does produce true clones in bypass. The illustration on the next page (courtesy of Jim Johnston) shows a series of FFT plots of a sine wave, illustrating the type of non-linear distortion products produced by truncation without dithering. The top row is an undithered 16 bit sinewave. Note the distortion products (vertical spikes at regular intervals, not harmonically related to the source wave). The second row is that sinewave with uniform dither. Note how the distortion products are now gone. The bottom row is the formerly dithered sinewave, going through a popular model of digital processor with a defective BYPASS switch, and truncated to 16 bits. This is what would happen if a (16-bit) CD was fed through this processor in so-called BYPASS mode, and dubbed to a CDR!

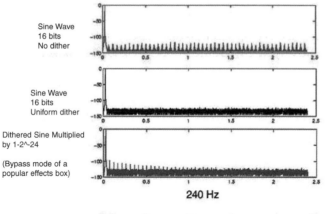

Sine Wave
16 bits
No dither

Sine Wave
16 bits
Uniform dither

Dithered Sine Multiplied
by 1-2^-24

(Bypass mode of a
popular effects box)

240 Hz

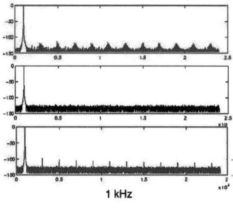

1 kHz

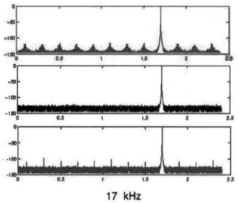

17 kHz

This is why every processor should be tested for bit transparency before attempting to make master-quality work with those processors patched into the signal chain.

IV. Some Practical Dithering Examples and Guidelines

1) When reducing wordlength you must add dither. Example: From a 24-bit processor to a 16-bit DAT.

2) Avoid dithering to 16 bits more than once on any project. Example: Use 24-bit intermediate storage, do not store intermediate products on 16-bit recorders.

3) Wordlength increases with almost any DSP calculation. Example: The outputs of digital consoles, DAWs and processors will be 24-bit even if you start with a 16-bit source.

4) Every "flavor" of dither and noise-shaping type sounds different. It is necessary to audition any "flavor" of dither to determine which is more appropriate for a given type of music. The most transparent-sounding dither may not be appropriate for "grungy" rock.

5) In any project, sample rate conversion should be the next-to-last operation, and dithering to the shortest wordlength must be last. Intermediate dithering may occur "behind the scenes," e.g. from 48 to 24 bits prior to feeding out of a processor. Truncation (without dithering) to 24 bits sounds far less bothersome to the ear[*][7] than truncating to 16 bits.

---

* Often, barely audible.

6) When bouncing tracks with a digital console to a digital multitrack, dither the mix bus to the wordlength of the multitrack. If the multitrack is 16-bit digital, that's a violation of #2 above, so try to avoid bounces unless the multitrack is 20-bit (or better). Example: You have four tracks of guitars on tracks 5 through 8, which you want to bounce in stereo to tracks 9 and 10. You have a 20-bit digital multitrack. You must dither the console outputs 9/10 to 20 bits. If you want to insert a processor directly patched to tracks 9 and 10, don't dither the console, just dither the processor to 20 bits.

One complication: The ADAT chips on certain console interface cards are limited to only 20 bits. Consult your console manufacturer. If the processor has a true 24-bit interface, but the console's is only 20 bits, then you need to dither the console feed to the processor to 20 bits and once again dither the processor output to 20 bits to feed the multitrack! The result will sound slightly warmer, wider, fuller.

V. Managing Wordlengths

Many engineers believe that expanding the wordlength of the existing samples in a workstation improves the sound. This is incorrect. The sound can never get more resolved than what was originally encoded. Regardless of the source sample's wordlength, the workstation will always calculate to its highest precision, effectively adding zeros to the tail of any shorter words to facilitate the calculation (the padded zeros do not change the original value). In other words, 16, 24 and 32-bit samples can coexist in a well-designed workstation,

and when calculations take place, all samples will be multiplied to the longer wordlength. Thus, there is even an advantage to bouncing a 16-bit session down to 24 bits, even though all the sources were 16 bit. The sonic difference may be subtle to significant depending on the quality of the sources. At the time of this writing, two workstations (Pro Tools and Digital Performer) do not allow using different source wordlengths in the same playlist, due to some kind of architecture limitation. This is a great inconvenience, time and space-waster, because all they do to convert the files is add padding zeros. Perhaps because of this inconvenience, neither of those workstations is commonly used by mastering engineers, who regularly mix wordlengths in the same session.

Auto-Dither

We often have to combine previously-mastered and dithered music with new material. If possible, we try to avoid cumulative dithering to 16-bit by passing the already-mastered source unmodified to the output medium. There are a couple of ways to accomplish this. The first is by using **auto-dither by source wordlength.** The Sonic Solutions workstations prior to HD had this useful facility built-in; in other words, if the source wordlength is equal to or shorter than the destination wordlength, then the dither generator shuts off automatically. At this time, I know of only one model of external dither processor that has this facility: the **Prism AD-2**. In the absence of the Prism, or if we prefer another type of dither, then we can route the already-mastered material to another DAW stream, direct to the output and bypassing the dither

generator. There are other kinds of **auto-dither**, including **auto-black** which turns off the dither if the source audio level goes below a certain threshold for a period of time, useful if the producer insists on total silence between pieces.

---

1   In practice, it's more than just the LSB which is exercised. It can be all the bits. In base 10, if we add two numbers, and the sum is greater than 9, we have to carry. In base 2, we also have to carry and if the next significant digit to the left is not a zero, we have to keep on carrying until the next digit up is a zero and turn it into a 1. In 2's complement, the addition of dither at the LSB level will affect the values of many digits, including the MSB, as the number changes polarity between negative and positive. You can see this on a bitscope, which seems to show two values at once because the numbers are always toggling with the addition of dither.

2   More exactly, below the coding floor of any particular wordlength. In other words, if we dither to 20 bits, whose coded range is 120 dB, we can encode low level signals below the -120 dB limit. Or if we dither to 8 bits, we can encode low level signals below 8-bit's normal limit of -48 dBFS.

3   The noise floor is raised 4.77 dB to be exact. This is the least amount of noise necessary to properly dither a digital audio signal and eliminate all possible distortion. The statistical distribution of the noise must be triangular probability. You can read about the math behind this in Lipshitz and Vanderkooy's papers as well as works by Bart Locanthi.

4   When I wrote an article about dithering around 1993, the situation was much worse. Today, only the most stubborn, ignorant, or simply cheap console manufacturers ignore the need for redithering in their products. And the more aware manufacturers have begun to dither the internal longword (e.g. 48 bits) up to 24 instead of truncating at the 24th bit, which produces an extremely subtle sonic improvement.

5   For signals which are correlated, the formula is dB change = 20 * log (ratio). For example, if we drop the level by a ratio of 1/2.... whose log is -.3010, then multiply by 20, the approximate result is -6 dB (6 dB down), to the nearest decibel. Note the use of the word approximate, and yes, the degree of accuracy used in such calculations affects the quality of our audio.

6   To be exact, the low level (ambience) information that was present in the

original wordlength is now spread proportionally over a much longer wordlength.

7   To put it another way, dither noise at −139 dBFS accumulates very slowly before it could become audible, or interfere with audible ambience. At this subtle a level, it's about the cumulative effect of multiple dithers (or lack of same) when processes are chained. I recommend that all wordlength reductions be dithered, even intermediate reductions from 48 to 24, for example, because as the material is further processed, previous distortions due to truncation start to be amplified and become audible as an edginess to the sound. This is why I insert a 24-bit dither generator into my SADiE workstation, when feeding external processors at 24 bits. Sonic Solutions workstations perform this chore automatically, transparent to the user. Z Systems Equalizers provide optional dither at the 24th bit, which should be engaged when processing. Weiss processors always dither when set to 24-bit output wordlength; it is not a user-settable option.

8   According to Jim Johnston, there are several forms of truncation, depending on the computer and the language in use, and none of them is good!

9   Chesky JD111, available at major record chains or through Chesky Records, Box 1268, Radio City Station, New York, NY 10101; 212-586-7799 (I produced this disc). The hard-to-find CBS CD-1, track 20, also contains a fade to noise test.

10 You may use a DAT machine on E-E (Electronics to Electronics) to truncate the signal, but be careful, some models of DAT machines actually pass 24 bits through on E-E!

11 Unless you are specifically looking for grunge, and a particular type of grunge at that. For the inharmonic distortion caused by quantization is very unmusical to the ear. Very different-sounding than turning a Marshall amplifier up to 11, for example. I'll take my grunge the old-fashioned analog way, if you please! In other words, if a particular type of music is designed be aggressive, *in your face*, it still sounds better to me if that aggression is obtained with a combination of high-resolution, pure sounding (analog-like) dither, and distortion-generating circuitry that produces musically-harmonic distortion. See Chapter 16 for more on this topic.

12 Since analog tape's noise floor is much higher than that of dither, many would argue that several generations of 16 bit dither circa -91 dB FS should be insignificant. I think it depends on the material. Pristine, digitally-recorded material can sound veiled when "over dithered." But some rock and roll sounds better with lots of noise, or with flat dither instead of noise-shaped dither. And the psychoacoustic argument goes on, which is why we have ears to make judgments!

CHAPTER 5

Decibels For Dummies

I. Introduction

This chapter summarizes the late 20th century approach to metering and leveling; it can be read as a preface to Chapter 15 in which we take these concepts into the 21st century. In the 20th century, because of their use of recording media with poor signal-to-noise ratios (SNR) engineers were often concerned with the signal peaks and with maintaining quality by maximizing the levels. With the advent of 24-bit recording, the SNR of our media is no longer an issue, but it is still crucially important for us to understand what the decibel scales on our meters are really telling us.

So many of us take our meters for granted—after all, recording is so simple: *all you do is peak to 0 dB and never go over!* But things only appear that simple until you discover one machine that says a recording peaks to -1 dB while another machine shows an OVER level, and yet your workstation tells you it just reaches 0 dB! We need to explore the concepts of the digital OVER, analog and digital headroom, machine meters, gain staging, loudness, signal-to-noise ratio and take a fresh look at the common practices of dubbing and level calibration.

II. Digital Meters and OVER Indicators

Recorder manufacturers pack a lot in a little box, often compromising on meter design to cut production costs. A few machines even have meters which are driven from analog circuitry—a definite source of inaccuracy. Even manufacturers who drive their meters digitally (by the values of the sample numbers) cut costs by putting large gaps on the meter scale (avoiding expensive illuminated segments). The result is that there may be a -3 point and a o dB point, with a large unhelpful no man's land in between. The manufacturer may feel they're doing you a favor by making the meter read o if the actual level is between -1 and o, but even if the meter has a segment at every decibel, when it comes to playback, the machine can't tell the difference between a level of o dBFS (*FS = Full Scale)* and an OVER. That's because once signal has been recorded, it cannot exceed full scale again, as illustrated below.

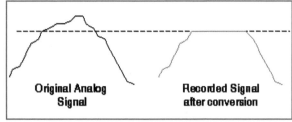

Original Analog Signal   **Recorded Signal after conversion**

While an original analog signal can exceed the amplitude of 0 dB, when that recording is reproduced, there will be no level above 0, yielding a distorted square wave. This diagram shows a positive-going signal, but the same is true on the negative-going end.

One way a signal can go OVER is during recording from an analog source. An early-warning indicator is a level sensor in an A/D converter, driven by the analog portion of the signal, which causes the OVER indicator to illuminate if the analog level is greater than the voltage equivalent to o dBFS. If the analog record level is not reduced, then a maximum level of o dB will be recorded for the duration of the overload, producing a distorted square wave.

---

* Contributed by Lynn Fuston.

MYTH:

*The red light came on while I was recording, but when I played it back, there weren't any overs, so I thought it was OK.**

After the signal has been recorded, distinguishing between a full scale recording and one that actually went OVER requires more meter intelligence than I've ever seen on a typical machine or DAW. I would question the machine's manufacturer if the OVER indicator lights on playback; it's probably a simple o dB detector rather than an OVER indicator. There are more sophisticated, calibrated digital peak meters such as those from Dorrough, DK, Mytek, NTT, Pinguin, RTW, Sony, and others, each with unique features (including custom decay times and meter scales), but all the good meters agree on one thing: the definition of the highest measured digital audio level. A true digital audio meter reads the numeric code of the digital audio, and converts that to an accurate reading.[1]

The Paradox of the Digital OVER

A well-designed digital audio meter can actually distinguish between o dBFS and an OVER. But if the digital levels on the medium cannot exceed o dB, how can the meter distinguish an OVER **after** the recording has been made? The answer is that a specialized digital meter determines an OVER by counting the number of samples in a row at o dB. The Sony 1630 OVER standard is three contiguous samples, because it's fair to assume that the analog audio level must have exceeded o dB somewhere between sample number one and three. Three samples is a conservative standard—most authorities consider distortion lasting only 33 **microseconds** (three samples at 44.1 kHz) to be inaudible. Depending on the nature of the music, distortion lasting as long as one or two **milliseconds** is likely inaudible. Thus, at higher sample rates, where many more samples go by in a short time, a case can be made to count many more contiguous full scale samples before warning the operator. Manufacturers of digital meters often provide a choice of setting the OVER threshold to 4, 5, or 6 contiguous samples, but it's better to err on the conservative side, to let the meter warn you before a problem could occur. If you stick with the 3-sample standard, you'll probably catch audible OVERs. But stand by, I'm about to recommend why you should mix at even lower peak levels!

Using External A/D Converters or Processors

There is no standard for communicating OVERs on an AES/EBU or S/PDIF line. So if you're using an external A/D converter, the recorder's OVER indicator will probably not function properly, if at all. Some external A/D converters do not have OVER indicators, so in this case, there's no substitute for an accurate external meter; without one I would advise not exceeding -1 dB. I've already received several overloaded tapes which were traced to an external A/D converter that wasn't equipped with an overload indicator.

When making a digital dub through a digital processor you'll find that most do not have accurate metering. Equalizer or filter sections can cause OVERs even when dipping levels! Contrary to popular belief, an OVER can be generated even if a filter is set for attenuation instead of boost, because filters can ring; they also can change the peak level as the frequency balance is skewed. Digital processors can also overload internally in a fashion undetectable by a digital meter. Internal stages may "wrap around" when they overload, without transferring OVERs to the output. In those cases, a digital meter is not a foolproof OVER detector, and there's no substitute for the ear, but a good digital meter will catch most other transgressions. When you hear or detect an overload from a digital processor, try using the processor's digital input attenuator, or simply attenuate its output if you are sure the processor has sufficient internal headroom, explained later in this chapter.

Oversampled Meters: Even More Sophisticated

Reading the simple numeric code from the digital stream may not be enough to detect OVERs in the converters that reproduce that signal. During the conversion from PCM digital to analog, built-in low-pass filtering causes occasional peaks **between the samples** that are higher than the digital stream's measured level, or even higher than full scale. **Digital designers have known for years that the actual output level of audio from a D/A converter can exceed 0 dBFS** but very few have taken this into account in the design. TC Electronic has performed tests on typical consumer D/A converters,[3] showing that many of them distort severely since their digital filters and analog output stages do not have the headroom to accommodate levels which exceed 0 dBFS! Besides D/As, certain processing elements of the signal chain can distort with intersample peaks, including sample rate converters and digital equalizers as we just explained. 0 dBFS+ peaks may reach as much as +3 dBFS with certain types of signals; what this means is that to make the cleanest recordings and to be perfectly safe, you should never exceed −3 dBFS on a simple (non-oversampling) digital meter! To demonstrate the problem and since this goes against typical *wisdom*, TC have developed an oversampling limiter and special oversampling peak meter in the System 6000.

Practice Safe Levels

Although there have been no psychoacoustic studies on their adversity, intersample 0 dBFS+ peaks cause some following processing circuits to linger and extend the distortion, which makes post-processing and broadcasting seriously problematic.[4] And some critical listeners report improvements when measured intersample OVERs are eliminated. It makes sense for production engineers to **practice safe levels during recording and mixing** by staying well away from 0 dBFS on a standard peak meter and leaving the decision on whether and how to raise levels to the mastering suite, where we make an educated decision. Mastering engineers, if maximizing levels, should at least use an over-counting meter, plus a digital limiter whose ceiling is set to −0.2 dB (see Chapter 10)[2] but preferably an oversampling limiter and oversampled meter (to prevent downstream problems with DACs and radio processing). Clipping of any type is to be avoided especially if a recording is to undergo further processing, as demonstrated in Appendix 1.[5]

MYTH OF THE MAGIC CLIP REMOVAL:

Turn it down after clipping and the clip will go away.

> "You would have to lower the peak level of a 24-bit recording by 48 dB to yield an effective 16-bit recording!"

The Myth of the Magic Clip Removal

If the level is turned down by as little as 0.1 dB, then a recording which may be full of OVERs will no longer measure any overs. But this does not get rid of the clipping or the distortion, it merely prevents it from triggering the meter. Some mastering engineers deliberately severely clip the signal, and then drop the level slightly, so that the meters will not show any OVERs. This practice, known as **SHRED**, produces very fatiguing (and potentially boringly similar) recordings.[6]

Peak Level Practice for Good 24-bit Recording

Even though 24-bit recording is now the norm, some engineers retain the habit of trying to hit the top of the meters, which is totally unnecessary as illustrated at left. Note that a 16-bit recording fits entirely in the bottom 91 dB of the 24-bit. You would have to lower the peak level of a 24-bit recording by 48

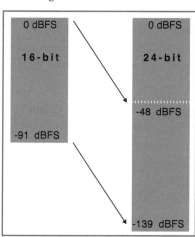

A 24-bit recording would have to be lowered in level by 48 dB in order to reduce it to the SNR of 16-bit. The noise floors shown are with flat dither.

dB to yield an effective 16-bit recording! So there is a lot of room at the bottom, and you won't lose any dynamic range if you peak to −3 dBFS or even as low as −10 dBFS; you'll end up with a cleaner recording. Distortion accumulates,[7] and at the mastering studio, a digital recording which is too hot can cause a digital EQ or sample rate converter to overload. A digital mix that peaks to −3 dBFS or lower makes it easier to equalize and otherwise process without needing an extra stage of attenuation in the mastering.

A number of 24-bit A/Ds advertise *additional headroom* by employing a built-in compressor at the top of the scale. As we have seen, there is no audible improvement in SNR by maximizing a 24-bit recording and no SNR advantage to compressing levels with a good 24-bit A/D.

How Loud is It?

Contrary to popular belief, the levels on a digital peak meter have (almost) nothing to do with loudness. For example, you're doing a direct to two-track recording (some engineers still work that way!) and you've found the perfect mix. Now, keep your hands off the faders, and let the musicians make a perfect take. During take one, the performance reached -4 dB on the meter; and in take two, it reached 0 dB for a brief moment during a snare drum hit. Does that mean that take two is louder? If you answered "both takes are about the same loudness," you're probably right, **because in general, the ear responds to average levels, not peak levels when judging loudness**. If you raise the master gain of take one by 4 dB so that it, too

reaches 0 dBFS peak, it will now sound 4 dB louder than take two, even though they both now measure **the same** on the peak meter.

Do not confuse the peak-reading meters on digital recorders with VU meters. Besides having a different scale, a VU meter has a much slower attack time than a digital peak meter. In Chapter 15 we will discuss loudness in more detail, but we can summarize now by saying that the VU meter responds more closely to the response of the ear. For loudness judgment, if all you have is a peak meter, use your ears. If you have a VU, use it as a guide, not an absolute, because it is still fairly inaccurate.

Did you know that an analog tape and digital recording of the same source sound very different in terms of loudness? Make an analog tape recording and a digital recording of the same music. Dub the analog recording to digital, peaking at the same peak level as the digital recording. The analog dub will sound about 6 dB louder than the all-digital recording, which is quite a difference! This is because the peak-to-average ratio of an analog recording can be as much as 12-14 dB, compared with as much as 20 dB for an uncompressed digital recording. Analog tape's built-in compressor is a means of getting recordings to sound louder (oops, did I just reveal a secret?).[8] That's why pop producers who record digitally may have to compress or limit to compete with the loudness of their analog counterparts.

MYTH:
*Normalization
Makes the Song
Levels Correct*

The Myths of *Normalization*

The Esthetic Myth: Digital audio editing programs have a feature called **Normalization**, a semi-automatic method of adjusting levels. The engineer selects all the segments (songs), and the computer grinds away, searching for the highest peak on the album. Then the computer adjusts the level of all the material until the highest peak reaches 0 dBFS. **If all the material is group-normalized at once**, this is not a serious esthetic problem, as long as all the songs have been raised or lowered by the same amount. But it is also possible to select each song and *normalize* it individually, which is part of the esthetic mythology—it's a real no-no. If you're making an album, never normalize individual songs, since the ear responds to average levels, and normalization measures peak levels, the result can totally distort musical values. A compressed ballad will end up louder than a rock piece! In short, **normalization should not be used to regulate song levels in an album**. There's no substitute for the human ear, and currently there is no artificial intelligence that does as well.*

The Technical Myth: It's also a myth that normalization improves sound quality of a recording; in fact, it can only degrade it. Technically speaking, normalization only adds one more degrading calculation and resulting quantization distortion. And since the material has already been mixed, it has already been quantized, which predetermines its signal to noise ratio—SNR of the recording cannot be further improved by raising it. Let me repeat: Raising the level of the material will not change its inherent signal to noise ratio but will only add more quantization distortion in an unnecessary step. **If the material is going to be mastered, do not normalize** since the mastering engineer will be performing further processing anyway.[9]

Judging Loudness the Right Way

Since the ear is the only judge of loudness, is there any objective way to determine how loud your CD will sound? The first key is to use a single D/A converter to reproduce all your digital sources and maintain a fixed setting on your monitor gain. That way you can compare your *CD in the making* against other CDs, in the digital domain. Judge DATs, CDs, workstations, and digital processors through this single converter.

III. Calibrating Studio Levels: Headroom and Cushion

Protecting your A/D and mix from clipping does no good if your analog console, preamplifiers or processors are distorting in front of the A/D! Since mastering engineers usually chain multiple pieces of gear, it's important to understand how to optimize analog levels, distortion and noise when making signal chains in front of your A/D converter. Ostensibly, typical balanced analog gear has a *nominal* level of +4 dBu (reference .775 volts[10], yielding 1.23 volts with sinewave. Unfortunately however, not all analog gear is created equal, and +4 dBu may be a bad choice of reference level. I use the term *nominal* to mean the *average* voltage level that corresponds with 0 VU, typically 20 dB below full scale digital (0 dBFS). We need to examine some

---

* When a client asks me if I *normalize* I reply that I never use the computer's automatic *normalization* method, but rather songs are leveled by ear. I avoid the term normalization because it has been misused.

easily overlooked factors when deciding on an in-house standard analog (voltage) level.

One factor is the clipping point of consoles and outboard gear. Before the advent of inexpensive 8-buss consoles, most professional consoles' clipping points were +24 dBu or higher. But a frequent compromise in low-priced console design is to use internal circuits that clip earlier, around +20 dBu (7.75 volts). This can be a big impediment to clean audio, especially when cascading amplifiers. To avoid the *solid-state edginess* that plagues a lot of modern equipment, the *minimum* clip level of every amplifier in a system should be 6 dB above the potential peak level of the music. The reason: Many opamps and other solid state circuits exhibit an extreme distortion increase long before they reach the actual clipping point, as they change from class A to class AB operation. This means clipping point should be at least +30 dBu (24.5 volts RMS) if 0 VU is +4 dBu!

You Can Never Have Enough Headroom!

A lot of solid-state designs start to sound pretty nasty when used near their clip point.[11] All other things being equal, the amplifier with the higher clipping point will sound better. Perhaps that's why tube equipment (with its 300 volt B+ supplies and headroom 30 dB or greater) often has a *good* name and solid state equipment with inadequate power supplies or headroom has a *bad* name. Most of the *robust-sounding* solid-state equipment I know uses very high power (but very expensive) supply rails.

Traditionally, the difference between average level and clip point has been called the *headroom*,

but in order to emphasize the need for even more than the traditional amount of headroom, I'll call the space between the peak level of the music and the amplifier clip point a *cushion*. With analog tape, a 0 VU reference of +4 dBu with a clipping point of +20 dBu provided reasonable amplifier headroom, because musical peak-to-average ratios were reduced to the compression point of the tape, which maxes out at around 14 dB over 0 VU. Instead of clipping, analog tape's gradual saturation curve produces 3rd and 2nd harmonics, much gentler on the ear than the higher order distortions of solid state amplifier clipping.

But it's a different story when the peak-to-average ratio of raw, unprocessed digital audio tracks can be 20 dB. Adding 20 dB to a reference of +4 dBu results in +24 dBu, which is beyond the clipping point of many so-called *professional* pieces of gear, and so doesn't leave any room at all for a *cushion*. If you adapt an active balanced output to an unbalanced input, the clipping point reduces by 6 dB, so the situation becomes proportionally worse.[12] Dual-output consoles that are designed to work at either professional or semi-pro levels can be particularly problematic. To meet price goals, manufacturers often compromise on headroom in professional mode, making the so-called semi-pro mode sound cleaner! It is an unpleasant surprise to discover that many consoles clip at +20 dBu, meaning they should not be using a professional reference level of +4 dBu (headroom of only 16 dB and no cushion). Even if the console clips at +30 dBu (the minimum clipping point I recommend), that only leaves a 6 dB cushion when reproducing

music with 20 dB peak-to-average ratio. That's why more and more high-end professional equipment have clipping points as high as +37 dBu (55 volts!). To obtain that specification, an amplifier must use very high output devices and high-voltage power supplies. Translation—better sound (all other things being equal), and also higher cost due to the need for more robust power supplies and devices.

These robust output drivers that have this kind of headroom sound better if they can deliver a clean high level into a 600 ohm load, which means they can probably handle long cable runs with their high capacitive loads. Long runs should probably be balanced, but since many mastering studios have small ground-loop areas, we often use custom-made unbalanced equipment, which often has simpler, quieter circuitry.

One of the most common mistakes made by digital equipment manufacturers is to assume that, if the digital signal *clips* at 0 dBFS, then it's OK to install a (cheap) analog output stage that would clip at a voltage equivalent to, say, 1 dB higher. This almost guarantees a nasty-sounding converter or recorder, because of the lack of cushion in its analog output section and the potential for 0 dBFS+ levels.

How can we increase the cushion in our system, short of replacing all our distribution amplifiers and consoles with new ones? One way to solve the problem is to recalibrate all the VU meters. SNR will not be significantly lost if we set 0 VU = 0 dBu or even -4 dBu (not an international standard, but a decent compromise if we don't want to throw out equipment), and things will sound cleaner in the studio. Once we've decided on a standard analog reference level, we calibrate all analog-driven VU meters to this level. At left is a diagram describing the concept of *cushion*.

IV. Gain Staging—Analog and Digital

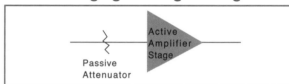

In the top device, signal enters a passive attenuator and exits through an active amplifier stage. This circuit effectively has infinite input headroom. The bottom device's input headroom is determined by the headroom of the input amplifier.

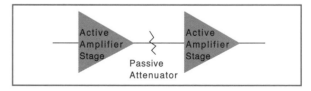

Analog Signal Chains

Now that we know how to choose an analog level, it's time to chain our equipment together. To really get a handle on our equipment, we should determine its internal structure. The above figures represent two possible internal structures. All structures are variations on this theme.

Clipping level of Analog Amplifiers= at least 26 dB over 0 VU

6 dB (minimum) cushion

Peak Program level= 20 dB over 0 VU= 0 dBFS

Actual Analog Headroom Needed (26dB or greater)

Traditional Headroom (20 dB)

Average level= 0 VU= -20 dBFS

Making a Cushion for Good Analog Sound with Digital Recording

To properly test analog devices and determine their internal makeup, use a good clean monitor system, an oscilloscope, a digital voltmeter and a sine wave generator that can deliver a clean +24 dBu or higher (a tough requirement in itself). The first type of device has a passive attenuator on its input, which means that we can feed it any reasonable source signal without fear of overload. We can prove this by turning the generator up and attenuator down; if the output never clips within a reasonable range of the generator, then the device must have a passive attenuator on its input. Then, we disconnect the generator and listen to the output of the device as we raise and lower the attenuator. There should be no change in noise or hiss, and the output noise should be well below −70 dBu unweighted, preferably below −90 dBu A-Weighted. This also is an indication that the device has a passive attenuator on its input. If the output noise changes significantly at intermediate positions of the attenuator, then the internal impedances of the circuit are in question, or there may be some DC offset. The output noise of this device will be limited by the noise floor of its output amplifier. We determine the best **nominal operating level** of this device by taking the output clip point and subtract at least 26 dB for headroom and cushion.

The second type of device's input is an active amplifier stage, whose design is much more critical. It is very rare to find a solid state device built this way which that won't clip with >+24 dBu input. While raising the signal generator, turn down the attenuator to keep the output from overloading. If we hear clipping prior to the generator reaching +24

dBu, then the device has a weak internal signal path. The clip point determines the nominal analog input level, which should be at least 26 dB below this clip point. Then, to check if the device's internal gain structure is well balanced, we see if the output stage clips at the same point as the input stage or at a higher level.

When cascading analog gear, the signal-to-noise ratio and headroom of the cascade is determined by the weakest link, but by studying the internal structure of each piece, it may be possible to increase SNR of the chain by running higher levels at points in the chain that have higher clipping levels. With test tone and then music, listen closely to the noise floor and high level sound quality at the last device in the chain; if the output of the chain sounds good and reasonably quiet, then I don't worry about tweaking the chain. I was able to improve the signal to noise ratio of a tube-based tape recorder whose gain structure resembles the second device. The original manufacturer's conservative schematic specified nominal internal levels of −10 dBu at the output of the second active stage. But since the tubes distort at well above +30 dBu (headroom of 40 dB), I decided to run the attenuator higher and run levels of 0 dBu in the second stage. This improved amplifier signal to noise ratio from the second stage on, by 10 dB, without endangering distortion. The tube tape recorder still has 30 dB of internal headroom.

In an analog signal chain, raising the music signal level as high as practical as early as possible (within the limits imposed by headroom and clipping point of A/D converters) will improve the signal to noise ratio of the entire chain. Then, later in the mastering, we will reduce the signal level digitally in the digital chain that follows.

Digital Signal Chains

Headroom of the Chain: It's a lot harder to grasp what's going on inside a digital signal chain, but we can test digital performance for headroom, clipping, and noise. Suppose we have a digital equalizer with several gain controls and equalization; we feed it a 1 kHz sine wave test tone at about −6 dBFS and turn up the 1 kHz equalization by 10 dB, observing that the output clips. Then we turn down the output gain control until the output is below 0 dBFS and verify by listening or FFT measurements that the internal clipping goes away. If not, then the internal gain structure of the equalizer does not have enough headroom to handle wide range inputs. We may be able to get away with turning down an input attenuator, but the early clipping indicates that this equalizer is not state-of-the-art. It is probably a first-generation fixed point unit and should be replaced. Modern-day digital processors have enough internal headroom to sustain considerable boost in early stages without needing an input attenuator, and clipping can be removed solely by turning down the output attenuator. The internal structure could be double-precision fixed point or floating point (see Glossary, Appendix 13); it's not easy to tell without asking the manufacturer. It is easy to be impressed by floating-point manufacturers' claims of hundreds of dB of headroom above 0 dBFS, but 24 dB or so internal headroom above 0 dBFS is probably enough; most well-designed fixed-point products have 24 or more dB internal headroom.

Distortion of the Chain and Individual Processor Levels: With a digital chain, we no longer have to consider the audio signal level between the various items of equipment; raising the source signal in a 24-bit digital signal chain does not make a meaningful SNR difference, considering the inaudible (approximately −139 dBFS) noise of the chain.[*] No longer should we get hung up on having a low signal level; instead, consider every calculation as a source of quantization distortion. Instead of *optimizing levels*, what matters most in a 24-bit digital chain is to reduce the number of total calculations; give the job of gain changes and other calculations to the components with the highest internal resolution (e.g., those which would introduce the least quantization distortion or *grunge*). In fact, we should avoid raising the signal until it reaches a device with the cleanest-sounding gain control, even if the source audio level is very low. For example, if the workstation has lower resolution, we try to hold everything at unity gain in the DAW and reserve the gain changes or EQ for higher-precision devices later in the signal chain. In other words, pass a perfect clone (bit-transparent copy) of the source from the DAW onto the next device in line to do processing.

Noise of the Chain: The only significant noise floors in a 24-bit chain are not from the chain itself but from the original sources, including mike preamp noise. We are primarily concerned with the

---

[*] Each processor does add its own quiescent or idle noise, which is cumulative, but in a good chain rarely adds more than 3 to 6 dB to the −139 dBFS RMS noise floor.

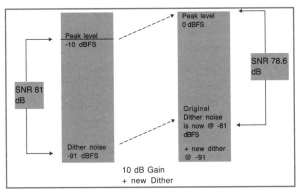

A 16-bit recording with peak level low at −10 dBFS. When gain is raised 10 dB and redither is added, the original 81 dB signal to noise ratio is reduced by about 2.4 dB.

impact of the summing of the higher level noises, and summing a new 16-bit dither with the source's dither noise can add a veil if the original was 16-bit.

Let's take an example of a 16-bit recording whose peak level is 10 dB low, as in the above figure. In mastering we may choose to raise its level by 10 dB and add 16-bit dither before turning it into a 16-bit CDR. This 16-bit recording's original 81 dB SNR is the difference between signal at −10 dBFS and dither noise at −91 dBFS.[13] When we raise the signal by 10 dB, both the original signal and the noise are raised equally, so the original signal to noise ratio is almost unchanged. However, the total SNR is the sum of the original dither which is now at −81 dBFS and the new dither which is at −91 dBFS. We ignore the insignificant noise of the gain processing, well below −130 dBFS, so the total is −78.6 dBFS, and the SNR of the source has been deteriorated by (81−78.6) or 2.4 dB. The more gain we apply to the source, the more distant the old noise will be above the added dither noise, and the smaller the new

dither will seem when the two noises are summed. So, reconsider doing anything if you have to raise a signal by only a few dB, because the new dither will be very close to the old; if we perform no gain change and just add dither, the noise floor is raised by 3 dB. If we lower the gain, the new dither predominates over the old. Despite this degradation, many times we have to live with compromises in mastering, since we still receive 16-bit sources; and we are forced to adjust the level according to the esthetics of the album. I've had considerable luck reducing cumulative sonic veiling by using noise-shaped dither.[14]

The manufacturers of the Waves L2 claim that peak limiting allows raising level enough to be significantly above the dither noise, and thus increases the signal-to-dither ratio and resolution. But exercise caution, because to my ears the apparent noise improvement is more than offset by the degradation of sound quality (the limiter reduces transient clarity).

If we could avoid 16-bit dither, by producing an output at 24-bit that the consumer could use, then mastering processing and gain-changing can be performed with no significant penalty, with noise floor 48 dB below the noise of 16-bit. This is the promise of delivering higher wordlengths to the consumer and another reason to record in 24-bit in the first place.

V. Analog to Digital Dubbing and Transfers

Dubbing and Copying—Translating between analog and digital points in the system

Let's discuss the interfacing of analog devices equipped with VU meters and digital devices equipped with digital (peak) meters. When you calibrate a system with sine wave tone, what translation level should you use? There are several de facto standards. Common choices have been -20 dBFS, -18 dBFS, and -14 dBFS translating to 0 VU. That's why some DAT machines have marks at -18 dB or -14 dB. I'd like to see accurate calibration marks on digital recorders at -12, -14, -18, and -20 dB, which covers most bases. Most of the external digital meters provide means to accurately calibrate at any of these levels.

How do you decide which standard to use? Is it possible to have only one standard? What are the compromises of each? To make an educated decision, ask yourself: What is my system philosophy? Am I interested in maintaining headroom and avoiding peak clipping or do I want the highest possible signal-to-noise ratio at all times? Am I interested in consistent loudness? Do I need to simplify dubbing practices or am I willing to require constant supervision during dubbing (operator checks levels before each dub, finds the peaks, and so on)? Am I adjusting levels or processing dynamics—mastering for loudness and consistency with only secondary regard for the peak level?

Consider that pure, unprocessed digital sources, particularly uncompressed individual tracks on a multitrack, will have peak levels 18 to 20 dB above 0 VU. Whereas typical mixdowns will have peak-to-average ratios of 14 to 18 dB (rarely up to 20). Analog tapes will have peak levels up to 14 dB, almost never greater. And that's how the three most common choices of translation numbers (18, 20, and 14) were derived. That's also why each manufacturer's DAT recorder has a different analog output level, which makes it a pain to interface in a fixed installation.

Broadcast Studios

In Broadcast, *speed and practicality* is our object, simplifying day-to-day operation, especially if the consoles are equipped with VU meters and recorders are digital. In broadcast studios, it is desirable to use fixed, calibrated input and output gains on all equipment. My personal recommendation for the vast majority of broadcast studios is to standardize on reference levels of -20 dBFS ~0 VU, particularly when mixing to 2-track digital from live sources or tracking live to multitrack digital. With a −20 dBFS reference, you will probably never clip a digital tape if you watch the VU. If the sources are compressed, the peak level may never reach full scale, but the SNR losses are insignificant with 24–bit recording. Use the top of the peak scale for headroom.

When dubbing from analog tape to digital, consider the analog tape to be a compressed source, and retain the VU reference at -20 dBFS, even if the digital never peaks above -6 dBFS. This will result in more consistent levels throughout the plant. When dubbing from digital to analog, optionally consider a −14 reference to avoid saturating the analog tape, or use a high headroom analog tape at high speed, or simply accept the 6 dB or so analog tape compression that we've been enjoying for years. For the major A/D/A converters in the complex, European broadcasters have settled on a -18 reference, since most of the material will have 18 dB or lower peak-to average ratio, and occasional clipping may be tolerated. I prefer the 20 dB choice to reduce clipping.

Recording Studios

For a busy recording studio that does most of its mixing, recording and dubbing to digital tape, standardizing on -20 dBFS will simplify the process and avoid clipping when watching VUs. When making dubs to analog tape for archival purposes, choose a tape with more headroom, or use a custom reference point (e.g. -14 instead of −20), as the goal is to preserve transients on the analog tape for the enjoyment of future listeners. For archival purposes, I prefer to use the headroom of the new high-output tapes for transient clarity, rather than to jack up the flux level for a better signal-to-hiss ratio.

One of the biggest problems in the contemporary recording studio is dealing with playback of CDs and the VU meter on the console, because many contemporary CDs have loudness levels that would damage a mechanical VU meter by *pinning* it, no matter what standard level you decide to calibrate the meter. Some recording studios solve this problem by switching the bus meter off when playing back commercial CDs, or by adding in a

variable meter attenuator, which I think is dangerous because they may forget to return the attenuator to normal. The K-System Meter (See Chapter 15) is the 21st century approach to the problem.

Mastering Studios

Mastering studios are working more frequently in 20-bit or 24-bit. And we can engage in a custom dubbing level for each analog tape, optimizing the level of the transfer according to sound quality, so fixed reference levels or calibration points for transfer are less important to us.

Analog PPMs

Analog PPMs have a slower attack time than digital PPMs, 6 to 10 ms instead of 1 sample (22 µS at 44.1 kHz). When working with a digital recorder, a live source, and desk equipped with analog PPM, I suggest a 5 dB "lead." In other words, align the highest peak level on the analog PPM to -5 dBFS (true peak) with sine wave tone.

In Conclusion

With this firm decibel foundation, we're now ready to begin discussing our mastering tools and techniques.

---

1 Ironically, there's still a tiny disagreement as to **which numeric code to read**, depending on the wordlength involved. Fortunately, a gentleman's agreement has been to use only the top 16 bits to determine level. Full scale 16 bits (positive going, 2's complement) is represented by the number 0111 1111 1111 1111. However, this number is infinitesimally smaller than full scale (positive) 24 bits, 0111 1111 1111 1111 1111 1111. To be exact, the difference is an error of (only) 0.0001 dB, and most people have agreed to ignore the discrepancy!

2 The manufacturers of the Benchmark A/D converter believe that counting contiguous samples is not a good idea, and they apply an even more conservative standard of any sample hitting 0 dBFS being considered an OVER, since an over-counting meter will never detect multiple contiguous high frequency signals at 0 dBFS because they're faster than the sample rate. I retort with the

psychoacoustic argument that: a) high frequency signals (e.g. 10 kHz) at full scale do not occur in real music and b) the ear is far less sensitive to short-duration high-frequency overloads. But still, there's nothing wrong with being conservative, especially during initial A/D conversion and especially with 24-bit recording!

3   Nielsen, Soren & Lund, Thomas (2000) 0 dBFS+ Levels in Digital Mastering. *AES 109th Convention, Preprint #5251.*

4   Jim Johnston (in correspondence) points out that processors such as MPEG coders (MP3), Dolby Digital encoders (AC3), WMA, Real, etc. **will add noise to your signal.** If you get too close to the edge, they will distort badly unless the input level is first reduced. The moral of the story is **do not get too close to digital max!** JJ recommends a maximum peak level at or lower than −0.2 dBFS for the benefit of post-processing.

5   Thomas Lund of TC has investigated a number of modern-day pop albums with the oversampled peak meter. He observes that most CD players are still in a distorted mode 200-700 ms after being hit by such peaks, as are radio processors because of SRC on their inputs, phase rotators, and other generally applied tricks.

6   Glenn Meadows and others discuss **shred**, on the Mastering Webboard:

Glenn: "Here's where I think all this is coming from, and it's kids oriented. Ever pull up to a stop light, and get blasted from the car next to you? (I assume the answer is yes). Well, besides being aggravated, actually listen to what's going on. ALL of the audio is clipped and distorted on the high end. THAT's what people THINK things sound like, and are SUPPOSED to sound like.

So, for the artists and producers, who are used to "cranking it up in their cars," and having the top and transients clipped/distorted, if they DON'T hear that in their offices, then the mastering is just plain wrong. So, it's once again filtering back to the mix engineers, to provide that hash in the mix to satisfy their clients (remember, we ALL have to satisfy our clients first and foremost), so instead of losing the gig to someone else who WILL provide that edge, everyone is doing the same thing.

[Unknown respondent:] In other words, you are stating that the music business is currently conducted by people who don't know what a record should sound like.

Glenn: "You got it. Clean is OUT, distorted is in. If it's clean, it's not right. Unfortunately, I've had too many sessions go that way in the past few months."

Chris Johnson: "There's no future in that... clipping causes ear fatigue. Ear fatigue means listeners listen less before ceasing the listening. These people are only committing commercial suicide by going for stuff with no longterm sales capacity. It's just the same as if you put everything through an Aural Exciter turned up so far it really HURT, only this time around it's distortion."

7   You don't always get the best Telco engineers on broadcast remotes. During a TV outside broadcast, I once complained to Telco, and he replied, "The distortion is leaving here ok!" Another time, during level testing, Telco asked me to "send me another one of those cycles."

8   As much of the "compression" of analog tape comes from the generation of additional harmonics as from the level saturation effect. A harmonic generator will reduce the peak to average ratio of a recording.

9   If perchance you decide to do a remix, and your previous mix revision was mixed at a low level, then by all means remix at a higher level. This is a good thing. Since the mixing process is a necessary (re)quantization step, this sort of "normalization" will raise the signal to noise ratio of the material, especially if you are mixing via analog console. With an analog mix, raising the level of the mix increases SNR by raising the level of the mix signal above the noise floor of the mixdown analog electronics and A/D. If you are mixing digitally, raising the

signal level increases the signal above the quantization distortion of the digital mixing DSP. But since the quantization distortion in a state-of-the-art DSP mixer will be around −139 dBFS, don't worry about raising the mix level unless it is significantly low (let's say, -10 dBFS to be conservative), for there will be no audible SNR improvement.

10   The origin of using **+4 dBu** as a reference for analog audio instead of a more convenient number like 0 goes back to the earliest days of the telephone company. The decibel is a relative measurement, but the reference used by the telephone company was based on **power.** And the telephone company's standard reference for 0 dB is one milliwatt, which across their standard impedance of 600 ohms yields 0.775 volts. This reference is commonly abbreviated as **0 dBm.** The VU meter then came along; it is calibrated to produce a level of 0 VU with 0 dBm, but if put across the 600 ohm line directly it would load it down and cause distortion, so the standard circuit included a 3600 ohm resistor in series with the VU meter. The 3600 ohm resistor attenuates the meter by 4 dB, so the circuit level has to be raised to +4 **dBm** in order to make the meter read 0 VU.

Nowadays, modern-day equipment generally has low impedance outputs (sometimes as low as 10 ohms or less), and high impedance inputs (greater than 10 k ohms), so there is no meaningful power transferred from gear to gear. Instead, a voltage reference is the only thing that is meaningful. And to keep using the same decibel levels we used for telephony, we kept the historical reference of 0.775 volts instead of a more convenient number like 1 volt! Now when the dB is referred to a voltage of 0.775 volts, we call that **0 dBu.** And to make a VU meter read 0 in a modern low impedance circuit with the right resistors, we have to feed in +4 dBu, or 1.23 volts. Also see Appendix 5, which is a short table of decibels.

The equations are:
If 0 dBu is 0.775 volts, then +4 dBu is 1.23 volts. 20 * log (1.23/.775) = 4.
I thank Mike Collins for reminding me to include this explanation.

11   This is of course dependent on the skill of the designer. Some IC operational amplifiers change from class A to class AB as they approach their clipping point, which can explain the sonic "nasties." However, many Mosfet power amplifier designs clip gracefully. Similarly, power supply design and regulation has a lot to say about sound quality near the clipping point. To avoid those nasties, measure and listen to be safe.

12   To be more exact, headroom is reduced 6 dB if you unbalance a transformerless amplifier's output. Transformer-coupled amplifiers retain their headroom even if unbalanced.

13   Simplifying the arithmetic, we assume the peak level is at −10 dBFS RMS and the dither noise is wideband and also RMS-measured at −91 dBFS (rounded from 96-4.77=-91.2). Anyway, chances are the music and room noise on the DAT are much higher than this dither noise, but the dither noise is the absolute minimum noise floor to consider. And many mastering engineers claim we can hear the degradation of dithering, even at as low a noise floor as −91 dB and even under music levels which are much higher!

14   You may ask: Other than the esthetic job of matching one song to another, why are we bothering to raise the level of the recording if the SNR of the source is worsened by the added dither? We also have to consider the noise floor of the final output electronics and D/A converter, and it is possible that by peaking closer to full scale we may overcome some of the weaknesses of the reproduction system's noisy analog outputs. It's a matter of finding the right balance and compromise amongst these several factors.

CHAPTER 6
Monitoring

I. Philosophy of Accurate Monitoring

The major goal of a professional mastering studio is to make subjective judgments as objectively as possible. You cannot afford to make mistakes when a record is released to thousands of listeners. Many of my clients are surprised to learn that a well-mastered CD can sound warm and clear on a wide range of systems, from low-end to high-end. How can this be done without compromising the integrity of the sound? Perhaps surprisingly, the answer lies less in using the right processing and EQ techniques (though these are the key), and more in the intelligent use of an accurate, high resolution monitoring system.

Elements of a High-Resolution Monitor System

A high-resolution monitor system is the mastering engineer's audio microscope, without which subtle processing decisions cannot even begin to be made. The monitor system permits hearing inner details in the music that otherwise might be missed, and might then cause problems for the end listener.

> *"The mastering engineer's monitor system is an audio microscope"*

The recipe for constructing a high-resolution monitor system probably hasn't been written, but we can describe some of the general elements:

1. With few exceptions, near-field monitors will not be found in a professional mastering room.[1] There are no little speakers, no representative cheap

speakers, no *alternative monitors*. Instead, there is a single pair of high quality loudspeakers (for stereo work), with which the mastering engineer is intimately familiar. He knows exactly how their performance will translate to the real world, and please the maximum number of listeners.

2. The mastering room is extremely quiet, with all noise-producing equipment banished to the machine room. Noise floor must be better than NC 30,[2] preferably NC 20 or less in the exceptional facility.

3. There are no significant obstacles between the monitors and the listener within the standard equilateral monitoring triangle.

4. The electronic chain is designed for maximum transparency. Often specialized or customized components are built which incorporate a bare minimum of active stages.

5. Monitor loudspeakers and amplifiers have wide bandwidth, high-headroom, and extremely flat frequency response. Sources of diffraction[3] are minimized. Cabinets are solid and non-resonant, as is the room, free of sympathetic vibrations and resonances.

6. Monitors and listener are in a reflection-free zone,[4] which means that reflections from nearby surfaces arrive at the listener at least 20 ms later than the direct sound (preferably >30 ms) and at least 15 dB down (preferably >20). This specification can be determined by time-delay spectrometry.[5]

The room is large enough to permit even, extended bass response, with no significant standing waves. Any remaining standing waves are controlled using techniques including Helmholz resonators or specialized diffusers. Room length should be at least 20 feet long for stereo, and in a critical mastering room, at least 30 feet long for multichannel, so that all speakers can be far enough from the walls to avoid the bass-resonance proximity effect.[6] The room should be wide enough so that first reflections from the side walls are insignificant, and/or the side walls are treated to minimize reflections. Dimensions should be symmetrical from left to right and a ceiling sloping upwards from the speaker end (cathedral ceiling) is a plus.

Acoustical design and electrical layout are accomplished by experienced and trained professionals.

Subwoofers and bass response

Stereo subwoofers, or prime loudspeakers whose response extends to the infrasonic, are essential for a good mastering studio. Vocal P pops, subway rumble, microphone vibrations, and other distortions will be missed without subwoofers, not just the lowest notes of the bass. Proper subwoofer setup requires knowledge and specialized test equipment (see Chapter 14). If subwoofers are inaccurately adjusted (e.g., "too hot," in a vain attempt to impress the client) then the results won't translate well to other systems.

Accurate subs are especially important in the hip-hop and reggae genres, but serve well to put rock and roll in perspective. By having accurately-calibrated subwoofers, we master a record that plays well on both boomy and thin systems.

Apparent bass response is also greatly affected by monitor level. The equal loudness contours (originally studied by Fletcher, Harvey and Munson) dictate that a recording which is mixed at too high a monitor level will seem bass-shy when auditioned at a lower level in a typical home environment. Thus, mixing and mastering at too loud a level is a conceit which we can ill-afford (see Chapter 15).

Monitor Equalization—by ear or by machine?

An inaccurate or unrefined monitor system not only causes incorrect equalization, it can also result in too much equalization. We must use our ear/brain in conjunction with test instruments to ensure monitor accuracy. Test equipment alone is not sufficient — for example, although some degree of measured high-frequency rolloff usually sounds best (due to losses in the air) there is no objective measurement that says, "this rolloff measures right," only an approximation. Different size rooms, monitor distances and monitor dispersions change the rolloff required to make the high end sound right.

Thus, for the high frequencies, the ultimate monitor tweak must be done by ear. But this leads to the chicken and egg problem: "If you use recordings to judge monitors, how do you know that the recording was done right?" The answer is to use the finest reference recordings (at least 25 to 50) to judge the monitors, and take an average. The highs will vary from a touch dull to a touch bright, but the majority will be right on if the monitor system is accurate. I try to avoid adding monitor correction equalizers; I prefer first to fix the room or replace the loudspeakers; my techniques include tweaks on

speaker crossover components until the monitors fall precisely in the middle of the "acceptance curve" of all 50 reference recordings.

Note however that a variety of factors - the number of people in the room, interconnect cable capacitance, power amplifiers, D/A converters, and preamplifiers — can all affect low and high frequency response, so if there are any changes to these, I immediately reevaluate the monitors' response with the known 25-best recordings!

Why Accurate Monitors Are Needed

Here is my *bell-curve theory*: Work to the middle of the curve, and you'll satisfy the maximum number of listeners. The mastering engineer strives to create a recording which will play well on the maximum number of reproduction systems. If you skew a recording in the bright direction, it will not play well on a lot of small systems that already have too much treble; conversely, if you skew it in the duller or heavier direction, with too much bass, it will not play well on systems that have too much bass. Thus, a recording which is well-balanced will satisfy the maximum number of listeners, as illustrated with the bell curve in this figure:

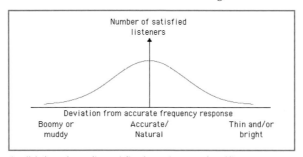

A well-balanced recording satisfies the maximum number of listeners.

The closer we can make the recording reach the middle of the curve, the more listeners we will satisfy. An accurate monitor system allows us to produce recordings which are in the middle of the curve. We pride ourselves on knowing just how much bass is going to be right, so that the recording will play well in a club, or in a small home system. This doesn't mean that we're home-free as soon as we construct the room with perfect response. For example, there will always be home and car systems that distort when certain bass frequencies are excessive. In this case, experience is the best teacher; there are bass changes we can make that will not skew a recording away from the middle of the bell curve. We always check references on various example systems. Usually the recording translates to all of them. Or if not, a small tweak will fix the problem, at the frequency that we've identified causes problems on the problem system. We engineer the change while listening on the accurate mastering system to confirm we are not skewing the recording away from the middle, or listeners with other systems will likely have the opposite problem. We also keep in mind that the ear hears peaks much easier than dips, so we can get away with some dips if necessary to please a recalcitrant client who judges everything on a problem system.

> *"A monitor which makes everything sound beautiful must not be accurate."*

II. Debunking Monitor myths

There is some resistance to the theory that you need accurate monitors, but certainly not among the majority of mastering engineers.[7] You can't argue with success—the most successful mastering engineers work with wide-range, flat-response monitor systems.

Myth #1: You must mix (master) with *real-world* monitors to make a recording for the real world

Here's a recent post from a mix engineer on Lynn Fuston's Internet bulletin board:

> **In Reply to: Best near-field monitors**
>
> **Frankly, I am at the point where I don't like to mix using reference monitors anymore. My monitors are so nice to listen to, but they are just too unreal.**
>
> **They are perfect for critical listening, but it is easier to make a real-world mix using an old receiver and a pair of old JBL home speakers and a boom box.**
>
> **I don't care how well you think you know your souped-up monitors. They will convincingly reproduce low frequencies that will distort like crazy on your neighbor's home stereo, and they will produce sparkling high end that completely disappears on the stereo at your mom's house.**

Beauty versus accuracy? First of all, I doubt the correspondent was describing *accurate* monitors. It sounds like he was describing *beautiful* monitors, because they are "so nice to listen to," the polar opposite of monitors which are "perfect for critical listening." There are speakers which are non-discriminatory; you know them well—everything sounds beautiful on them. A monitor which makes everything sound beautiful or which masks the fine differences between sources, must not be accurate. *Beautiful* systems are loudspeakers which are voiced, or which have faults that always make them sound "beautiful" (such as horn resonances, smeared imaging, diffraction, and dispersion qualities that emphasize the ambience in a source). On the contrary, an *accurate* monitor is merciless, revealing all distortions or frequency anomalies. On my mastering system, excellent recordings sound wonderful and beautiful, but inferior recordings do not sound very pleasant. That's a characteristic of a monitor system which is "perfect for critical listening."

Good monitors sound sparkly? This is not true. Accurate monitors do not sound *sparkly*. The poster remarked, "the [good monitors] will convincingly reproduce low frequencies that will distort like crazy on your neighbor's home stereo, and they will produce sparkling high end that completely disappears on the stereo at your mom's house." I think he must be describing someone else's *good* monitors, because a mastering engineer listening to accurate monitors will not be tempted to turn the bass up too far or cut the treble too much. The poster's conclusions have to be based on working with inaccurate, low resolution monitors.

Typical Monitor Speakers? There is no such thing as a *typical* or *representative* small monitor. Just

like the bell curve pictured above, mini-monitors' frequency responses vary all over the place.[8] Mixing engineers who believe that their particular flavor of colored mini-monitor is accurate will produce mixes with faults that sound bad on other monitors. Only a few mix engineers with a strong adaptive ability have learned how to work with small and near-field monitors and mentally compensate for their weaknesses. Though this minority of well-trained mix engineers can get excellent results with mini-monitors, mastering engineers should never depend on them.

Most times I can tell an NS-10/nearfield mix when it arrives for mastering. The bass drum is far too boomy (a particular problem with NS-10 mixes), the vocal is often too low (probably caused by center buildup in the nearfield environment), the reverb is sometimes too low (the *headphone* effect enhances inner details), the midbass of the bass instrument is depressed (caused by resonances or comb-filtering artifacts from console surface), the stereo separation is very small (imagine a big pair of headphones), and the high end is, well, unpredictable. But one time out of ten, I am shocked and pleased to learn that a mix engineer got good results with colored mini-monitors. But there's more to this story than meets the ear! That mixing engineer made it a point to take references of the recording to various places to see how it was translating, and then made adjustments before committing the mix. Not all models of nearfield-monitors are tonally colored, so what remains to conquer are the problems due to their position and proximity to reflecting surfaces.

When I mix, I choose monitors that are as accurate as I can obtain for the mixing space. I go to great effort to locate the monitors on solid stands, far from obstructions like consoles and racks or reflections like the control room glass. I may even move the loudspeakers off the console to the left or right, which makes producers think I'm crazy until they sit down and listen. It seems weird to be moving faders and looking to the side, but not in the name of getting a great mix. It's demonstrable that mix engineers use much less EQ when the monitors are accurate.

Myth #2: Adding high end helps inferior monitors that are weak in the highs.

This is an untruth. Firstly, the recording will sound sharp, tinny and fatiguing on any monitor that has adequate highs, and there are plenty of such representatives along the bell curve of inferior monitors. Next, radio play will suffer, because as mentioned, the radio limiters will just cut back the highs that you have added. But most important...

The midrange is the key. As described in the chapter on Equalization, adding too much high end depresses the lower midrange. You may end up with a vocal which has no power when reproduced on a limited bandwidth system; for example, adding highs actually reduces the strength of the male vocal in the mix when auditioned on a limited bandwidth

{ *"The Midrange is the Key"* }

system. The major power of a male vocal is in the fundamental range circa 250 Hz. If that range is depressed, the recording runs the risk of having a vocal which will not translate over the widest variety of systems. Try this: Take a great recording. Play it, go into the next room and listen. The information still comes through despite the filtering of the doorway, carpets and obstacles. Then try filtering the recording severely below 200 and above 5 kHz (like the sound of an old, bad cinema loudspeaker). A good recording will still translate. This tells you that the midrange is the key. If you lose the midrange, you lose it all. I am reminded of my first experience with my audiophile album of Paquito D'Rivera *Tico Tico*. This recording was made with minimalist miking, no equalization, nor compression. It has a very natural tonal balance, yet it plays well everywhere. Why? Because the midrange is right.

Myth #3: Heavy compression is necessary to prevent small monitor systems from overloading.

I have found the opposite to be true, with few exceptions. When I take my dynamic, impacting masters to a little Aiwa 3-piece system, they sound (comparatively) compressed, with fewer transients and less impact. If I reduced the transient clarity in the mastering, it would only sound worse on the smaller system, which does its own compressing!

I believe that high-quality monitoring is nearly as important for mix engineers as for mastering, because the mini-monitors don't reveal the damage of all those tempting low-resolution plugins and overcompression, and then it's too late to fix.

III. Refinements

Alternate Monitoring Systems

Mastering engineers use alternate loudspeakers as a double-check, not as a benchmark. I place all alternate monitoring systems outside the mastering room. Having an alternate system in the mastering room wastes time, and confuses the client. Furthermore, the alternate loudspeakers are likely to interfere acoustically with the main system. It is better to focus on a single monitor system that will not fool you into making wrong judgments. At Digital Domain, I have a second system in a separate room that I can feed "live" from the mastering room. This system has very large, "loose-sounding" woofers, and represents one extreme in the acceptance bell curve. It is fairly representative of what may happen to the bottom end of the

Here's what we're up against.

recording in a club, and somewhat helps interpret what may happen in a car. Though cars are so unpredictable, about all we can say is they will have very uneven bass response and a resonance at one or more low bass frequencies. Plus user controls that we often find in the smile shape (as in this photo).

I've learned to watch out for recordings where the client is looking for very hot bass or bass drum, and I use the "extreme" system to demonstrate what could happen if they push things too far! Because if we boost the record's bass in the mastering room to get *that sound*, it won't sound right *anywhere else*. It'll actually overdrive a typical car system. Many clients are not used to a neutral reproduction system; the hip-hop or reggae client may want it to sound like it does in his car in the mastering room. The boomy alternate listening room does the trick.

One mastering studio has a radio station transmitter and processor in their machine room, and invites the client out to their car to hear what it 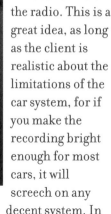 will sound like on the radio. This is a great idea, as long as the client is realistic about the limitations of the car system, for if you make the recording bright enough for most cars, it will screech on any decent system. In other words, use the car system as an example of an extreme, not the least common denominator.

Narrowcasting

There are boombox systems, club systems and car systems especially engineered for music such as hip hop whose bass response/resonance is extremely exaggerated. Properly-engineered recordings sound so thick on these systems that the vocals are almost completely lost! It is almost impossible to make a master that plays well on such an extreme system that doesn't sound thin and lifeless on all the others. We cannot include these extreme systems when making a master if we want to please the maximum number of listeners. Instead, the best solution is to make a separate (dedicated) master for the club(s) or venues.

IV. In Summary

The major goal of the mastering studio is to make subjective judgments as objectively as possible. Mastering engineers confirm that accurate monitoring is essential to making a recording that will translate to the real world. The fallacy of depending on an inaccurate "real-world-monitor" can only result in a recording that is bound to sound bad on a different "real world monitor."

Even the best master will sound different everywhere, but it will sound most correct on an accurate monitor system. Which leads us to this comment from a good client:

I listened to the master on half a dozen systems and took copious notes. All the notes cancelled out, so the master must be just right!

1 See the section on comb-filtering in Chapter 3. Jim Johnston (private correspondence) notes that nearfields use a completely different listening method than what almost anyone uses in the real world, i.e. most real world listeners, other than boombox and headphone listeners are well into the diffuse field of the room.

2 NC 30. Noise criterion 30 decibels, follows an attenuation curve whereby at 1 kHz noise level is 30 dB, and at lower frequencies is permitted to rise.

3 Very few near-field monitors pass the "bandwidth and compression test." Almost none have sufficient low frequency response to judge bass and subsonic problems, and very few can tolerate the instantaneous transients and power levels of music without monitor compression. If your monitors are already compressing, how can you judge your own use of compression?
Diffraction is the bounce of an acoustic wavefront from cabinet edges, causing a "smearing" of the sound quality. This can be reduced by using round instead of sharp cabinet edges, and soft materials on the edge instead of hard.

4 This term was coined by Dr. Peter D'Antonio of RPG.

5 The advantage of this monitoring environment is that time-domain errors in the musical material will be more audible, since they will not be masked or smeared by the monitoring room itself. Time delay-based measurement used to be extremely expensive, but has reached affordability with the advent of fast personal computers and decent audio software. In the absence of TDS equipment, an objective-subjective test called the LEDR test can help determine if nearby reflections are interfering with the monitoring. LEDR (Listening Environment Diagnostic Recording) is available from Chesky Records, (http://www.chesky/com) on JD37. First play the announce track and confirm that the announcer's positions are correct. If not, then adjust speaker separation and angle. Then play the LEDR test. The beyond signal should extend about 1 foot to the left and right of the speakers. If not, then look for side wall reflections. Similarly, the up signal should rise straight up, 3 to 6 feet, and the over signal should be a rainbow rising at least as high as the up. If not, look for interfering objects above and between the speakers, or defective drivers or crossovers. Frequency response of left/right pairs must be well-matched for a perfect LEDR score.

6 Unless the speakers are placed in soffits within the wall structure, which requires considerable acoustical expertise. It's much easier to design a room with free-standing loudspeakers.

7 There are a few major mastering engineers left who use non-standard, non-flat monitors, but like the best mix engineers, they have learned their faults and know how to make a master translate to the world. However, I would not advise that a new mastering engineer start out this way. Very few people have the ability to adjust their inner hearing this well. Similarly, some mastering engineers skew their monitor systems by using underpowered tube power amplifiers to make their judgments, which I feel is dangerous, as the natural compression of tubes may prevent them from knowing if a recording needs some compression or may mask overcompression. See Chapter 10—Compression and Monitoring. Tubes can work in a high-powered monitor amp, and hundreds of watts are required to keep tube amplifiers feeding typical inefficient loudspeakers from skewing in the overcompressed direction.

8 The LSR series from JBL accomplishes the most appropriate best compromise in monitor accuracy. Each smaller monitor in the series has a strong family resemblance to its larger cousins, with very linear frequency response down to its bandwidth limit. Which means that when placed in a linear environment (rarely encountered on the top of a console), the smaller LSRs will only be missing the extended portion of the low range, with the rest being pretty accurate.

"**we'll**
FIX IT
IN THE
MIX."

—Anon

"

IT'S NOT HOW LOUD YOU make IT. IT'S HOW YOU make IT LOUD.

"

—BOB KATZ

CHAPTER 7

Putting The Album Together

Introduction

Sergeant Pepper is often cited as the first rock and roll *concept album*, i.e. an elaborately-designed album organized around a central theme that allegedly makes the music more than a simple collection of songs. This started a trend in the 70's that many assume has more or less died. But is the concept album really dead? I'm not so sure; I treat *every* album that comes for mastering as a *concept album*, even if it doesn't have a fancy theme, artwork or gatefold. The way the songs are spaced and leveled contributes greatly to the listener's emotional response and overall enjoyment of the album. It is possible to turn a good album into a *great album* just by choosing the right song order, though, unfortunately, the converse is also true.

I. Sequencing: How to Put an Album in Order

Sequencing is an art. Sometimes, the musicians making an album have a good idea of the song order they'd like to use, but many people need help with this tricky chore. Traditionally, the label's A&R person would help put the album in order, but in today's world of independent productions that service is not always available. This is frequently the producer's job, or clearly someone experienced, politically "neutral"* and esthetically inclined. A mastering engineer bridges the nebulous division between artist, producer, and engineer—having heard thousands of albums and being *au courant*, he may provide useful guidance during this process.

---

* Albums produced by a band member(s) sometimes suffer from the *more me* syndrome, where each musician wants to hear his or her instrument louder. The only way to avoid more me is to use a producer/engineer who has no "political" alliances and is working for the concept of the album as a whole.

This is my approach: First let me tell you what usually does not work—Don't try to respond intellectually. One musician thought it would be a good idea to order his album by the themes presented in the lyrics; he started with all the songs about love, followed by the songs about hate, and finally the songs about reconciliation. It was a musical disaster. The beginning of his album sounded musically repetitive, because all his love songs tended to use the same style, and furthermore, the progression of intellectual ideas simply was not obvious to the average listener, who primarily reacted to the musical changes. Even when the listener got the intellectual point, it didn't contribute much to the enjoyment of the album. Listening to music is first and foremost an emotional experience. If we were dealing here with lyrics (poetry) without music, perhaps the intellectual order would be best, but the intellectual point of the album will still come through, even if the songs are organized for primarily musical reasons.

Before proceeding to order the album, it's important to have its gestalt in mind: its sound, its feel, its ups and downs. I like to think of an album in terms of a concert. Concerts are usually organized into *sets*, with pauses between the sets when the artist can catch her breath, talk briefly to the audience, and prepare the audience for the mood of the next set. On an album, a *set* can consist of only one song, but most often is three or four. There are no strict rules, but usually the space between sets is a little greater than the typical space between the songs of a set, in order to establish a breather, or mood change.* Sometimes there can be a long segue

(crossfade) between the last song of a set and the first of the next. These basic principles apply to all kinds of music, vocal and instrumentals.

Now comes the job of organizing the sets. To make it easier, I usually prepare a rough CD of all the songs, or a playlist on a DAW (my favorite) to allow instant play of all the candidates. This is a lot easier than it was in the days of analog tape. Then I make a simple list, describing each song's characteristics in one or two words or symbols, such as *uptempo*, *midtempo*, *ballad*. Sometimes I'll give letter grades to indicate which songs are the most exciting or interesting, trying to place some of the highest grade songs early in the order.† I may note the key of the song, although this is usually secondary compared to its mood and how it kicks off. If there's a bothersome clash in keys, sometimes more spacing helps to clear the ear, or else I exchange that song with one that has a similar feel and compatible key.

The opening track is the most important; it sets the tone for the whole album and must favorably prejudice the listener. It doesn't have to be the *hit* or the *single*, but almost always should be up-tempo and establish the excitement of the album. Even if it's an album of ballads, the first song should be the one that hits the listener's heart and soul.

If the first song was (hopefully) exciting, we usually try to extend the mood, keep things moving just like a concert, by a short space, followed by an

---

* Similarly, classical albums have shorter spaces between movements than between the major numbers.

† That's life. Not every song is a masterpiece, but it's important to give your best impression as early as possible.

up- or mid-tempo follow-up. Then, it's a matter of deciding when to take the audience down for a breather. Shall it be a three- or four-song set? I examine the other available songs, then decide if it will be a progression of a mid-tempo or fast third song followed by a relaxed fourth, or end with a nice relaxed third song.

At this point, there are track numbers penciled next to the candidates for the first set of the album. I play the beginning of the first song to see how it works as an opener, then skip to the last 30 or 40 seconds, play it out and jump to the start of the second song to see if that works. The listener actually reacts more to the musical transition than to the entire feel of the previous song. This is how to join different musical feels; an up tempo song that comes down gently at the end can easily lead to a ballad. If the set doesn't flow, I substitute songs until it works.

Then, I check off the songs already used on the list, and pick candidates for the second set, usually starting with an up-tempo in a similar "concert" pattern. This can be reversed, of course; some sets may begin with a ballad and end with a rip-roaring number, largely depending on the ending mood from the previous set. A set can also be a roller coaster ride, depending on the mood we want to create. Regardless, when you consider the album in terms of sets, it becomes a lot easier to organize. By the way, the ultimate listener doesn't usually realize that there are sets; our work ends up as only a subliminal contribution to the feel. As the set list gets filled up, it becomes a jigsaw puzzle to make the remaining pieces fit. Perhaps the third or fourth set

doesn't work quite as well as the first. Perhaps one of the songs just doesn't transition into the other. At that point I try a one-song set, or see if this problem song works better in an earlier set, either replacing a song, or adding to the earlier set. It can get frustrating, but it will all come together in time.

The Odd Man Out

One song may just not fit well musically with the rest. For a Brazilian samba album which I was mastering, the artist also recorded a semi-rock blues number. She said everyone loved this song in Brazil, so we couldn't excise it from the album, but stylistically it did not seem to gel as a part of any set. At first I suggested putting it last as a "bonus track," but this ruined the feel of the original album ending, which was a beautiful, introspective song that really did belong at the end. Eventually, we found a place for the offender near the middle of the sequence, as a one-song-set, with a long-enough pause before and after. It served as a bridge between the two halves of the album.

The Right Kind of Ending

So, how to end the album? What is the final encore in a concert? It's almost never a big, uptempo number, because the audience always cries "more, more, more." You've got to leave them in a relaxed, comfortable "goodbye mood," otherwise you'll be playing encores forever. That's why the last encore is usually an intimate number, or a solo, with fewer members of the band. The same principle applies with the record album. I usually try to create a climax, followed by a dénouement. The climax is obviously an exciting song that ends with a nice peak. This, followed by one or two *easy-going* songs to close out the album. When I find the perfect sequence, it's a real treat!

II. Spacing The Album

The first thing to remember is never to count the seconds between songs. Experienced producers know that the old "4 second" "3 second" or "2 second" rule really does not apply, although it is clear that album track spacing has gotten shorter over the past 50 years, along with the increased pace of daily life. The correct space between songs can never accurately be estimated or counted, so putting an exact number on it is probably meaningless. Different people start counting at different times; the last few moments of a decay often signal the feel of the space between the tunes. The computer may objectively say that a space is only 1 second, but the ear may feel it's closer to 2.5. So I've stopped counting seconds, and just go by the feel. As a general rule, the space between two fast songs is usually short, the space between a fast and a slow song is medium length, and the space between a slow and a fast song is usually long. The space following a fadeout is usually very short, because the listener in a noisy room or car doesn't notice the tail of a fadeout. Often we have to shorten fadeouts and make segues* or the space will seem like forever at home and especially in the car. Spacing is also dependent on the mood of the producer and time of day. If you space an album in the morning when you're relaxed, it almost always sounds more leisurely than one which has been paced in the afternoon, when hearts are beating faster. The solution is to be aware of your inner self and not make too short a space when you're in a fast mood, or too long a space when you're very relaxed; the result will probably average out for the listener.

Consider the *pace* of an album, which is affected by intertrack spacing. As described above, we often want the first set to be exciting, so you may want to control the pace by using shorter spaces within the first set and then slightly longer spaces thereafter. Tricks like these have some psychological power over the listener. An interesting observation is that if you start with tight spaces and then make the rest of the spaces "normal," the normal spaces seem too long, because your internal sense of timing has been altered by the pace of the first section. Manipulate spaces to produce special effects—surprises, super-quick and super-long pauses make great effects. One client wanted to have a long space in the middle of his CD, about 8-10 seconds, to simulate the change of sides of an LP. Rather than rejecting his idea out of hand (always respect the input of creative individuals), I tried the super-long space, and it worked! This was largely due to his choices of songs and the order. The set which began side two had a significantly different feel, and the long space helped to set it off, like a concert intermission.

Some engineers like to think of spaces as punctuation marks. There's a **comma space**, a **semicolon**, and a **period**. Never judge a space by *dropping the needle on the record*, that is, by auditioning 30 seconds or so of one tune's tail followed by the beginning of the next. Inevitably the listener will need a bit more of a breath before starting the next, especially if it's the space between

---

* *Segue* (pronounced seg-way)—a crossfade or overlap of two elements. Webster's: proceed without interruption. Italian: *seguire*, to follow.

two sets. That **period space** won't feel like a period when you've heard the entire song, or the whole set in context. Experience teaches us to anticipate these effects, so we add more of a breath after an exciting song and we know to preview far enough back to get more of the holistic feel. Still, sometimes the first CD reference needs spacing adjustments.

For a fast-paced pop album, if in doubt I prefer to make a space too short rather than too long. I sometimes will cut a space shorter and shorter until it is obviously too short and then add just the soupçon necessary to make it sound "just right," especially knowing that it always seems longer at home. Then there's the question of the *ideal space*, when the rhythm of the previous song leads very well into the attack of the next, where we count beats, and make the following song land on the beat. Finally, there's the mystery space, where it's not obvious what will work best. So, I try both long and short spaces, inching them up or down until it's obvious which approach is best.

We didn't have this kind of luxury in the days of analog tape, and it's interesting to note that when an LP master comes in for conversion to CD the spaces always seem too long. One reason, as I've said before, is the current quicker pace of life, but the other is that vinyl noise acts as a filler. When there's dead silence between tracks, spaces always seem longer. I may remove 2 or more seconds out of an LP space and it will feel just fine on CD.

III. PQ Coding

Spaces and PQ (Track) Coding

The CD Redbook standard does not permit official pauses shorter than 2 seconds between tracks. This doesn't mean you cannot have a one second or shorter space between songs, it only means that there will be no official pause between tracks, where the CD player would be counting backwards (officially, this is called Index Zero). Instead, the next track mark also functions as the end mark of the previous track.

When two songs segue into one another, the placement of the next track mark is critical, because CD players take finite time to cue—up to about 5 SMPTE frames, for older players. So if there is an overlap where the previous song is fading out on top of the next, the track mark has to be placed extremely close to the top of the next song, or slow-cuing CD players will reveal a piece of the previous sound.* Sometimes this cannot be avoided, but many times an experienced mastering engineer will find a solution. Live albums with applause require special attention to both editing and PQ coding; fading up and down between songs is very disconcerting to the listener. I prefer a delicately-edited album that sounds like a continuous concert. But then comes the decision of where to put the track marks, because there are no dead spaces. For track beginnings, I keep in mind that the fastest CD players take 1 SMPTE frame to cue and the slowest about 5 frames, and try to find a track position that

---

* Conversely, there are one or two slow CD players that cue too late, missing the downbeat if the track mark is too close.

doesn't reveal the previous noise, or up-cut the downbeat of the track. It's an art and a science, and often a compromise

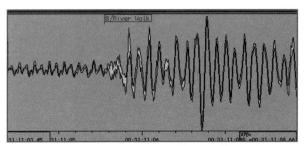

when a previous noise comes very close to the downbeat, illustrated here.

Track mark placed very tight to the downbeat with no offset to avoid hearing talking which comes before the mark.

Hiding Information in the Gap

When a cut from a concert album is played on the radio, it's often desirable to start the tune on the downbeat, but the listener at home wants to hear the atmosphere between cuts and the artists' charismatic introductions. To accomplish this dual feat, the creative mastering engineer takes advantage of the compact disc's Index 0 and Index 1 time, as in the following figure.

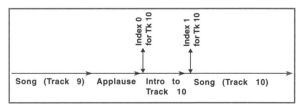

In this example, the song for track 9 ends with applause, and the official end of song 9 is at the Index 0. The time between Index 0 and Index 1 is called the *pause* or *gap time*, during which the CD player counts backwards to zero, but in this case there is sound in the gap. This permits the CD player's **random play** function to ignore the boring or irrelevant parts. Similarly, the introductions,

count offs, sticks, and so on, for songs on any album can be placed *in the gap* so they will not be heard on the radio or in random play. Note that by putting the speeches into the pause time, they do not increase the official length of either track. Unfortunately, the most primitive CD players only respect Index 1, so the introduction would be treated as the end of the previous track, producing some incongruous results in random play. Furthermore, many current computer (software-based) CD players and many modern-day DVD players also ignore Index 0, which is destroying a critical part of the artistry of the Compact disc.[*] To top it off, most DVD players cue CDs very loosely, revealing unintended material. Alert your congressman, err, rather, licensors Sony and Philips that the CD standard is rapidly eroding, hindering the artistry that we have enjoyed for over 20 years. Regardless, I always PQ code masters assuming they will be played on CD players that respect the standard; there is little other choice.

In this vein, it pays to be vigilant for many CDR duplicators will mute the pause audio, sometimes even taking many seconds OUT and putting just 2 blank seconds IN (the minimum pause length in the CD standard). Imagine your classic Pink Floyd The Wall, which has continuous sound, being gapped by accident at the plant. These copiers were found to be copying in **Track At Once** Mode, rather than **Disc at Once**, instead of simply cloning the disc.[†] Certainly frustrating.

---

[*] The second disk of a multi-disk set that has a start id higher than 1 will crash many computers, according to Bob Olhsson, in correspondence.

[†] Thanks to Dan Stout for this information, as viewed on the excellent Mastering Webboard.

PQ Offsets

Since CD players can vary in their reaction times, the editing program can apply typical offsets, or show the PQ codes exactly as they will appear on the disc. For example, a start time offset of 12 CD frames* means that the actual track mark will be 12 frames (160 ms) in front of its visual location on the screen if you choose to display the mark without the offset. Sophisticated DAWs let you rehearse the effect of cuing with or without the offsets.

Redbook† Limits

The Redbook specifies the Compact Disc. A CD may have up to 99 tracks and each of these tracks may have up to 99 indexes (AKA subindices). Rarely do we code CDs with indexes since many players do not support them and most people don't know how to use them. Classical engineers used to code each major piece with a track mark and the movements within via indexes. But today most classical CDs place a track mark for each succeeding piece.

The minimum CD track length is 4 seconds. Mastering engineers have been known to create a hidden track by inserting many short, blank 4-second "tracks" at the end of the CD prior to the "hidden" one.

Disc-At-Once, Track-At-Once and Standalone CD Recorders

I would never use a standalone CD recorder to make CDRs for replication. There is no provision for Index 0, and the location of Index 1 (the track mark) can only be as accurate as a manual button push. Plus, when recording one track at a time, these standalone recorders work in **Track-At-Once**

mode, which puts an E32 error onto the disc wherever the laser stops recording. Computer-based machines should be set to work in **Disc-At-Once** mode, which means that the CD must be written in one continuous pass.

PQs and Processor Latency

Since I like to master on loadout, with all processors in line, I have to consider the latency (delay) of all the processors, which I have seen up to 12 SMPTE frames with a full chain including up- and down- sampling and the linear-phase equalizer, which has a tremendous processor latency. The trick is to measure the delay and slide the PQ marks by this amount.

Hidden Tracks in Pregap

Some CD players have the ability to rewind in front of track one; this is called the pregap or first Index 0. One company claimed to have the rights to putting hidden tracks in that position, but it's not even permitted in the Redbook standard, and many plants will not press CDs with a hidden track in the pregap. To the best of my knowledge, there is no way to produce a DDP with this feature, so only CDR masters can be produced in this way if the DAW allows it.

---

* There are 75 CD Frames in a second, as opposed to SMPTE frames, 30 per second.

† The **Redbook** defines the standards for the audio CD as defined by Sony and Philips.

IV. Editing

I love the art of editing, because it gives instant gratification. There's nothing like generating a hundred smiles in a day, one after each successful edit! I think a whole book should be written on editing techniques, but ultimately the skill of fine editing can only be learned through guided experience: the school of hard knocks, and an apprenticeship. A good mastering engineer has a well-developed **editing esthetic**, which helps us turn a rough-hewn work into an audio masterpiece.

The purpose of this short section is to discuss some of what is possible in digital audio editing, and what is expected of a good audio master. Using sophisticated workstations, we can perform edits that were impossible in the days of analog tape and the razor blade. I once spent 30 hours painstakingly editing a spoken-word version of a novel, a task which now might be accomplished in a single day. SADiE's playlist-editing mode makes this real easy.

The Tale of the Head and Tail

Editing heads and tails is an important skill born of experience and musical knowledge.

Head noise cleanup. Because mechanical artifacts can easily distract the listener's attention from the emotional feel and involvement in the music, a mastered work should feel consistent and *smooth* (unless a jarring, jumpy style is intended). For example, mastering workstations allow us to edit the beginning of a song with a careful fade-up. Sometimes this fade-up is made fast (equivalent to a 90 degree cut), because for some music the downbeat is king. But a fast fade-up often sounds wrong with soft music, especially pieces that begin with solo vocal or acoustic instruments. A delicate acoustic guitar solo can sound abrupt if the noise of the room and preamp noise is suddenly brought up from silence. Unless we perform just the right speed and shape of fade-up the *air* (roomtone) noise will call attention to itself.

Natural Anticipation. We also have to be aware of the important role played by *natural anticipation:* the human breath before the vocal; or the movement of the guitarist's hand before a strum; or the movement of the fingers and keys prior to hearing a piano downbeat. Often it sounds unnatural to cut off these kinds of anticipation; I dislike openings of songs that sound choked because the recording engineer has cut off the air or space or breath or even subtle movements of the musicians. If the breath is better included, but sounds a bit loud, then a gentle fade-up can produce just the right esthetic. I advise mixing engineers not to cut off the tops when sending songs for mastering, for the mastering engineer probably has better tools to fix these, and a quiet, meditative environment to make these artistic decisions properly. 60% of the time, I'll remove these extra noises, but use the rest to good advantage to help the subliminal feel and pace of the album.

Tail Noise Cleanup. Sometimes the tail end of a song contains noise from musicians or equipment, which draws attention to itself by the transition from noise to the silence between pieces. The simplest and most common solution is called a *follow fade*, which is usually a cosine or S-shaped fade to silence. A good mastering engineer may spend a minute or more on such a fade to ensure that the tail ambience or reverberation does not feel cut off, whilst at the same time, the hiss or noise is brought to silence at just the right speed so that it isn't noticed. We can take advantage of the fact that hiss and noise are masked by signal of the right amplitude, so the follow fade can and should be slightly slower than the natural decay. The delicate decay of a piano chord at the end of a tune should feel like it's ending naturally, even while avoiding the thump of the release of the pedal. Some sophisticated mastering workstations contain reverse S curves, allowing us to raise the gain at the tail, after having previously lowered it, in order to hear some fine inner detail.

Fadeouts. I think a good-sounding musical fadeout is one that makes us think the music is still going on; we're still tapping our feet even after the sound has ceased. Although we can apply the same cosine shape we use for tails, fadeouts are a distinct art in themselves. Typically, a fadeout will start slowly, and then taper off rapidly, mimicking the natural hand movement on a fader because most people don't like to sit and listen too long to a fade that lingers. On the other hand, a fadeout should not sound like it fell off a cliff, and often in mastering we get material that has to be repaired because the mix engineer dropped the tail of the fade too fast. Since editing is like whittling soap, I recommend that mix engineers send unfaded material so it can be refined in the mastering. It is difficult to satisfactorily repair a fade that was too fast at the

end; sometimes an S-shape helps, and sometimes we can apply a taper on top of the original taper.

Adding tails. Although editing is like whittling soap, sometimes we're called upon to make more soap. And the soap we create can sound more authentic than what had to be cut away! If the musicians or instruments make a distracting noise during the ambient decay, the ambience will sound cheated or cut off if we perform a follow fade to remove the noises. In the figure below is a fadeout, to the right of which you can see the noise made by the musicians. Unfortunately, these noises occurred during the reverberant tail, so the ambience sounds cut off. The trick is to feed just the tail of the music into a high-quality artificial reverb and capture that in the workstation, which you can see in the bottom panel. Also notice that the predelay of the reverb

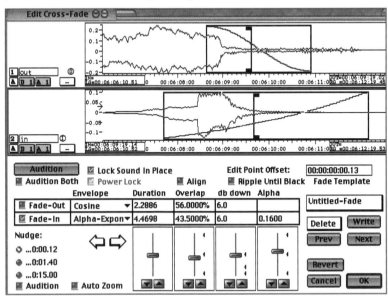

Adding a tail via a crossfade to artificial reverb.

postpones its onset. This can be adjusted in the mastering DAW's crossfade window which allows us to carefully shape, time, and adjust the level of the transition to this artificial reverb in a manner that can sound completely seamless. Thus we have performed the impossible: putting the soap back on the sculpture!

Sometimes an analog tape may have a lot of echoey print through or hiss noticeable at the tail of the tune. If adding tails with reverb does not work well, in this case it is advisable to edit to the digital safety version of the mix, so I advise clients to send both versions.

Adding Room Tone

Room tone is essential between tracks of much natural acoustic and classical music. Recording engineers should bring samples of room tone to an editing session. Room tone is usually not necessary for pop productions, but if a recording gets very soft and you can hear the noise of the room, going sharply to *audio black* can be disconcerting. The object is not to draw the listener's attention to the onslaught or removal of noise, as illustrated in the figure at left.

Room tone should be recorded in advance as a separate "silent take" with no musicians in the room. If the room tone was not supplied in a separate take by the mixing engineer (at least 4, preferably 10 seconds or more), it is almost impossible for us to manufacture a convincing transition and we have to be satisfied with a fade to/from silence. In stubborn cases I have manufactured a matched room tone by shaping pink noise, but it can be a very time-consuming (thus expensive) process.

Repairing Bad Edits. One type of bad edit is where the reverberation of one take has been cut off by the insertion of a new one. This is a classic error caused by the producer instructing the musicians to begin the retake exactly at the intended edit point, instead of a few bars earlier, a much better practice which would not only give the musicians a running start, but also generate the reverberant decay of the preceding note for the editor to work with. Because the producer did not record the reverberation, the ear notices the cutoff of the reverb, which is not masked by the transient attack of the next downbeat. Luckily, when it comes to mastering, we can repair some of these bad edits even if the original takes are not available. The trick is to separate the original take and the insert at the edit point, use an artificial reverb chamber to re-create the missing tail as above, then join the edit back together. Since this would involve mixing more than two elements, sometimes more than one (stereo or surround) track is necessary for the brief mix.

Editing and assembling concert albums can be a great pleasure. The edited concert album is the perfect example of the principle of willful

Labels: Decay of previous / Follow fadeout to remove musician's noise / Fadeup on Roomtone / Roomtone / Edit within the Roomtone / Medium fast fadeup on breath of next track and slow fadeout of roomtone

5:08:13.40   .00   00:15:15:00.00   OUT=00:26 =00:15:22:02.78

Editing room tone in an acoustic work requires considerable artistry. An edit must not call attention to itself.

suspension of disbelief because real-life applause is almost never as short as 15 or 20 seconds, and real-life artists have to stop to tune their instruments. The object is to prune the concert down to its essence so that the home listener is never bored on replay. Editing applause is an art; you have to be familiar with the feel of natural applause. Cutting applause and ambience between different performances exercises the power of the workstation's crossfades. There can never be silence between numbers, there must be some degree of room tone (audience ambience). The room tone which precedes a quiet number has a very different feel than the sound of the audience at the end of a loud one, and it is necessary to create an imperceptible transition between the two. My approach is to do the major cutting on one pair of tracks (for stereo), and wherever it needs transitional help, mix in a bed of compensating ambience on another track pair. I once put an audience ambience loop under the only studio cut on a live album, and to this day no one has been able to figure out which track is the ringer!

V. Leveling The Album

The greater a recording's dynamic range, the harder it is to judge "average level" and you have to listen in several spots. I usually start with the loudest song on the album and find its highest point. I then engineer the processing to create the impact I'm looking for, hold the monitor at the predetermined gain, and make the rest of the songs work together at that monitor gain. The rest of the album falls in line once the loudest song has its

proper level and impact. During the processing of this loudest song, it's important to ensure the chain of processors are in their optimum gain without overload; this is the test for the rest of the album. These days, digital limiters keep from going "over level" (distorting the digital system), although a limiter pushed too hard produces a squashed and unpleasant sound (see Chapters 9-11 on dynamics).

The ear judges level by comparison to the surroundings, and adapts to loud and soft passages by lowering and raising its *human gain*. Thus, a soft beginning may seem too soft following a loud climax, but the same level would be fine in the context of the middle of a song. And a loud passage following a silence seems even louder. That's why you have to pay attention to context when judging apparent levels. Leveling and dynamics processing are inseparable, for the output (makeup gain) of the processors also determines the song's loudness compared to the others (see Chapter 10). A more compressed song may sound louder than another even if its peaks don't hit full scale (0 dBFS). If you change the processing, you have also changed its level, so it's all done by ear. After working on the loudest song and saving the settings, I usually go to the first song and work in sequence. Then the second song, and next I check the transition between the first and second. In a good mastering room, this transition will usually work without any fine-tuning because we've been monitoring at a consistent gain while doing our decision-making. If one song appears too loud or soft in context, I make a slight adjustment in level until they work together, or sometimes increase the spacing to "clear the ear." If the first song is hot and up-tempo and the second begins quietly, it is sometimes necessary to turn up the intro of the second song so it will work in context. So you can see why it's important to have the album in proper order before mastering!

Extra-soft beginnings, endings or even middle spots require special attention. Meter readings are fairly useless in this regard; only experience will tell us when something is too soft and has to be raised. In Chapters 9 thru 11, we'll get into some manual and automatic techniques for altering internal dynamic range.

Ear Fatigue? After leveling and processing the last song, I always review song numbers one and two, to make sure they still fit well into the context. There may be a tweak that can further optimize the first couple of songs. Or, I might find that the album has been growing in amplitude due to ear fatigue and the latter songs may need to be lowered.

The Domino Effect

Overzealous leveling practice (where the engineer or producer is trying to make every song super-hot) can produce a *Domino Effect*. Suddenly, the song which used to be the loudest, doesn't sound as loud as it did before. This is psychoacoustics at work, or possibly listening fatigue. Not every song can be the loudest! If the loudest song was good enough before, the problem may be the unintentional escalation. Instead of trying to push the loudest song further, thereby squashing it with the limiter, I try to lower the previous song by even a few tenths of a dB, which will restore the impact of the next song by use of contrast.

CHAPTER 8

Equalization Techniques

I. Introduction

Interaction

Mastering is the art of compromise. It is the art of knowing what is sonically possible, and then making informed decisions about what is most important for the music. The first principle of mastering is this: *Every action affects everything else.* This principle means that we cannot just import practices from elsewhere into the mastering room. Equalization practice is an especially clear case of where a technique used in mastering is crucially different from an apparently similar technique used in mixing. For example, when mastering, adjusting the low bass of a stereo mix will affect the perception of the extreme highs. Similarly, if a snare drum sounds dull but the vocal sounds good, then nine times out of ten, the voice will suffer when you try to equalize for the snare.[1] These problems occur even between elements in the same frequency range: when you work on the bass drum, for example, the bass guitar will more than likely be affected, sometimes for the better, sometimes worse. If the bass drum needs EQ but the bass instrument is correct, it may be possible with careful, selective equalization to "get under the bass" at the fundamental of the drum, somewhere under 60 Hz. But just as often a bass drum exhibits problems in its harmonics, which overlap with the range of the bass instrument. A resonance problem in the bass instrument may be counteracted by

> "Mastering is the art of compromise"

dipping around 80, 90, 100 Hz ... but this can easily affect the low end of the vocal or the piano or the guitar. Sometimes we can't tell if a problem can be fixed until we try. We should never promise a client miracles—that way they're delirious when we can deliver them!

II. What is a Good Tonal Balance?

Perhaps the prime reason clients come to us is to verify and obtain an accurate tonal balance. The output of the major mastering studios is remarkably consistent, pointing to their very accurate monitoring. While it is possible to help certain individual instruments, most of the time our goal is to produce a good spectral balance. But exactly what is a "good" tonal balance? The ear fancies the tonality of a symphony orchestra. On a spectrum analyser, the symphony always shows a gradual high frequency rolloff, and so will most good pop music masters. The amount of this rolloff varies consid-erably depending on the musical style and even the moment in the music, so mastering engineers rarely* use the spectrum analyser display to make EQ judgments.

{ *"Practice is the best of all instructions"* — CHINESE FORTUNE COOKIE }

Everything starts with the midrange. If the mid-frequency range is lacking in a rock recording, it's just like leaving the violas or the woodwinds out of the symphony. The fundamentals of the vocal, guitar, piano and other instruments must be

---

* We don't use the spectrum analyser to judge musical balance, but it's useful to have around to reveal problems, e.g. identify noises at discrete frequencies or ultra high or low frequency noise.

correct, or nothing else can be made right. The mastering engineer's job is to make sure that the tonal balance is well within the acceptable range, that things don't *stick out* inappropriately, that the sound is pleasant, warm and clear, and is correct for the song and the genre. Some pieces of music require laid-back cymbals, others are just crying out for an *in your face* treatment; with the right monitors and experience it is possible to know that the EQ is just right.

While we always seek an absolute standard in EQ, a recording can have an intentional color, for example, a brighter, thinner sound, and the ear will "train" itself and learn to accept a slight deviation from neutral.[2] Once the ear has been "trained," if you throw a naturally EQ'd song in the middle of this, it will seem fat and muddy by comparison. The mastering engineer is there to ensure that the deviation from neutral is not excessive because if it is then the sound will not translate adequately on the widest variety of playback systems. We must recognize when a sibilant vocal is acceptable, or must be controlled, for esthetic and technical reasons.[3]

Specialized Music Genres

I try to keep the symphonic tonal balance in my head as a basic reference for most rock, pop, jazz, world music, and folk music, especially in the mid to high frequency balance. This works most of the time. But some specialized music genres deliberately utilize very different frequency balances, and for them the *symphony ideal* is not appropriate. For example, in some styles of music, 'too much' (or 'too little') bass is just right. You could think of Reggae as a symphony with lots more bass instruments whereas punk rock is often extremely aggressive, thin, loud and bright. Punk voices can be thin and tinny over a fat musical background, with the natural fundamental-harmonic relationships completely strained. When this is done for a whole record it can be fatiguing, but it can be interesting and musically special when it's part of the artistic variety of the record.*

Be aware of the intentions of the mix

Equalization (and other processing) affects more than just tonality—it can affect the internal balance of a mix. So a good mastering engineer must be capable of evaluating the mix intentions of the producer/engineer/musicians and be sensitive to the needs of the production team. We must not unintentionally alter carefully-constructed instrumental interrelationships. For example, raising the bass level to get a warmer tonality will inevitably raise the level of, say, the bass instrument compared to, say, the vocalist. Sometimes this is exactly what the producer intended, because it is possible that the lack of warmth will be traced to a monitoring issue in the mix environment, and the same issues that caused a lack of warmth could also be reducing the bass instrument level on an absolute basis. Regardless, when I feel that I am affecting a balance, I always discuss my feelings with the producer to make sure that the balance "fault" which I perceive was not intentional.

---

* Yes, there are artistic punk rock records! I believe that the musical integrity of the artist determines the worth of a recording, not the style they work in.

III. Equalization Techniques

Parametric Equalizers

There are two basic types of equalizers — **parametric** and **shelving**— named for the shape of their characteristic curve. Parametric EQ is favoured in recording and mixing. Invented by George Massenburg circa 1967[4], the parametric is the most flexible curve, providing three controls: center frequency, bandwidth, and level of boost or cut. Mix engineers like to use parametrics on individual instruments, either boosting to bring out their clarity or salient characteristic, or selectively dipping to eliminate problems, or by virtue of the dip, to exaggerate the other ranges. The parametric is also the most popular equalizer in mastering since it can be used surgically to remove certain defects, such as overly-resonant bass instruments. A simpler (non-parametric) equalizer has fixed frequency and bandwidth and only the level is adjustable per band.

Q's and Bandwidth

Equalizer Q is defined mathematically as the product of the center frequency divided by the bandwidth in Hertz at the 3 dB down (up) points measured from the peak (dip) of the curve. A low Q means a high bandwidth, and vice versa. The first figure on the next page shows two parametric equalizers with extreme levels for purposes of illustration: On the left, a 17 dB cut at 50 Hz with a very narrow Q of 4, which is 0.36 octaves. The bandwidth is 12.5 Hz. On the right, a 17 dB boost centered at 2 kHz, with a fairly wide (gentle) Q of 0.86, which is 1.6 octaves. The bandwidth is 2325

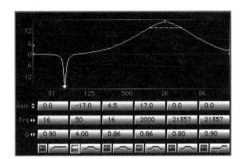

Parametric equalizer with +17 dB boost centered at 2 kHz with a fairly wide bandwidth of 1.60 oct (Q = 0.86), indicated by the dashed white line at the 3 dB down points. A cut of −17 dB at 50 Hz with a very narrow bandwidth of 0.36 octaves (Q = 4).

Hz, represented by the dashed white line.[*]

The choice of high or low Q depends on the situation. Gentle equalizer slopes almost always sound more natural than sharp ones, so Q's of 0.6 and 0.7 are therefore very popular. Use the higher (sharper) Q's (greater than 2) when you need to be surgical, such as dealing with narrow-band resonances or discrete-frequency noises. It is possible to work on just one note with a sufficiently narrow-band equalizer. I also use higher Q's when I want to emphasize an instrument with minimal effect on another instrument. For example, a poorly-mixed program may have a very weak bass instrument; boosting the bass circa 80 Hz may help the bass instrument but muddy the vocal, in which case I narrow the bandwidth of the bass boost until it stops affecting the vocal. The classic technique for finding a resonance is to **focus the equalizer:** start with a large boost (instead of a cut) to exaggerate the unwanted resonance, and fairly wide (low value) Q, then sweep through the frequencies until the resonance is most exaggerated, then narrow the Q to be surgical, and finally, dip the EQ the amount desired.

Shelving Equalizers

A shelving equalizer affects the level of the entire low frequency or high frequency range below or above a specified frequency. For example, a 1.5 kHz high shelf affects all the frequencies above 1.5 kHz. In mastering, **shelving equalizers** take on an

increased role, because we're dealing with overall program material. One interesting variant on the standard shelf shape can be found in the Waves Renaissance EQ and Manley's Massive Passive, very useful mastering equalizers. This *resonant shelf* is based on research from psychoacoustician Michael Gerzon, who believed it to be a very desirable shape. I like to think of it as a combination of a shelving boost and a parametric dip (or vice versa). In the top figure, a low Q (0.71) bass shelf of 11.7 dB below 178 Hz is mollified by a gentle parametric dip above 178 Hz, all controlled by a single band of the equalizer. This is an extreme boost for illustration, but this type of curve can be useful to keep a vocal from sounding thick while implementing a bass boost.

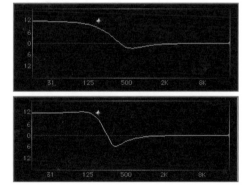

Top: Gerzon resonant shelf with a low Q. Bottom: The same with a high Q. The dip just past the shelving boost frequency is characteristic of the Gerzon resonant shelf.

The bottom figure shows the same boost with a high Q of 1.41.

Shelving equalizers can have low or high Q, with Q defined as the slope of the shelf at its 3 dB up or down point.

Using Baxandall for air

As I mentioned in Chapter 3, the *air band* is the range of frequencies between about 15-20 kHz, the

---

[*] Many equalizers define bandwidth in octaves instead of Q. Appendix 6 contains a convenient table for converting between Q and bandwidth.

highest frequencies we can hear. An accurate monitoring system will indicate whether these frequencies need help. An *air boost* is contraindicated if it makes the sound harsh or unintentionally brings instruments like the cymbals forward in the depth picture. Very few people know of a third and important curve that's extremely useful in mastering: the Baxandall curve, named after Peter Baxandall (pictured at right). Hi-Fi tone controls are usually modelled around the Baxandall curve. Like shelving equalizers, a Baxandall curve is applied to low or high frequency boost/cuts. Instead of reaching a plateau (shelf), the Baxandall continues to rise (or dip, if cutting instead of boosting). Think of the spread wings of a butterfly, but with a gentle curve applied. You can simulate a Baxandall high frequency boost by placing a parametric equalizer (Q= approximately 1) at the high-frequency limit (approximately 20 kHz). The portion of the bell curve above 20 k is ignored, and the result is a gradual rise starting at about 10 k and reaching its extreme at 20 k (see fig). This shape often corresponds better to the ear's desires than any standard shelf and a Baxandall high frequency boost makes a great *air eq.*

Be careful when making high frequency boosts (adding *sparklies*). They are initially seductive, but can easily become fatiguing. In addition, the ear often treats a high frequency boost as a thinning of the lower midrange, which completely changes intended program balance or the mix that was intended. The highs come up, but for example, the cymbals, triangle and tambourine also become louder. Is this consonant with the musical intent? In

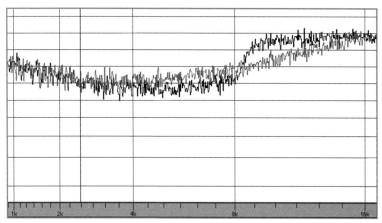

Gentle Baxandall curve (pink) vs. sharp Q shelf (black). Many shelving equalizers have gentler curves and may approach the shape of the Baxandall. Try a shelf with 3 dB per octave slope for this purpose.

accordance with the first principle of mastering, you must pay attention to the instrumental and vocal balance as well as the tonal balance whenever making changes in any EQ range.

High-Pass and Low-Pass Filters

On the left of the figure on the next page is a sharp high-pass (low cut) filter at 61 Hz, and on the right, a gentle low-pass (high cut) filter at 3364 Hz. The frequencies are defined as the points where the filter is 3 dB down. High-pass and low-pass filters are used to solve noise problems in mastering but they can make their own problems as we shall soon see. They're hard to use surgically because they affect everything above or below a certain frequency. High-pass filters are used to reduce rumble, thumps, p-pops and other noises. Low-pass filters are sometimes used to reduce hiss, though since the ear is most sensitive to hiss in the 3 kHz range, a parametric dip may be more surgical than the radical pass-filter solution. I rarely apply a

At left: Sharp high-pass filter at 61 Hz. At right: Gentle low pass filter at 3364 Hz.

standard filter to reduce hiss except for short passages, preferring specialized noise-reduction solutions instead (see Chapter 12).

EQ Yin and Yang

Remember the yin and the yang: Contrasting ranges have an interactive effect. For example…

- A slight dip in the lower midrange (~250 Hz) can have a similar effect to a boost in the presence range (~5 kHz).
- Adding bass will make the highs seem duller and reducing bass will make the sound seem brighter.
- Adding extreme highs between 15-20 kHz will make the sound seem thinner in the bass/lower midrange.
- Warming up a vocal will reduce its presence.

{ *"Remember the yin and the yang: Contrasting ranges have an interactive effect"* }

Yin and yang considerations imply that you are likely to be working in two contrasting ranges at once to assure that the sound is both warm and clear. Harness the yin and yang when the level is too high—pick the frequency band which you can reduce in level. Harshness in the upper midrange/lower highs can be combated in several ways. For example, a harsh-sounding trumpet-section can be improved by dipping around 6-8 kHz, and/or by boosting circa 250 Hz. Either way produces a warmer (sweeter) presentation, and your choice of which frequency range to work on will be influenced partly by what other instruments are playing at the same time as the trumpets. The next trick is how to restore the sense of *air* which can be lost by even a 1/2 dB cut at 7 kHz, and this can often be accomplished by raising the 15 to 20 kHz range, often only 1/4 dB can do the trick.[5] Never forget the first principle; it's easy to fall into the trap of concentrating on one element while forgetting how it is affecting the rest.

One channel or both (all)?

Most times making the same EQ adjustment in both (all) channels is the best way to proceed as it maintains the stereo (surround) balance and the relative phase between channels. But sometimes it is essential to be able to alter only one channel's EQ. For example, with a too-bright high-hat on the right side, a good vocal in the middle and proper crash cymbal on the left, the best solution is to work on the right channel's high frequencies.

Start subtly first

Sometimes important instruments need help, though, ideally, they should have been fixed in the mix. The best repair approach is to start subtly and advance to severity only if subtlety doesn't work. For example, if the piano solo is weak, we try to make the changes surgically:

- only during the solo
- only on the channel where the piano is primarily located, if that sounds less obtrusive
- only in the frequency range(s) that help, fundamental, harmonic, or both

- only as a last resort by raising the entire level, because a keen ear may notice a change when the gain is brought up

Realize the limitations of the recording

There is only so much that can be accomplished in the mastering and waiting until the mastering stage to fix certain problems usually produces compromise. There is little we can do to fix a recording where one instrument or voice requires one type of equalization and the rest requires another.[*] For example, rolling off the low end to correct a heavy synth bass is sure to lose the punch of the bass drum. Or brightening a vocal can make the tambourine sound fatiguing. In these cases I often recommend a remix. If a remix is not possible, then we resort to specialized techniques such as M/S equalization or multiband dynamics (compression/expansion) to bring out a weak instrument or hide another, which can produce fabulous results, sometimes indistinguishable from a remix (we explain M/S in Chapter 13). But the better the mix we get, the better the master we can make, which implies that a perfect mix needs no mastering at all! Even so, it is worth the time to get the approval of an experienced mastering engineer working in a neutral monitoring environment, even if she decides that no mastering or polishing is needed.

Instant A/B's?

With good monitoring, equalization changes of less than 1/2 dB are audible. I believe that instant A/B comparisons deceivingly hide the fact that a subtle change has been made, as the change will only be noticed over time.[†] I will take an equalizer in and out to confirm initial settings, but **I never make instant EQ judgments**. Music is so fluid from moment to moment that changes in the music will be confused with EQ changes. I usually play a passage for a reasonable time with setting "A" (sometimes 30 seconds, sometimes several minutes), then play it again with setting "B." Or, I play a continuous passage, listening to "A" for a reasonable time before switching to "B." For example, over time it will become clear whether a subtle high frequency boost is helping or hurting the music.

{ *"The perfect mix may need no mastering at all!* }

Fundamental or Harmonic?

The extreme treble range mostly contains instrumental harmonics. Surprisingly, the fundamental of some crash cymbals can be as low as 1.5 kHz or below. When equalizing or processing bass frequencies, it is easy to confuse the fundamental with the second harmonic. The detail shot of a SpectraFoo™ Spectragram in **Color Plate *Figure C8-01*** illustrates the importance of the harmonics of a bass instrument. High amplitudes are indicated in red, descending levels in orange, yellow, green, then blue.

Notice the parallel run of the bass instrument's fundamental from 62-125 Hz and its second and third harmonics from 125-250 and up. Should we

---

[*] Bernie Grundman calls this a recording which is "not uniform," as quoted in *The Mastering Engineer's Handbook* (see Appendix 10).

[†] This is a fundamental part of the see-saw arguments for and against blind testing methods, something which we will not cover in this book.

equalize the bass instrument's fundamental or the harmonic? It's easy to be fooled by the octave relationship; the answer has to be determined by ear—sometimes one, the other or both. To find out which is most important, I use the focusing technique, sweeping the equalizer from the fundamental to the harmonic. But in mastering we may not have the liberty of choice, since the equalizer may simultaneously affect the bass instrument, bass drum, and the low end of the piano, guitar, vocals, etc. It might be necessary to choose the frequency which has the least effect on other instruments rather than the ideal one for the focal instrument. It's also a matter of feel; in a rhythm piece, we can forgo delicacy and make it kick with a general bass boost.*

Bass boosts can create serious problems

Since the ear is significantly less sensitive to bass energy, bass information eats up lots more power (6 to 10 dB) for equal sonic impact below about 50 Hz, and requires about 3-5 dB more between 50 and 100 Hz.[6] This means that our low frequency equalization practice may use up so much energy that it affects the loudest clean level we can give to a song. It also explains why bass instruments often have to be compressed to sound even. Historically, the high pass filter was our best friend when we made LPs, to prevent excess groove excursion and obtain more time per LP side. Digital media do not have this physical problem, but the psychoacoustic problem of the ear's low frequency insensitivity still exists.

One possible way to save "energy" is to use a fairly sharp high pass (low cut) filter somewhere below 40 Hz, which does not significantly affect the energy of the bass drum or the low notes of the bass. I do not make this decision lightly as many recordings sound better flat; the monitor system's woofers must have calibrated, extended response for this judgment. The high pass filter must be extremely transparent and have low distortion. During mastering, I listen carefully, switching a filter in and out to determine if it is helping or hurting. Sometimes a gentle filter is a better choice than a steep one, as when dealing with a boomy bass drum or bass. But subsonic energy, rumble or thumps require a steep filter to have minimal effect on the instruments. When "uncoloring" a resonance, a fairly narrow parametric filter tuned to the offending frequency is also a good choice.

Mix engineers working with limited bandwidth monitors run the risk of producing an inferior product. Subwoofers permit you to hear low frequency leakage problems that tend to muddy up the mix, for example, bass drum leakage in vocal and piano mikes. It's much better to apply selective high pass filtering during the mixing process because mastering filters will affect all the instruments in a frequency range. For example, mix engineers can usually get away with a steep 80 Hz filter on an isolated vocal, but it's extremely rare to see a mastering engineer use one on a whole mix. A mixing engineer should form an alliance with a mastering engineer, who can review her first mix and alert her to potential problems before they get to the mastering stage.

---

* If that's what the piece needs. I shudder to think that readers may take each recommendation in this chapter literally, and apply it to their work. Mastering engineers do not automatically equalize; we always listen and evaluate first. Many pieces leave mastering with no equalization at all.

IV. Other refinements

Linear-phase Equalizers

All current analog equalizer designs and nearly all current digital equalizers produce phase shift when boosted or cut; that is, signal delay varies with frequency and the length of the delay changes with the amount of boost or cut. Hi-Q filters produce the most phase shift. This kind of filter will always alter the musical timing and wave shape, also known as *phase distortion*. Daniel Weiss says,

> [In contrast] a particular type of digital filter, called the **Symmetric FIR Filter**, is inherently linear-phase.[7] This means that the delay induced by processing is constant across the whole spectrum, unconstrained by eq settings.[*]

Since FIR filters are expensive to implement in real time, linear-phase equalizers have only recently appeared. Rather than FIR filters, the Weiss uses a complementary IIR technique to obtain linear-phase. This technique seems to avoid one of the downsides of the FIR approach, which can produce weird results at certain frequencies unless they use extreme computing power (MIPS).

John Watkinson believes that much of the audible difference between EQs comes down to the phase response.[†] I don't think engineers have a good handle on the sonic deteriorations of phase-shift in equalizers; after my first linear-phase experience, it was hard to go back. To my ears, the linear-phase sounds more analog-like than even

analog! The Weiss has a very pure tone and seems to boost and cut frequencies without introducing obvious artifacts. Ironically, while mastering a punk rock recording, it proved too *sweet* in linear-phase mode so I had to return to normal mode to give the sound some *grunge*. So clearly much of the qualities we've grown accustomed to in standard equalizers must be due to their phase shift.

Most times I choose linear-phase mode. But both filter designs have their Achilles' heels.

> Whenever you have to equalize, you will alter the signal in both the time and frequency domains (as mathematics requires); there will always be a **time artifact**. In the *analog* style equalizer, which is usually mathematically termed *minimum-phase*, the alteration will be primarily to spread the signal **downstream**, i.e. does not lead the original signal by much, if any. A **downstream modification** translates into different delays at different frequencies dispersing the original signal. In some cases this effect is quite audible. If one uses a *digital approach*, one can either mimic the analog behavior, or use a *linear-phase*, aka *constant delay* filter. This filter will **equally precede and follow the signal;** part of the filter may create a pre-echo effect, modifying the

---

[*]   Described by Daniel Weiss at the Weiss website, http://www.weiss.ch.
[†]   **Studio Sound** Magazine, 9/97.

leading edge of transients and signal changes. A high Q linear-phase filter can introduce audible pre-echo in the short millisecond range; it's exactly like a floor bounce but without the comb-filtering. Any time that a high Q filter is used, careful listening with both types of equalization may be necessary to decide which choice is best.[8]

Neither approach is fundamentally better. The minimum phase (analog-style) equalizer tends to smear the depth and imaging, and occasionally that artificial smearing produces a pleasantly vague image. The linear phase equalizer can subtly deteriorate transient response. It might be a good idea for manufacturers to allow us to select filter types per band; I might choose minimum-phase for a steep high pass, and linear phase for a gentle presence boost.

Dynamic Equalization

Multiband dynamics processing can also be treated as dynamic equalization, where the time constants or thresholds have little effect on the actual dynamics but rather more on the tonal balance at different amplitudes. Dynamic equalizers emphasize or cut low, mid or high frequencies selectively at either low levels or at high levels. These can be used as noise or hiss gates, rumble filters that only work at low levels (especially useful for traffic control in a delicate classical piece), sibilance controllers, or ambience enhancers. They can enhance inner details of high or low frequencies

at low levels, where details are often lost. They can be used to reduce harshness, enhance clarity at high levels or for other purposes, as described in detail in Chapter 10.

---

1   We're always seeking techniques (beyond simple equalization) to isolate one instrument from another, and it is possible to greatly improve the impact and clarity of the snare and other percussion instruments without changing the tonality of the vocal, using upward expansion with just the right attack and release times. It's frequently possible to enhance or punch a bass drum without significantly affecting the bass instrument, by using selective-frequency dynamics processing. And so on. See Chapters 10-11.

2   We all believe we have "the absolute sound" in our heads, but are surprised to learn how much tonal variance is tolerable as the ear/brain accomodates. Similarly, the eye accustoms itself to varying color temperatures, which only call attention to themselves when they change. A good photographer can usually identify Ektachrome from Kodachrome, but both look good on their own, and their color difference primarily shows up when you place two slides side by side.

3   Technically, **sibilance** can wreak havoc with the high frequency limiters in FM radio which are there to handle a preemphasis boost. An over sibilant vocal can cause the radio limiters to clamp down and lose definition, in extreme, the sound will bounce and words will be lost at the rate of the radio limiter's recovery time. Thus, overly bright records can sound dull on the air; brightness is self-defeating when it comes to radio processing.

4   In 1967, young George Massenburg began the search for a circuit which would be able to independently adjust an equalizer's gain, bandwidth and frequency. The key word is *independent*, for most analog circuits fail in this regard and the frequency, Q, and gain controls interact with each other. He called this circuit a **parametric equalizer** and his circuit remains proprietary today.

5   Moving coil cartridges sometimes have a dip in the 8 kHz range and a rise from 10 to 20 kHz, which gives them a *sweet sound*, amounting to a tone control in the reproduction system. I prefer my reproduction system to be neutral and to correct problems in the program material itself. But since a lot of older program material was equalized on lower resolution monitor systems, it makes sense to have a tone control in your home playback system.

6   This is dictated by the psychoacoustic *equal loudness curves*, first researched by Fletcher, Harvey and Munson in the 1930's.

7   **FIR** stands for Finite Inpulse Response, and **IIR** for Infinite Impulse Response. Readers interested in a detailed theoretical explanation of the difference between FIR and IIR filters should invest a little time in John Watkinson's **The Art of Digital Audio**.

8   Jim Johnston, in correspondence.

CHAPTER 9

How To Manipulate Dynamic Range for Fun and Profit

PART ONE:
MACRODYNAMICS

I. The Art of Dynamic Range

Dynamic Range is defined as the *ratio* between the loudest and softest passages of the body of the music; hence it should not be confused with loudness or *absolute* level; the term **dynamic range** is only concerned with **differences**. For popular music, this is typically only 6 to 10 dB, but for some musical forms it can be as little as a single dB or as great as 15 (very rare). In typical pop music, soft passages 8 to 15 dB below the highest level are effective only for brief periods, but in classical, jazz and many other acoustic forms, soft passages can last several minutes.

Microdynamics and Macrodynamics

The art of manipulating dynamics may be divided into **Macrodynamics** and **Microdynamics**. I call music's rhythmic expression, integrity or bounce, the *microdynamics* of the music. I call *macrodynamics* the loudness differences between sections of a song or song-cycle. Usually dynamics processors (such as compressors, expanders) are best for *microdynamic manipulation*, and manual gain riding is best for *macrodynamic manipulation*. The micro- and macro- work hand in hand, and many good compositions incorporate both microdynamic changes (e.g. percussive hits or instantaneous changes) as well as macrodynamic (e.g., crescendos and decrescendos). If you think of a music album as a full-course meal, then the progression from soup to appetizer to main course and dessert is the macrodynamics. The spicy impact

MYTH:
"Of course I've got dynamic range. I'm playing as loudly as I can!" *

---

* A common misconception. Thanks to Gordon Reid of Cedar for contributing this audio myth.

of each morsel, is the microdynamics. In this chapter we concentrate on macrodynamics.

Dynamics in Musical History

Dynamic changes became very important to western music sometime between the medieval Gregorian chants and the classical period, when composer Franz Josef Haydn surprised us with perhaps the first example of simultaneous micro- and macrodynamics.[1] Since ancient times, many "non-western" styles, such as African, Afro-Caribbean, Eastern, Indian, Balinese and other Oriental music forms, have stressed rhythm (microdynamics, especially in the form of percussion) as much as melody, and in the twentieth century of integration, heavy percussive rhythm became extremely important to western musical forms as well.[2]

Any genre that does not grow in musicality will quickly die, and dynamic contrast plays a big musical role. Today's Rap and Hip-hop music has taken a 250-year-old lesson from classical composition, by beginning to incorporate a melodic and harmonic structure. The genre can further grow and avoid sounding tiresome by expanding its dynamic range, adding surprises. Silence and low level material creates suspense that makes the loud parts sound even more exciting. Five big firecrackers in a row just don't sound as exciting as four little cherry bombs followed by an M80. Radio,

{ *The soundtrack for the movie* **The Fugitive** *is mixed like a relentless, fatiguing music single.* **Titanic** *was mixed like a beautiful record album.* }

TV and Internet distribution are currently too compressed to transmit the joy of wide dynamic range, but it sure turns people on at home, and also in the motion picture theater.

Films provide an ideal framework to study the creative use of dynamic range. The public is usually not consciously aware of the effect of sound, but it can play a role in a film's success. I think the movie *The Fugitive* succeeded because of its drama, but despite an aggressive, compressed, fatiguing sound mix. From the beginning bus ride, with its super-hot dialog and effects, all the crashes were constantly loud and overstated, completely destroying the impact of the big train crash. I can hear the director shouting, "more more more" to the mix engineers. Haven't they heard of the term *suspense?* Because when everything is loud, then really, nothing is loud. In contrast, the sound mix of '97's biggest movie, *Titanic*, is a masterpiece of natural dynamic range. The dialog and effects at the beginning of the movie are played at natural levels, truly enhancing the beauty, drama and suspense for the big thrills at the end. Kudos to director James Cameron and the Skywalker Sound mix team for their restraint and incredible use of dynamic range. That's where the excitement lies for me.

Life Imitates Art?

Clearly, modern recording techniques and equipment have aided in the creation of whole new musical styles, for example, hip hop, which uses digital editing and processing to create the beats of the music in a highly compressed, often low-dynamic-range style.[3] This is basically an extension of a trend in popular music that began many years ago with the invention of electric instruments and amplifiers, and has accelerated exponentially with modern recording techniques and powerful digital processors. Successive styles have incorporated less and less dynamic range, both macrodynamics and microdynamics. Going hand in hand with this trend is an exponential increase in distortion from style to style and year to year. This may very well be due to a vicious circle that is centered in the mastering engineer's hands, for inevitably, most masters tend to be more compressed than the sources[4]—and what sources do recording engineers listen to for inspiration? Mastered records! We may have bred the very disease which we seek to eliminate!

While I find the current high-distortion trend very fatiguing and unlistenable after short periods of time, we must remember that one man's meat is another man's poison—never more true in the case of popular music. Musical and sound styles have been created out of the very results of pushing digital compressors beyond their usual settings, for example, sound qualities such as *squashing* and *shred*. Which is why the successful mastering engineer must be familiar with and enjoy listening to many musical styles and sounds, including perhaps those sound qualities that would not normally be considered *clean* by practicing engineers. I simply hope that the cycle has reached its peak, since there's nowhere to go but back down, when music has dynamic range of 3 dB and distortion that tears the hair out of one's ears. In due time, these new styles will become assimilated into the larger musical vocabulary, and we can hope that decent and exciting dynamics will return as a rule rather than the exception.

The Art of Decreasing Dynamic Range

The dynamics of a song or song cycle are critical to creative musicians and composers. As engineers, our internal sound quality reference should be the sound quality of a live performance; we should be able to tell by listening if a recording will be helped or hurt by modifying its dynamics. Many recordings have already gone through several stages of transient-destroying degradation, and indiscriminate or further dynamic reduction can easily take the clarity and the quality downhill. However, usually the recording medium and intended listening environment simply cannot keep up with the full dynamic range of real life, so the mastering engineer is often called upon to raise the level of soft passages, and/or to reduce loud passages, which is a form of **manual compression**.[5] We may reduce dynamic range (compress) when the original range is too large for the typical home environment, or to help make the mix sound more exciting, *fatter*, more coherent, to bring out inner details, or to even out dynamic changes within a song if they sound excessive.[6]

Experience tells us when a passage is too soft. The context of the soft passage also determines whether it has to be raised. For example, a soft introduction immediately after a loud song may have to be raised, but a similar soft passage in the middle of a piece may be just fine. This is because the ears self-adjust their sensitivity over a medium time period, and may not be prepared for an instantaneous soft level after a very loud one. Thus, meter readings are fairly useless in this regard. How soft is too soft? The engineers at Lucasfilm discovered that having a calibrated monitor gain and a dubbing stage with NC-30* noise floor do not guarantee that a film mix will translate to the theatre. During theatre test screenings, some very delicate dialogue scenes were "eaten up" by the air conditioning rumble and audience noise in a real theatre. So they created a specially-calibrated noise generator, added to the mixing studio's monitor system, labeled "popcorn noise," which could be switched on whenever they wanted to check a particularly soft passage. For similar purposes, the "typical" (alternate) listening room we have at Digital Domain has a ceiling fan and other noisemakers. Whenever I have a concern, I start the DAW playing a loud passage just before the soft one, and take a walk to the noisy listening room. If the soft passage seems a bit too soft in comparison to the loud one, it will be obvious in there.

The Art of Increasing Dynamic Range...

...can also make a song sound more exciting, by using the art of contrast or by increasing the intensity of a peak, for much of the impact of a song comes from its internal dynamics and transients. The trick is to recognize when an enhancement has become a defect—musical interest can be enhanced by variety, but too much variety is just as bad as too much similarity. Musical taste, experience and a great monitor system are required to make these judgments. Increasing dynamic range is known as **expansion.** Another reason to expand is to restore, or attempt to restore the excitement of dynamics

which had been lost due to multiple generations of compression or tape saturation; in this case we are increasing the recorded range.

The Four Varieties of Dynamic Range Modification

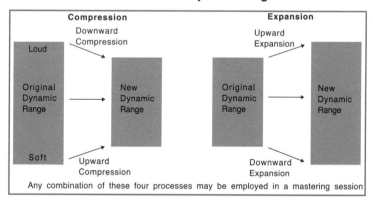

We always use the term **Compression** for the reduction of dynamic range and **Expansion** for its increase. There are two varieties of each: **upward compression, downward compression, upward expansion, and downward expansion**, as illustrated in the above figure.

Downward compression is the most popular form of dynamic modification, taking high level passages and bringing them down. Limiting is a special case—downward compression with a very high ratio (to be explained in Chapter 10). Examples include just about every compressor or limiter you have ever used. For clarity in this book, we will always use the short term **compressor** to mean **downward compressor** unless we need to distinguish it from **upward compressor.**

Upward compression takes low level passages and brings them up. Examples include the encode

---

* A room with an NC-30 rating is very quiet.

side of a Dolby® or other noise reduction system, the AGC[7] which radio stations use to make soft things louder, and the type of compressor frequently used in inexpensive video cameras and consumer VCRs. In Chapter 11 we will introduce you to a powerful upward compression technique that is extremely transparent to the ear.

Upward expansion takes high level passages and brings them up even further. Upward expanders are very rare and very precious, for in skilled hands they can be used to enhance dynamics, increase musical excitement, or restore lost dynamics. Examples include the peak restoration process in the playback side of a Dolby SR, the DBX Quantum Processor, the various Waves brand dynamics processors, and the Weiss DS1-MK2 when used with ratios less than 1:1 (to be explained).

Downward expansion is the most common type of expansion: it takes low level passages and brings them down further. Most downward expanders are used to reduce noise, hiss, or leakage. A dedicated noise gate is a special case—downward expansion with a very high ratio (to be explained). Examples of downward expanders include the classic Kepex and Drawmer gates, Dolby and similar noise reduction systems in playback mode, expander functions in multi-function boxes (e.g., Finalizer), and the gates on recording consoles. For clarity in this book, we will use the simple term **expander** to mean the downward type unless we need to distinguish it from the upward type.

II. The Art of Manual Gain-Riding: Macrodynamic Manipulation

In General

Level changes need to be made in the most musical way. To this end, internal level changes are least intrusive when performed manually (by raising or lowering the fader), as little as a 1/4 dB at a time, as opposed to using processors such as compressors or expanders, which tend to be more aggressive.

When gain riding, rock the boat the right way; try to go with the waves, don't fight them. If the musicians are trying for upward impact, pulling the fader back during a crescendo can be devastating since taking the fader down during a peak diminishes the intended impact. If you're doing a live recording and you sense the musicians are going to overload the recorder, you're already too late. The best case scenario is to use your sixth sense as early as possible, and lower the fader as slowly as possible, and only enough to fix the anticipated problem. An experienced live recording engineer will log where she made such changes, so that the original dynamic range may be restored by reversing the moves in post-production. Another trick is to measure peak levels during rehearsal, and assume the concert will have a peak at least 3 dB hotter! Having calibrated faders makes that adjustment easier. The art of manual leveling can really improve a production. We can enhance a great rock or pop mix during mastering, first by discovering any inappropriate level changes that the mix engineer may have missed, and by reversing them we can restore or enhance where the music is trying to go. I've heard many a rock piece where the climax was emasculated because the mix engineer kept on dropping the master fader to keep from overloading. In mastering we can correct for this unintentional error with delicate changes; it's amazing what a dB here or there can accomplish. It's also our responsibility to check with the client in case their level change was intentional! A great rock and roll mix is extremely rare; during mixing it's really hard to simultaneously pay attention to the internal balances as well as the dynamic movement of the music between, for example, verse and chorus. A sensitive mastering engineer will take a well-balanced mix the rest of the way; you may not even realize what was missing or how much it can be enhanced until you hear the mastered version. We try to enhance those moments where it should have swelled or dipped, for this is where some of the excitement of the song can be generated.

How and When to Move the Fader

Extra-soft beginnings, endings or even middle spots require special attention. If the highest point in the song sounds "just right" after processing, but the intro sounds too soft, it's best to simply raise the intro, finding just the right editing method to restore the gain to normal after the intro using one or more of these approaches:

- Sometimes a long, gradual decrescendo is the solution, which might occur at the end of the intro, or slowly during the first verse of the body.
- Sometimes a series of 1/4 or 1/2 dB edits, taking the sound down step by step at critical moments. This is useful when you don't want the listener to note that you're cheating the gain back down and

you may be forced to work against the natural dynamics.

- Sometimes a quick edit and level change at the transition between the raised-level intro and the normal-level body creates a nice effect and is the least intrusive.

The reverse approach, that is, purposely creating a softer intro so that the body of the song seems louder and has impact on the entrance can also work. In this case, the quick edit (gain change) between intro and body provides dramatic impact.

The Art of Changing Internal Levels of a Song

Some soft passages must be raised. But if the musicians are trying to play something delicately, pushing the fader too far can ruin the effect of the soft passage. The art is to know how far to raise it without losing the feeling of being soft, and the ideal speed to move the fader without being noticed. In a DAW, physical fader moves are replaced by commands, crossfades, or by drawing on a volume/time line. The true magic of the mastering engineer is to be so invisible that no one knows you have anything up your sleeve; if they think the sound is being manipulated, you haven't done your job.[8] Here's a technique for decreasing the dynamic range in the least damaging and most helpful way. I learned this over 30 years ago from Alec Nesbitt's book **The Technique of the Sound Studio** (see Appendix 10). When doing it live, you must know the score, to anticipate the moves of the musicians. But after the fact, on a digital audio workstation it's real easy, for the waveform is the score. Supposing that you must take a loud passage down. The best place to take the level down is at the end of the preceding soft passage before the loud part begins. Look for a natural dip or decrease in energy prior to the beginning of the crescendo, and apply the gain drop during the end of the soft passage before the crescendo begins. That way, the loud passage will not lose its comparative impact, for the ear judges loud passages in the context of the soft ones.

The figure at right from a Sonic Solutions workstation illustrates the technique. The gain change is accomplished through a crossfade from one gain to another.

The producer and I decided that the *shout chorus* of this jazz piece was a bit overplayed and had to be brought down from triple to double forte (which amounted to a dB or so).[9] To retain the contrast, the trick is to drop the level during the soft passage just before the drum hit announcing the shout chorus. You'll see this in the 12 second crossfade from unity gain (top panel) to -1.5 dB gain (bottom panel); the drum hit is just to the right of the crossfade box. If done right, you'll still feel goose bumps as the musicians make a delicate soft move (now enhanced with a further decrescendo by the mastering engineer), and then hit you with the chorus.

Some songs start with a very soft introduction, and this may have to be raised. Other songs start softly and build to a big climax. I like to start mastering by going directly to the climax. After I get a great sound with the necessary processing, I return to the beginning and if there's room, I may **lower** the gentle introduction, which will enhance the body that follows by contrast. This also reduces the temptation to raise the loud part so much that it

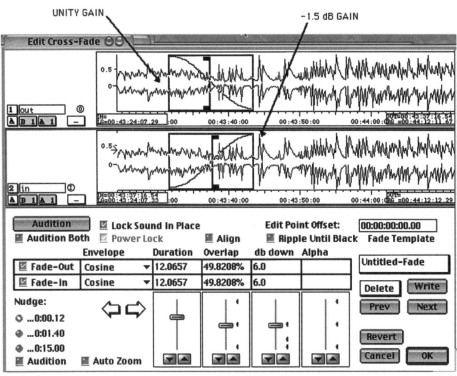

might be squashed by excessive processing. In the following figure, I've reduced the level of a song's introduction, and slowly introduce a crescendo (20 seconds long) that enhances the natural build of the song as it goes into the first chorus. The top panel is at −1 dB gain, bottom panel is at unity (0 dB) gain, achieved at the end of the crossfade.

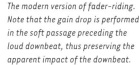

The modern version of fader-riding. Note that the gain drop is performed in the soft passage preceding the loud downbeat, thus preserving the apparent impact of the downbeat.

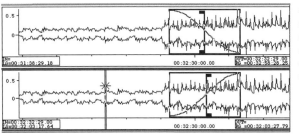

A soft introduction has been reduced even further, and the impact of the body of the song is enhanced by gradually increasing the gain during the beginning of the main part of the song.

Another trick is to increase the space before a song, which increases its dynamic impact by extending the tension caused by silence. Give the ear a chance to adjust to silence and then hit them with all you've got! The best musicians know how to use space within their music; they consider the rests to be as important as the notes.

In Conclusion

Macrodynamic manipulation is a sometimes overlooked but powerful tool in the mastering engineer's arsenal. In the next chapter we move on to the use of compressors, expanders and limiters to manipulate **microdynamics**.

1 Surprise Symphony, No. 94 in G, 1791, incorporated a mischievous drumbeat in the middle of a slow passage. This type of microdynamic instantaneous impact is often termed a sforzando in western music. To 20th century ears, Haydn's piece seems rather tame. Especially after you've been exposed to John Williams' quasi-classical **Suite from Close Encounters** reproduced on a decent Hi-Fi.

2 Especially with the influence of Afro-Caribbean musical forms on jazz (and eventually R&B, fusion, and rock) when in the 1940's Dizzy Gillespie brought percussionist Chano Pozo into his band.

3 Naturally with many exceptions. For example, I think The Geto Boys *Da Good Da Bad and Da Ugly*, one of the honor roll CDs (listed at www.digido.com), is a masterpiece of inventive musicality, dynamic range, depth, and tone on the same order as a good classical work.

4 It's hard for a mastering engineer to return a master to a producer that isn't louder than what was sent, even if the original recording was already too loud and compressed. But I find that producers like to receive recordings which are clearer and more impacting than what they sent in, even if the master is not quite as loud. Dare to try it!

5 Please do not confuse the term *dynamic range reduction* (compression) with *data rate reduction*. Digital Coding systems employ data rate reduction, so that the bit rate (measured in kilobits per second) is less. Examples include the MPEG (MP3) or Dolby AC-3 (now called *Dolby Digital*) systems.

   Since it's not good to refer to two different concepts with the same word, we should encourage people to use the term **Data Reduction System** or **Coding system** when referring to data and **Compression** only when referring to the reduction of dynamic range.

6 Excessive is definitely in the ear of the behearer! It's very important to develop an esthetic which appreciates the benefits of dynamic range, and which also knows when there is too much—or too little. This is clearly a matter of taste, as well as objective knowledge of the requirements of the medium and listening environment.

7 AGC (automatic gain control) has been given a bad name by its ubiquitous use in consumer and professional camcorders. Listen to the news reports on TV where a portable camera was used with AGC to see what I mean. You will hear severe hiss modulation in between syllables, and the transient syllabic impact is reduced.

8 This is true for most of the "natural" music genres, with some exceptions being hip-hop, psychedelic rock, performance art, etc., where the artists invite the engineer to contribute surprising or rococo dynamic effects.

9 Producers don't always use classical Italian dynamic terms to describe their needs. The mastering engineer should chose the bonding language which is best for the client—"Make it louder, man!"

CHAPTER 10

How to Manipulate Dynamic Range for Fun and Profit

PART TWO:
DOWNWARD
PROCESSORS

I. Compressors and Limiters: Objective Characteristics

Part two and Part three of this series are about microdynamic manipulation, which is primarily achieved through the use of dedicated **dynamics processors**. In this chapter (part two), we look at how *downward* processors work. Before we can learn how to use devices such as compressors and expanders, we must study the objective characteristics of the devices which perform the job.

Transfer Curves (Compressors and Limiters)

Let's begin with the **measurable characteristics** of processors which perform downward compression, simply called **compressors and limiters**.

A transfer curve is a picture of the input-to-output gain characteristic of an amplifier or processor. A straight wire or unity-gain* amplifier would yield a straight diagonal line across the middle at 45°, called the *unity gain line*. A family of **linear curves** can be drawn, as in these three figures:

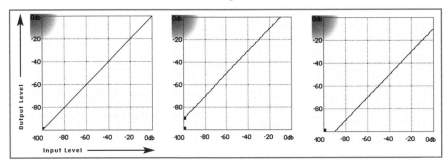

Three transfer curves. *At left, a Unity-Gain Amplifier, then an amplifier with 10 dB gain, then with 10 dB loss (attenuation).*

---

* Unity-gain means the ratio of output to input level is 1, or 0 dB.

Input level is plotted on the X axis, and output on the Y. At left is a unity gain amplifier, followed by one with 10 dB gain, and with 10 dB loss (attenuation). As long as there is a straight line (not a curve) at 45°, the amplifiers are linear. Notice that the middle plot would yield distortion for any input signals above −10 dBFS.

The threshold of a compressor is defined as the level above which gain reduction begins to occur. **Compression ratio is the ratio of input change to output change above the threshold.** At left in the following figure is a simple compressor with a fairly gentle 2.5:1 compression ratio, and a threshold at around −40 dBFS (which is quite low and would yield strong compression for loud signals). 2.5:1 means that for a level increase of the source of 2.5 dB, the output will only go up 1 dB, or for a rise of 5 dB, the output will only go up 2 dB, or as can be seen in the plot, an input change of 20 dB yields an output change of a little less than 10 dB (once the curve has reached its maximum slope). A compressor such as this would actually make loud passages softer, because the output is less than the input above threshold; this is always the case unless you follow the compressor with a gain makeup amplifier.

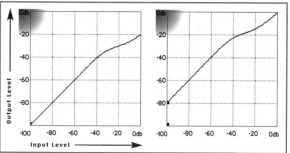

At left, Compressor with 2.5:1 ratio and −40 dBFS Threshold and no gain makeup. At right, the same compressor with 20 dB gain makeup.

At the right-hand side of the figure, by using gain makeup (a simple gain amplifier after the compression section), we can restore the gain such that a full level (0 dBFS) signal input will yield a full level signal output. In this illustration, the amplifier has an extreme amount of gain, 20 dB, which would considerably amplify soft passages (below the threshold). In typical use, makeup gains are rarely more than 3 or 4 dB. Loud input passages from about −40 to about −15 are still amplified in this figure, but above about −15 dBFS, the curve slopes back to unity gain and resembles that of a linear amplifier. Far below the threshold, it's a fairly linear 20 dB amplifier and can have pretty low distortion because there is no gain reduction action. At full scale, 20 dB of gain makeup is summed with 20 dB of gain reduction, yielding 0 dB total gain. This particular compressor model's curve levels off towards a straight line above a certain amount of compression, so the ratio only holds true for the first 15-20 dB above the threshold. Other compressor models continue their steep slope, thus maintaining their ratio far above the threshold. There are as many varieties of compression shapes as there are brands of compressors, and they all give different sounds. To get the greatest esthetic effect from any compressor, most of the music action must occur around the threshold point, where the curve's shape is changing; thus, it is likely a real-world compressor's threshold would be nearer −20 to −10 dBFS, where most of the musical movement takes place.

The following figure shows a very high ratio of 10:1, without gain makeup. Notice that the output is almost a horizontal line above the threshold. Most authorities call any compressor with a ratio of 10:1

or greater a limiter. There are very few analog compressors with greater ratios, however, some digital limiters have been built with ratios of 1000:1 in order to prevent even the minutest excursion or overload above full scale (0 dBFS). The portion of the curve at or near the threshold is called the **knee**, which is the transition between unity gain and compression. The shape of the knee can make the transition gentle, or hard. The term **soft knee** refers to a rounded knee shape, and **hard knee** to a sharp shape, where the compression or limiting

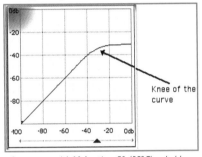

Compressor with 10:1 ratio, -32 dBFS Threshold, without gain makeup

kicks in quickly above the threshold. Conceivably, the change from unity to 10:1 could be instantaneous, in which case the knee

would be a sharp angle instead of round, producing a sharp sonic change, thus a limiting effect. The need for a gentle knee depends a lot on how much musical activity is occurring at the threshold. If there is a lot of musical activity or movement around the threshold, the knee shape can be critical. For those models of compressors that do not have knee adjustments, some of the effect of the knee can be accomplished by tweaking the ratio and/or threshold.

Attack and Release Times

Attack time is defined as the time between the onset of a signal that is above threshold and full gain reduction. It can be measured in micro or

milliseconds though it can be as long as a second or two. Typical compressor attacks used in music range from 50 ms to 300 ms, with the average used probably 100 ms. **Release time**, also known as **recovery time**, is defined as the time between when a signal drops below threshold and when the gain returns to unity. Typical compressor release times used in music range from 50 ms to 500 ms or as much as a second or two, with the average used probably 150-250 ms.* The terms **short** or **fast** with attack or release time may be used interchangeably, they mean the same thing. Similarly, **slow** and **long** attack and release times mean the same.

At the left side of the following figure is the envelope shape of a simple tone burst, from a high level to a low one and back again.

At left, a simple tone burst from high to low level and back. At right, the same tone burst passed through a compressor with very fast attack, high ratio, and fast release time

At the right side is the same tone burst passed through a compressor with a very fast attack, high ratio, and fast release, and whose threshold is midway between the loud and soft signals. Note that the loud passages are instantly brought down, the soft passages are instantly brought up and there is less total dynamic range, judging by the relative vertical heights (amplitudes).

---

* One manufacturer, DBX, measures release time in dB/second, which is probably more accurate, but I find hard to get used to.

At left in this next figure is the envelope of a compressor with a low ratio, slow attack time and a slow release time. Notice how the slow attack time of the compressor permits some of the original transient attack of the source to remain until the compressor kicks in, at which point, the gain reduction brings the level down. Then, when the signal drops below threshold, it takes a moment for the release time to take action, and the gain is still low, then slowly the gain comes back up. A lot of the compression effect (the "sound" of the compressor) occurs during the critical release period, since as you can see, except for the attack phase, the compressor has actually reduced gain of the high level signal.

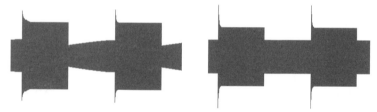

At left, a Compressor with a low ratio, slow attack time and slow release time. At right, higher ratio, faster attack and very fast release.

Contrast this with the compressor at the right, which has a much higher ratio, faster attack, and very fast release time. The higher ratio clamps the high signal down farther, and with the fast release, as soon as the signal goes below threshold, the release time aggressively brings the level up. This type of fast action can make music sound strongly compressed because it brings down the loud passages and quickly brings up the soft passages.

Here is another variation, a compressor with a release delay:

Output of a Compressor with a low ratio, slow attack time, slow release time plus release delay

A release delay control allows more flexibility in painting the sound character. Very few compressors provide this facility. It's useful when we want to retain more of the natural sound of the instrument(s), not exaggerate its sustain when the signal instantly goes soft, or reduce "breathing" or hissing effects when the source is noisy. The release delay is part of the subtle pastel color palette of the mastering artist.

The next figure illustrates what happens when the attack and release times are much too fast.

When the combination of attack and release times are extremely fast (typically <50 ms), a compressor can produce severe distortion, as it tries to follow the individual frequencies (waves) instead of the general envelope shape of the music

The distortion is caused by the compressor's action being so fast that it follows the shape of the low frequency waveform rather than the overall envelope of the music. This problem can occur with release times shorter than about 50 ms and correspondingly short attack times.

II. Microdynamic Manipulation: Adjusting the Impact of Music with a (downward) Compressor

The Mixing Engineer as Artist

Compressors, expanders and limiters form the foundation of modern-day recording, mixing and mastering. With the right device you can make a recording sound more percussive or less percussive, punchy or wimpy, smooth or bouncy, good or bad, mediocre or excellent.

When used by skilled hands, compression has produced some of the most beautiful recordings in the world, and a lot of contemporary music genres are based on the sound of compression, both in mixing and mastering, from Disco to Rap to Heavy Metal. A skilled engineer may intentionally use creative compression to paint a mix and form new special effects; this *intended distortion* has been used in every style of modern music. The key words here are *intent and skill*. Surprisingly, however, some engineer/artists don't know what uncompressed, natural-sounding audio sounds like. While more and more music is created in the control room, I think it's good to learn how to capture natural sound before moving into the abstract. Picasso was a creative genius, but he approached his art systematically, first mastering the natural plastic arts before moving into his cubist period. Similarly, it's good practice to know the real sound of instruments. Try recording a well-balanced group in a good acoustic space with just two mikes; it's a lot of work, and a lot of fun! Before multitracking was invented, there was much less need for compression, because close miking exaggerates the natural dynamics of instruments and vocals. At first, compressors were used to control those instruments whose dynamics were severely altered by close miking, e.g. vocals and acoustic bass. Later, when modern music began to emphasize rhythm, many instruments began to get lost under the energy, inspiring the creative possibilities of compressors and a totally new style of recording and mixing. Certainly the advent of the SSL console, with a compressor on every channel, changed the sound of recorded music forever.

Limiting Versus Compression In Mastering

Mastering requires new skills to be developed since we generally work on overall mixes instead of individual instruments. In mastering as well as mixing, compression and limiting change the peak to average ratio of music, and both tools reduce dynamic range. Most mastering engineers use compressors to intentionally change sound and limiters to change sound as little as possible, but simply enable it to be louder.* That's why limiters are used more often in mastering than in mixing. There is no perfectly invisible limiter, but compression changes the sound much more than limiting does. Think of compression as a tool to change the inner dynamics of music. While reducing dynamic range, it can "beef up" or add "punch" to low- and mid-level passages to make a stronger musical message. With limiting, however, with fast enough attack time (1 or 2 samples), and a

---

* As with compressors, it is the gain makeup process that permits the output of a limiter to be louder. When the peaks have been brought down, there is room to bring the average level up without overloading.

carefully-controlled fast release,* even several dB of limiting can be transparent to the ear. *Consider limiting* when you want to raise the apparent loudness of material without severely affecting its sound; *consider compression or upward expansion* (see next Chapter) when the material seems to lack punch or strength or rhythmic movement.

The BBC performed research in the 1940's demonstrating that distortion shorter than about 6-10 ms is fairly inaudible, which was the basis for the 6 ms integration time of the BBC PPM meter. In this modern solid-state world, some transient distortion as short as 1 ms will change the audible sound of the initial transient, particularly for instruments such as piano. So be sure to use your ears before limiting or reducing even short transients. With good equipment and mastering technique, wide range program material with a true peak to average ratio of 18 to 20 dB can often be reduced to about 14 dB with little effect on the clarity of the sound. That's one of the reasons 30 IPS analog tape is desirable as the medium to mix to: it has this limiting function built-in. A rule of thumb is that short duration (a few milliseconds) transients of **unprocessed digital sources** can be reduced by 4 to 6 dB with little effect on the sound; **however, this cannot be done with analog tape sources**, which have already lost the short duration transients. Any further transient reduction by

---

* The faster the release time, the greater the distortion, which is why the only successful limiters which use extra fast release times have **auto-release control**, which slows down the release time if the duration of the limiting is greater than a few milliseconds. The effective release time of an auto-release circuit can be as short as a couple of milliseconds, and as long as 50 to 150 milliseconds. If limiting a very short (invisible) transient, the release time can be made very short.

compression or limiting will not be transparent (though it may still be esthetically acceptable or even desirable).

All digital limiters affect the sound to some extent, softening the transients and even fattening the sound slightly, as they allow us to raise the average level and the loudness. The less limiting we use, the cleaner and more *snappy* the sound, unless we are looking for a sound with softer transients. In an ideal mastering session, the limiter should only be acting on occasional inaudible peaks. Limiting distortion is especially audible on material which already has little peak information because a limiter is not designed to work on the RMS portion of the music and limiters can sound pretty ratty when pushed into the RMS region. Watch out for severe bass distortion because the time constants of a limiter are too fast for optimal compression.

A manual for a certain digital limiter reads "For best results, start out with a threshold of −6 dBFS." This is like saying "always put a teaspoon of salt and pepper on your food before tasting it." Instead, mastering engineers should judge how much limiting to use based on the desired absolute loudness (compared with other CDs) and how much degradation we can accept. Some sources can tolerate 6 dB of limiting without significant degradation, others 1 or none.

The World's Most Transparent Digital Limiter

The most transparent limiter is to use no limiter at all! When we are trying to make a section louder, if there is a very short peak (transient) overload, for example, during a section of a drumbeat, a skilled mastering engineer can perform a short-duration gain drop that can be invisible to the ear, with the DAW's editor. This **manual limiting** technique allows us to raise a song's apparent loudness without the attendant distortion of a digital limiter, so it is the first process to consider when working with open-sounding music that can be ruined by too much processing. We can often get away with 1 to 3 dB manual limiting typically for a duration of less than 3 ms. But longer duration gain drops will affect the sound as much as or more than a good digital limiter. We use as little gain reduction as possible and when trying to make material louder, squeeze as much level as possible without clipping, for it helps keep the limiting invisible.

Equal-Loudness Comparisons

Since loudness has such an effect on judgment, it is very important to make comparisons at equal apparent loudness. During an instant A/B comparison the processed version may seem to sound better, if it is louder, but long-term listeners prefer a less fatiguing sound which "breathes." When you make comparisons at matched apparent loudness, you may be surprised to discover that the processing is making the sound worse, and it was all an illusion.

The Nitty-Gritty: Compression in Music Mastering

Consider this rhythmic passage, representing a piece of modern pop music:

shooby dooby doo **WOP**…
shooby dooby doo **WOP**…
shooby dooby doo **WOP**

The accent point in this rhythm comes on the backbeat (**WOP**), often a snare drum hit. If we strongly compress this music piece, it might change to:

SHOOBY DOOBY DOO WOP…
SHOOBY DOOBY DOO WOP…
SHOOBY DOOBY DOO WOP

This completely removes the accent feel from the music, which is probably counterproductive.

A light amount of compression might accomplish this…

shooby dooby doo **WOP**…
shooby dooby doo **WOP**…
shooby dooby doo **WOP**

…which could be just what the doctor ordered for this music because strengthening the sub accents may give the music even more interest. Unless we're trying for a special effect, and purposely creating an abstract composition it's wrong to go against the natural dynamics of music. (Like the TV weatherperson who puts an accent on the wrong syllable because they've been taught to "punch" every sentence: "The weather **FOR** tomorrow will be cloudy"). Much of hip hop music, for example, is intentionally abstract—anything goes, including any resemblance to the natural attacks and decays of musical instruments.

To manipulate the music requires careful adjustment of threshold, compressor attack and release times. If the attack time is too short, the snare drum's initial transient could be softened, losing the main accent and defeating the whole purpose of the compression. If the release time is too long, then the compressor won't recover fast enough from the gain reduction of the main accent to bring up the subaccent (listen and watch the bounce of the gain reduction meter). If the release time is too fast, the sound will begin to distort. If the combination of attack and release time is not ideal for the rhythm of the music, the sound will be "squashed," and louder than the source, but "wimpy loud" instead of "punchy loud." It's a delicate process, requiring time, experience, skill, and an excellent monitor system.

The best place to start adjusting a compressor is to find the approximate threshold first, with a fairly high ratio and fast release time. Adjust the threshold until the gain reduction meter bounces as the "syllables" you want to affect pass by. This ensures that the threshold is optimally placed around the musical accents you want to manipulate, the "action point" of the music. Then reduce the ratio to very low and put the release time to about 250 ms to start. From then on, it's a matter of fine tuning attack, release and ratio, with possibly a readjustment of the threshold. The object is to put the threshold in between the lower and higher dynamics, so there is a constant alternation between high and low (or no) compression with the music. Too low a threshold will defeat the purpose, which is to differentiate the "syllables" of the music; with too low a threshold **everything will be brought up to a constant level**.

Typical Ratios and Thresholds

When working on microdynamics in the above fashion, compression ratios most commonly used in music mastering are from about 1.5:1 through about 3:1, and typical thresholds in the −20 to −10 dBFS range. But there is no rule; some engineers get great results with ratios of 5:1, whereas a delicate *painting* might require a ratio as small as 1.01:1 or a threshold of −3 dBFS. Sometimes a recording requires the most gentle *invisible* compression without trying to alter its built-in dynamics. One trick to compress as invisibly as possible is to use an extremely light ratio, say 1.01 to 1.1 and a very low threshold, perhaps as low as −30 or −40 dBFS, starting well below where the action is. We may choose a low ratio to lightly control a recording that's too *jumpy* or to give a recording some needed *body*. It's unusual to see such low ratios used in tracking and mixing but very common in mastering of full program material, partly because with full program material, larger ratios may draw attention to the magic behind the curtain or reveal breathing, pumping or other artifacts.

We have noted before that every brand of processor (both compressors and expanders) has its own unique characteristics and sound. Part of the fun of mastering (and mixing) is discovering the special characteristics of different compressors. Even with the same settings, some are *smooth*, others are *punchy*, some bring out percussion better than others. This is not due to attack and release times per se, but rather to the curve or acceleration of the time constants, whether the device recovers linearly from gain reduction, whether the gain returns to unity quickly or slowly at the beginning. Design engineers spend much research time psyching out these particular characteristics, and the best we poor mortals can do is listen and see what we like.

Fancy Compressor Controls

Some compressors provide a **crest factor** control, usually expressed in decibels, or a range from RMS (or full average) to quasi-peak through to full peak. What this means is that the compressor acts on either the average parts of the music, the peak parts, or somewhere in between. Ostensibly, compressors with RMS characteristics sound more natural as they correspond with the ear's sense of loudness, but the best-sounding compressor I own is peak-sensing.

The Weiss model DS1-Mk2 is the first dynamics processor I've encountered with **two different release time** constants, *release fast* and *release slow*. The user sets a threshold of average transient duration, such as 80 ms, above which a sound movement is called *slow*, and below which it is called *fast*. Thus, instantaneous transients can be given a faster release time, but sustained sounds a slower one, which results in a more natural-sounding compression, especially with heavy compression. Indicator lights on the front panel aid in these adjustments.

Compression and Monitoring

I recall mixing a purist jazz recording using excellent powered monitors equipped with a driver protection circuit, which is ostensibly inactive except on peaks. However, when I arrived at my mastering room, I discovered that the recording "jumped out" too much, and required a bit of compression, a fact hidden during the mix and which I feel would have been similarly hidden had I monitored the mix with low-powered tube amplifiers (which self-compress).

As I mentioned in Chapter 6, it is a myth that you have to "precompress" for small systems. It's actually the converse. I made an excellent *snappy-sounding* master where we were concerned that the upper dynamics might have a bit too much upward impact. But when the recording was auditioned on a typical boom box or bookshelf system, the peaks were squashed compared to the mastering room audition and actually would have benefited from even more impact. Thus I have learned that if it "sticks out a little too much" on a high-headroom mastering system, then it's probably going to be fine when played on an inferior system. However, you'll never learn if something needs a bit more compression or is too compressed when listening on a monitor system that squashes the sound.

Multiband processing

Multiband compression is probably the most powerful and potentially deadly audio process that's ever been invented. Basically, a multiband processor splits the information into two, three or more frequency bands, so that the compression action in one band will not cause another band to be affected. For example, if the vocal causes a bit of gain reduction, it will not pull down the bass drum (or vice versa), which might occur if you used a full-band compressor. This is the virtue and the

MYTH:
Program Compression is required to protect small reproduction systems.

justification of splitting processing into multiple bands. However, multiband compression has been overused, and hyped in my opinion. It can easily produce very unmusical sound or take a mix where it doesn't want to be. This tool requires careful judgment on the part of the mastering engineer.

Multiband processing was probably first introduced by TC Electronic in their M5000, then in their ubiquitous Finalizer, and brought to great sophistication (and much better sound quality) in their System 6000. Tube-tech has produced a three-band tube compressor. But multiple bands are hardly needed; one or two bands are usually enough. Rarely do even hip-hop recordings need more than two bands to sound punchy and strong. I use more than two bands in my mastering no more than a few times a year, when multiple bands have been a lifesaver. I largely use multiband compression (and expansion) to fix bad mixes that could not be remixed, for one key to a great master is to start with a great mix!

{ *"One key to a great master is to start with a great mix."* }

When To Consider multiband processing

- When there is a heavy and somewhat isolated bass drum and/or bass, splitting the processing into two bands prevents the drumbeats from modulating the rest, or vice versa.
- When you want to let transients (percussive sounds) through while still *punching* the sustain of the sub accents or the continuous sounds. Transients contain more high frequency energy than continuous sounds, so splitting the processing

into a low and a high band permits using gentler compression or no compression at high frequencies (e.g., higher threshold, lower ratio).
- When there is too much sibilance. Sibilance can be controlled by using selective compression in the 3 through 9 kHz range (the actual frequency has to be tuned by listening to the vocalist). Try a very fast attack and medium release and a narrow bandwidth for the active band.
- When the mix is bad or certain elements appear to be weak in the mix, multiband processing can save the day, assuming a remix is not possible. I once received a rap project that was somehow mixed with very low vocal and extremely loud percussion and bass drum, and a remix was not possible. By compressing and then raising the level of the frequencies in the vocal range (circa 250 Hz) I was able to *remix the piece* and very nicely, turn the vocal up. Clearly, multiband compression is a power that should be used very wisely!

However, before trying multiband, first

- See if simply raising the attack time in a one-band compressor permits sufficient transient energy to come through. Or, try upward expansion (described in the next Chapter) instead.
- Try using few bands, only two if possible. This avoids potential phase shift and unnatural relationships between the mix elements of the mix, which can become the enemy of the mix engineer's delicate creation.

Equalization or Multiband Compression?

When multiband processing is available, the line between equalization and dynamics processing becomes nebulous, because the output levels of each band form a basic equalizer. Use plain equalization when instruments at all levels need alteration. Or consider multiband compression, to provide spectral balancing at different levels. For example, a song may get harsh-sounding when it gets loud, and it is possible to simulate the euphonic high-frequency saturation characteristics of analog tape by using a bit more compression at high frequencies.

If we're already using split dynamics, we make our first pass at equalization with the outputs (makeup gains) of each band. Multiband compression and equalization work hand-in-hand. Tonal balance will be affected by the crossover frequencies, the amount of compression, and the makeup gain of each band. In general, the more compression, the duller the sound, because of the loss of transients. I first try to solve this problem by using less compression, or altering the attack time of the high-frequency compressor, and as a last resort, I use the high frequency band's makeup gain or an equalizer to restore the high-frequency balance.

Clipping, Soft Clipping and Oversampled Clipping

Clipping is the result of attempting to raise the level higher than 0 dBFS, producing a square wave, a severe form of distortion. Clippers are devices which electronically cut momentary peaks out of the waveform to allow the overall level to be raised. Soft clipping attempts to do this with less distortion. I've decided that I don't like the quality of distortion produced by clipping or soft clipping, at least at 44.1 kHz SR (see Chapter 16). I believe there are better approaches. The first is not to raise the

level at all, for many CDs are already too hot for their own good. Or use a good limiter, which sounds better than clipping to my ears. In Appendix 1, radio gurus Bob Orban and Frank Foti explain why clipping is a severe problem for radio processors. The jury is still out when it comes to oversampled clipping, whose distortion artifacts can be reduced by half in the audible (20-20 kHz) range, but isn't that really like saying *she's a little bit pregnant?*

Compression, Stereo Image, and Depth

One sure way to destroy the depth in a recording is to compress it too much. Compression brings up the inner voices in musical material. Instruments that were in the back of the ensemble are brought forward, and the ambience, depth, width, and space are degraded. But not every instrument should be "up front". Pay attention to these effects when you compare processed vs. unprocessed and listen for a long enough time to absorb the subtle differences. Variety is the spice of life. As always, make sure the cure isn't worse than the disease.

The Mastering Engineer's Dilemma

Without compressors in CD changers and in cars, it is extremely difficult for the mastering engineer to fulfill the needs of both casual and critical listeners. It is our duty to satisfy the producer and the needs of the listeners, so we should continue to use the amount of compression necessary to make a recording sound good at home. But try to avoid using more compression than is

{ *"Not every instrument should be up front."* }

> *"Never in the history of mankind have humans listened to such compressed music as we listen to now."* — Bob Ludwig*

required for home listening. This approach will actually help radio play (see Appendix 1). If compromises have to be made for car or casual play, try to use transparent-sounding techniques such as parallel compression (see next Chapter), which satisfy even critical listeners. Audition test masters in all environments, hopefully arriving at a decent compromise.

III. For the Mixing Engineer: How To Avoid Hypercompression† during Mixing and Tracking

Letter from a DIGIDO.COM visitor:

> I found your site through a link. I was looking for information on how to use my compressors to make my music better. What I found was instruction on how not to use my compressors to make my music better. The quality of my recordings has gone up greatly since I read your articles.

How to Avoid making Hypercompressed Mixes

Hypercompression is a form of sound squashing, where everything has an unrelenting and fatiguing intensity, with lost transients and reduced definition. When overused, mastering

---

* In correspondence. A variation of this quote is in Owsinsky, Bobby. *Mastering Engineer's Handbook*.

† The expressive term **hypercompression** was coined by Lynn Fuston of 3D Audio.

---

tools can produce this result, though the tools to do it have migrated to the mixing studio, with a lot of unfortunate sonic results (and a few sonic gems), in my opinion. Hypercompression produces the reverse effect from the intent of a good mix— *boring, lifeless mush*. Perhaps the current slack in music sales is related to hypercompression and its tendency to give everything a monotonous sameness—is the public voting against compression with its pocketbook? Lately it seems about the only place we can enjoy good dynamic range and impact is in the motion picture theatre. This book is partly about how we can bring similar life to our music masters. In this chapter we concentrate on some advice for the mixing engineer.

Let me tell you a sad story. A pop-rock band once sent me a mix that they felt a bit uneasy about, though they could not exactly express why. When I received the DAT it was obvious why. Here's what I heard:

· there was absolutely no dynamic range left, it was "maxxed to the max."
· there was no transient information.
· the sound was grainy and literally lifeless (squashed)
· all the songs sounded continuously and fatiguingly loud. I couldn't listen for more than a couple of minutes at a time.
· although the obvious intent was to produce a hot, clear, punchy sound, the result was exactly the opposite.

No wonder the band felt uneasy, but still they couldn't put their finger on the problem. All the mix elements were there, and the tonality seemed fine. It was easy for me to tell: their engineer had **mixed directly from multitrack through a 3-band mastering compressor to DAT**. In a way I admired his work because he obviously had slaved for hours at the dials "perfecting" this most disappointing sound. Amazingly there were no intermodulation artifacts between the frequency bands, an example of the power of this box, for I was instantly able to identify the brand and type of processor he had used. I called the group and asked them to check if he had made an unprocessed mix as well. Unfortunately he had not. Sadly, I was unable to do anything to salvage this production. I tried a bit of upward expansion (to undo the damage), and the band felt it was an improvement, but an upward expander can only accomplish something when there is "movement" in the source to grab onto (to amplify). Why do you suppose he did this? The motivation was eventually traced to a misguided desire to make the recording "radio-ready" (see sidebar).

Here are some ways to avoid hypercompression during mixing, which easily occurs when consoles and DAWs have a compressor on every channel strip. Everyone has his own style of working with compressors and there are no rules. But I suggest that when learning or beginning a mix, start by working *without* any compressors! Then you'll discover the necessity which was the mother of its invention. The compressor will then become for you a tool to handle problems which cannot be handled with fader moves, not a crutch or substitute for good recording and mixing techniques. Learn about the natural dynamics and impact of musical instruments, then begin to alter them with

compressors (which can include using compression to create special effects). Every 5 years or so, give yourself a reality check...try making a recording or mix with little or no compression. You'll rediscover the parts of music that make it lively and aid in its clarity. It's a real challenge, but a refresher course may point out that *less compression* will buy you a more open, more musical sound than you've previously been getting.

Start mixing fresh each time— free yourself of preconceptions. Although you compressed the bass on 9 out of the last 10 albums, maybe this time you won't need a compressor. Each musician is an individual and their sound must be respected. In general, the better the bass player, the less compression will be needed, and the greater the chance that compression will "choke up" his sound. If you get to know the sound of your instrumentalists you can then ask yourself: are you trying to capture the sound of your instrumentalists or intentionally creating a new sound? Get a great mix that sounds **alive** and **clear** and **big** and then later see how much better it can be made in the mastering suite, for mixing and mastering are two different things. After mixing for a while, compare the mix to the raw, unaltered monitor mix (which can be a sobering experience); be honest, have you lost some of the magic that you captured on the recording day? Has the sound closed down instead of opening up?

> **The Real Recipe for Radio-Ready**
> The real recipe for Radio-Ready includes:
> 1) Write a great original song, use fabulous singers and wonderful arrangements.
> 2) Be innovative, not imitative.
> 3) Make sure the music sounds good at home. Keep the dynamics lively, interesting and unsquashed, and some of that virtue will make it through the radio processing.

---

* Not every piece of music should be *big-sounding*, but I think you get the idea.

The process of refining a mix should always include revisiting your compression (and EQ) settings and questioning your work. Compressors are often used to create a tighter band sound, making the rhythm instruments sit in a good, constant place in the mix. But the wrong compression setting can take away the sense of natural breathing and openness that makes music swing and sway. Thus, I recommend that during mixing, after you've inserted a few compressors on certain instruments (e.g., the bass, rhythm guitar, vocal) and listened for a while, try comparing with the compressors bypassed (total automation makes that process easy; store two fader snapshots so you can switch between them). If you've lost some of the swing, or the subtleties of the musician's performance, then try reducing or eliminating some compression.

I think some of today's mix engineers have to learn (or relearn) the ability to mix loudly and clearly. Rock and Roll music is often a casualty of compressor abuse. I receive rock mixes from well-meaning engineers that should be getting louder and louder and reach a climax, but which have lost their intensity, producing *wimpy loud sound.*[*] There is dynamic inversion; instead of a chorus sounding lively and dramatic, it's been pulled back. To make a better sound and ease the mastering engineer's job, check the climaxes; do they sound open, or squashed? Squashing is a common problem in rock mixes, for it is very difficult to maintain excitement all the way to the highest peaks, but squashing is very

hard to repair in mastering. One trick is to start mixing during the climax of the song, make the climax sing and swing, using just enough compression on individual instruments to do the trick; then, return to the beginning, work your butt off riding faders where necessary during the soft passages **but without changing the thresholds from the position used for the peak of the song.** This helps avoid overcompression on the loud passages and keeps the song sounding exciting. It's better to send material that's mixed well and powerfully at the mid levels but at the high levels is not squashed. Even if the climaxes don't sound loud enough to the mix engineer, he should consider it a *work in progress*, for the mastering engineer can take it to the next level of performance, with the punch it needs at mid levels and strength and volume at high levels.

I advise against mix engineers trying to mix through dedicated mastering processors unless you have the patience to refine the many parameters against the constantly-changing parameters of a mix in progress. Even bus compressors built into consoles are not usually optimized for processing overall music. A processor on the bus will change the mix in mysterious ways; it's not predictable whether the vocal or any instrument will stand out, and it can fight the mix instead of helping it. Wideband bus compression causes all the instruments to be modulated by the attack and transients of the loudest instrument. A rim shot or cymbal crash can take down the reverberation and the sound of all the other instruments. Any compressor on a mix bus can quickly become a

---

[*] "It's like there has been an unlearning curve. As flexibility has improved, respect for the integrity of the source has all but vanished as people become lost in the possibilities." Bob Olhsson, Mastering Engineer's Webboard.

crutch, a substitute for good mixing techniques. Some mix engineers add delicate bus compression *after the mix has been achieved*, to see if it fattens the sound without deterioration. And to keep the bus compressor from punching "holes" in your mix, they use a very slow attack/release and very little compression (e.g. 1 dB).

Hedge Your Bets. Many mix engineers will subvert **Murphy's Law of Experience** and print two versions to send to mastering, one with bus compression and one without. I often find the bus-compressed version has fatter bass (which the client likes) but wimpy highs and attacks (which the client doesn't like), but in mastering you can have your cake and eat it too: I can supply dynamics processing with carefully-applied multiple time constants, yielding a more impacting result that still has "fat bass." Of course, if the mix was made so aggressively through the bus compressor that removing it would change the mix, then there is no point in providing two versions; be aware that you are painting yourself into a corner, if a remix is not an option.

But what if you want to mix *aggressively*...

This should be the province of the experienced mixer who knows that this is the practice that works best for the particular music, client, or audience and who recognizes the fine subjective line between **aggressive bus** compression and hypercompression. In other words, some engineers mix aggressively on purpose with the bus compressor (or against it); which is only ok if:

- the music truly calls for it
- the experienced mix engineer is aware of all the effects of the bus compressor on the sound

But be careful how you make it loud, because if you deteriorate the clarity of the sound, there's little that can be done to fix it in the mastering. When mixing with aggressive bus compression, I advise you to ascertain the mastering engineer's opinion on this mix in progress. Recently I asked a client why he was using bus compression on his mix, and he replied, "because I think it doesn't sound loud enough without it." But through demonstration, we found out that his mix sounded *wimpy loud* **but not better** (e.g., fatter, punchier, clearer, fuller). I suggest that you concentrate on mixing and save the question of absolute loudness for the mastering; when mixing, go for **better** when auditioned at the same loudness (i.e. turn up the monitor gain until it sounds loud enough). I think Mastering engineers can do a better job and for much music would prefer not to receive bus-compressed mixes—we can stand back objectively, fine-tuning time constants and bandwidths, maximizing the sound quality (and level) without destroying the rhythm, melody or dynamics of the music. Each tune will be optimally and precisely adjusted in the context of the whole album. Attempting these sorts of decisions during mixing, without having the perspective of the entire album, is dangerous since it's irreversible.

> *Learning from your mistakes gives you room to make even bigger ones!*
> — Murphy's law of experience

If you wish to try your hand at mastering processing after mixing, by all means do so, perhaps as an example of the type of sound you are looking for, but also bring an unprocessed mix safety to the mastering session.

Monitor gain* has a tremendous effect on these matters of judgment. The higher you place the monitor gain, the less the chance of over-compressing. If the music mix sounds properly "punchy" at a higher monitor gain, then leave the rest of the magic for the mastering rather than add another DSP process or take the sound downhill. The VU meter (as opposed to the peak meter) is our friend. Have one hanging around, preferably calibrated to 0 on the VU meter = -20 dBFS on the peak meter with a sine wave, or if necessary, to as high as −14 dBFS peak. If the VU meter is reading hot, then the sound may be overcompressed.

Stop Emulating Squashed CDs

Many mixing engineers compare their mixes against already-pressed CDs, but be careful what you choose as a standard. Ironically, mastered CDs often do not sound like what comes out of the mix, so how can you emulate something which can only be done post-mix? And emulating aggressively-mastered CDs for a mix may contribute to the vicious circle of escalating loudness. What you really need is to hear the sound of a good mix *before* it was sent for mastering. But since that's not available, choose from the plenitude of pop records that have been well-mixed and conservatively mastered. Visit www.digido.com for *The Honor Roll*,

a listing of well-mixed and conservatively-mastered current CDs.

Avoiding Compression Problems during Tracking

When tracking vocalists (who have a habit of belting now and then), a well-adjusted compressor can sound reasonably transparent, and most engineers agree the cure is better than the disease. But watch out for a *closed-in* sound, clamping down when the vocalist gets loud (which reduces clarity and impact), which can be caused by improper time constants, too high a ratio, or using the wrong compressor. Compare IN versus BYPASS before committing to tape. Match levels to make a fair comparison. If you notice too much degradation, maybe it's time to consider a different compressor or change the settings you are using. The sound should be open and clear... remember that no amount of equalization in the mixdown can substitute for capturing a clear sound quality during tracking. This is true for all the lead instruments, including trumpets and electric guitars. If possible, put the uncompressed sound on a spare track—it may save your life. If there's any rule, nine out of ten engineers would prefer to save the decision on drum and percussion compression until mixing. There are always exceptions—every piece of music is unique.

---

* I prefer the term **monitor gain** to **volume control**. See Chapter 14

CHAPTER 11

How To Manipulate Dynamic Range for Fun and Profit

PART THREE: THE LOST PROCESSES

Introduction

This chapter introduces two processes which should be part of every audio engineer's vocabulary. To be successful with them, you have to learn to think like a contrarian, but it's well worth it.

I. Upward Compression

Over-concentration on the use of downward compressors—makes it easy to overlook the psychoacoustic fact that the ear is much more forgiving of the upward "cheating" of soft passages than of the awkward "pushing down" of loud passages. The latter feels like an artificial loss while the former can feel very natural.

Let me introduce you to a venerable compression technique which has finally come of age. Imagine compression that requires just a single knob—no need to adjust attack, threshold, release or ratio. The sound quality is so transparent* that careful listening is required to even know the circuit is in operation! A few years ago New Zealand radio engineer Richard Hulse discussed with me his practice of **parallel compression,**[1] which accomplishes upward compression. Richard was using analog components and got acceptable results, but he thought that a digital implementation could sound even better and suggested I try one. I found the digital version of this technique to be so successful that today I often use it to fatten sound and bring up soft passages in place of manual gain riding. The principle is quite simple: Take a source, and mix the output of a compressor with it. Many

---

* For me, the term **transparent** means the signal path sounds as clean as the source.

mix engineers have practiced this approach with their analog tools. In the digital domain, it is possible to sum the source with a compressor without any side effects, by using a precise time delay for the "dry" signal which exactly matches that of the compressor, as shown in this block diagram (one channel only of stereo shown):

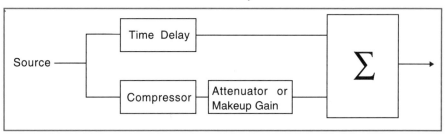

The Parallel Compression technique employs a matched time delay in the "dry" signal path to avoid phase shift or comb filtering. This yields very transparent-sounding upward compression.

In principle, the distortion of the parallel compression technique can be much lower than standard (downward) compression, since most of the signal has a linear path, and the non-linear path is added to the main path.[2] The amount of compression is controlled by the attenuator or makeup gain. The object of the technique is for the parallel compressor to contribute less and less to the total sound as the signal gets louder. This is accomplished by using a very low threshold, thereby putting the parallel compressor into gain reduction almost all the time.

Here are suggested optimal settings for the parallel compressor, derived from original experiments performed by Richard Hulse:

- Threshold −50 dBFS. A very low threshold ensures that the parallel compressor will be into extreme gain reduction during loud passages. Because the output of the parallel compressor has

been *pushed down* during loud passages, it will contribute only negligibly to the total level. In principle, if you add in a second signal that is 20 dB or more down, the second signal will not perceptibly contribute to the total level.

- Attack time as fast as possible. One millisecond or less if available. This ensures that the transient impact of the original sound will be preserved, for as soon as a loud transient hits, the compressor goes into gain reduction. It helps for this compressor to have **look ahead**, which means that it has a built-in time delay that permits it to look at the incoming signal levels and perform predictive gain reduction.

- Ratio 2:1 or 2.5:1 (I prefer 2.5). The net ratio of the sum of the parallel chain varies depending on how much of the parallel compressor is being added in. Richard has developed a chart so you can go by the numbers, but I find it unnecessary and simply go by ear.

- Release time medium length. Experiments show that 250-350 milliseconds works best to avoid breathing or pumping, although in cases where the reverberation is very exposed, particularly *a capella* music, as much as 500 ms. may be needed to avoid overemphasizing the reverb tails.

- Output level or makeup gain adjusted to taste. With the parallel compressor off (−∞ gain), there will be no compression. 0 dB or higher, compression will be very noticeable, with soft or even medium-level passages being raised in level. A nice subtle compression can be achieved with makeup settings of −5 through −15 dB (the lower the level of the compressor, the less total compression).

To determine the time delay needed to compensate for the compressor, adjust the parallel compressor to a 1:1 ratio and unity gain output. If possible, invert the polarity to either half of the chain. Then adjust the time delay until there is a complete null. Typical delays are 5 to 10 samples, but can be much more if there is considerable look-ahead delay in the parallel compressor. If a (non-delayed) polarity invert is not available, adjust the time delay until signal level is maximum (it will have 6 dB extra gain when the delay is correct) and check with pink noise to confirm there is no comb-filter effect.

Correspondents have told me they have successfully implemented this technique in Pro Tools, Digital Performer, and in SADiE. Every digital processor can easily include a parallel compression algorithm. Weiss has incorporated it in their DS1-MK2. I've adapted a single engine of the TC Electronic System 6000 to stereo parallel compression: Feed the signal into the 5.1 (surround) compressor, use the front L/R channels as the "dry" signal, bypassing the sidechain. Use the SL/SR channels as the compressed signal. The time delay is automatically taken care of as all channels of the 5.1 compressor have matched delay. I then assign the output level of the compressor to a fader and adjust to taste by listening. The fader level is a fair guide to how much compression is being applied; there is no need to look at a gain reduction meter. During operation, the *contrarian* engineer just looks for extreme low level passages, and adjusts the parallel compression until the level dips sound more natural or the sound gets a bit fatter and fuller if desired.

Parallel compression can also be used multiband, to separately *fatten* a bass instrument, or to give more presence to low level passages, which is more like dynamic equalization than compression. I assign the output level of each band to a fader, and adjust the sound to taste. The nice thing about the fattening qualities of this compression technique when helping the bass instrument is that the body of the sound gets fatter without destroying the transient impact. Or when increasing the presence frequencies at low levels, the sound can be clearer and better defined without becoming harsh at mid or loud levels.

Even at severe settings, parallel compression sounds much better to my ears than any squashing I've heard from severe downward compression. Unlike downward compression, this form of upward compression preserves the transients or initial attacks very well. In addition, there's room to be expressive at the top levels with upward expansion (see next section) if the original material was too compressed at high levels. Like any process, if upward compression is pushed too far, it will eventually call attention to itself. The first audible artifact will be increased sustains and emphasized reverberation, then, finally, breathing or pumping. These artifacts can sometimes be reduced by raising the release time of the parallel compressor. However, if the music is so open that the process continues to call attention to itself, the only solution is to abandon the processor and manually raise the passages which are too soft.

II. Upward Expansion

Another underused but incredibly useful processing technique is **upward expansion.** Some people think of an upward expander as the **uncompressor**, but it is far more than that (indeed there is a limit to how much a sound can be restored once it has been excessively compressed). Rather, upward expanders can be used to emphasize different parts of the dynamic rhythm from those affected by downward compressors, and the result is often more consonant with the natural movement of the music. For example, upward expansion is great for restoring the liveliness of typical uninteresting musical samples from samplers. It can also put the snap back into a slightly-squashed snare drum. Upward expansion is definitely a technique worth learning, and is no more difficult to use than a downward compressor, once you learn to think like a contrarian and use the threshold, ratio, and attack/release.

Historically, upward expanders were not easy to build until the advent of the VCA.[3] Once you have a VCA-based compressor, it's a simple matter to turn it into an upward expander by reversing the sign (polarity) of the sidechain signal. Probably the first commercial dedicated upward expander was in a device made by DBX called the model 117, circa 1971, designed to enhance dynamics in a hi-fi system. Another early upward expander was the Phase Linear Peak Unlimiter. The honor for the first digital upward expander goes to the Waves C1 (plug-in), algorithms designed by Michael Gerzon. The first stand-alone digital upward expander was in the DBX Quantum mastering unit, followed shortly by the Weiss DS1-MK2. The Waves C4 (plug-in) is the first single processor to perform all of the four dynamics processes, though it can perform only one of the four at a time on each band. It is very desirable to be able to do simultaneous upward compression, upward expansion, and limiting in a single box.

Ironically, downward compression doesn't make the loud parts louder, it makes them softer, pushing ascending passages downward. A loudness increase is obtained as the incoming level decreases and the compressor goes into the release phase, raising the gain. **In contrast,** when the parameters have been optimized, **upward expansion increases the loudness of passages that are ascending in volume**, in rhythm with the upward motion of the music. (Hence it may be necessary to use output attenuation instead of makeup gain to prevent the output from overloading.) There is a small increase in dynamic range, but if used delicately for microdynamic purposes, the upward expander becomes as valuable a production tool as the downward compressor.

This next figure shows an upward expander with a severe .75:1 ratio and threshold at −32 dBFS. Without attenuation it will overload with input levels exceeding about −10 dBFS. Note that the ratio of an upward expander can be expressed in decimal or fraction form depending on the manufacturer's preference. The Waves and DBX units use decimal form, while the Weiss unit expresses this in fraction form as 1:1.33. Typically, the range of ratios used in upward expansion is far smaller than those used when compressing. Commonly, from a very gentle 1:1.01 through about 1:1.2 (fraction); equivalent to from 0.99 through .83 (decimal). A common value used for music enhancement is around .91 decimal (1:1.10 fraction).

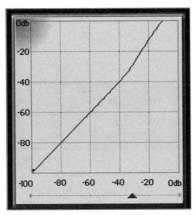

An upward expander with .75:1 ratio, expressed in decimal (1:1.33 expressed as a fraction). Threshold is −32 dBFS, and without adding loss, the output will overload if input exceeds approximately −10 dBFS.

At left, upward expander with fast attack, slow release. At right, slow attack, fast release.

The next figure contrasts fast and slow attack, and fast and slow release when used with an upward expander. As you can see, the dynamic characteristics are opposite from the compressor examples shown in the previous chapter.

The best way to learn how to use an upward expander is to compare it to a downward compressor, described in the chart on the next page (values given in the chart are only for general purpose guides).

Compromises When Making Hot Masters

Both Downward Compression and Upward Expansion result in compromises if you are trying to make a master *super-hot* (high absolute loudness). The problem with downward compression is that it is hard to avoid the squashing effect and loss of dynamics. By splitting the bands (multiband, see Chapter 10), you can slightly postpone the inevitable sonic degradation. The problem with upward expansion is that if you are trying to make a recording *hot*, you must follow the expander with a limiter to increase the level, but the limiter will fight the advantages of the expander, and soon becomes the limiting factor (oops!). When the limiter is used conservatively, it will not deteriorate the sharp transients, and the upward expander can do its job of making the upward-going dynamics more exciting. Prove it by bypassing the limiter at matched compare levels and see if it's hurting the sound of the music. If it is, and you cannot live with the degradation, the only solution is to master at a lower level.

DOWNWARD COMPRESSION	UPWARD EXPANSION
makes sound louder during the **descent** of the music (release phase).	makes sound louder during the **rise** of the music (attack phase).
tends to make sound fatter and exaggerate low frequencies (subject to time constants and threshold).	tends to exaggerate transients and high frequencies (subject to time constants and threshold).
Attacks that are too short (fast) cause transients to be lost.	Attacks as short as a few ms can restore and sharpen lost transients (e.g., from analog tape or overcompressed sources).
Typical attacks 100 ms through 300 ms. Less than 100 tends to blur transients.	Typical attacks 1 ms through 300 ms. If a transient still sounds too sharp and trying >150 ms attack, perhaps this is not the right process for this music, or consider a touch of limiting after the expansion.
tends to make things sound duller or warmer.	tends to make sounds brighter or sharper.
If sounds "jump out" too much, raise the ratio, shorten the attack, and/or speed up the release.	If sounds "jump out" too much, lower the ratio, lengthen the attack, and/or slow down the release.
If attacks seem too sharp, shorten the attack time.	If attacks seem too sharp, lengthen the attack time, or consider compression.
If sustains seem too long or too prominent, lengthen the release time.	If sustains seem too short, lengthen the release time.
If attacks seem too dull, lengthen the attack time.	If attacks need enhancement, shorten the attack time.
If you don't like the percussiveness (e.g., snare drum), speed up the attack. To increase the ratio of rhythm to melody, lengthen the attack. Downward compression is not good at helping the impact of percussion instruments.	If you don't like the percussiveness (e.g., snare), slow down the attack. To increase the ratio of rhythm to melody, shorten (speed up) the attack. Upward expansion is very good at helping the impact of percussion instruments, however, sometimes at the expense of the vocal balance because the percussion becomes more prominent.
	can work very well with upward compression, which fills in any perceived low level "holes" or lost sustain.
Very easy to degrade the liveliness or "bounce" of the music if time constants are not optimized or if overused.	Very easy to enhance the liveliness or "bounce" of the music, but watch out for too much "bounce" or exaggerated dynamics.
tends to go **against** the natural movement of the music, especially when the parameters are not optimized.	tends to work **with** the natural movement of the music, especially when the parameters have been optimized.
tends to de-emphasize musical accents and emphasize the sub accents and sustains in reverse proportion to their original movement.	tends to emphasize the hottest musical accents and to a lesser degree, the sub accents in increased proportion to their original movement.
	Very useful to follow with a limiter, as loud passages are being brought up by the expander. As long as the limiter is used to cheat down very short, momentary transients, it will not significantly diminish the effect of the upward expansion. The limiter's gain reduction meter should be moving very little and on brief occasions, while the expander's gain increase meter should be bouncing with the syllables of the music that's being enhanced. However, if the limiter's gain reduction meter starts to mirror the expander's gain increase meter, then the two processes are canceling each other out and there's too much limiting.
can decrease the overall dynamic range of the song (macrodynamics), in addition to affecting the microdynamic *bounce* of the music.	can increase the overall dynamic range of the song (macrodynamics), making a climax seem even more climactic, which can be very effective.

III. Changing Microdynamics Manually

It is possible to change musical microdynamics without using processors by doing manual edits and gain changes in a DAW. In this figure, I have artificially enhanced the attack of the first note of a song with very brief manual upward expansion (it's the brevity which makes it microdynamic):

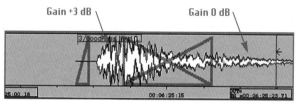

Creating an Artificial Sforzando

At left, the first few milliseconds of the note have a greater gain (in this case, 3 dB), and then there is a crossfade to a gain of 0 dB, resulting in a *sforzando*. An interesting story is that the producer was looking for a surprise when this track entered, and I initially had the beginning attack at +5 dB, but when he took the reference CD home, he reported the attack was too startling, so I took it back a bit for the final master.

This chapter completes our dynamics trilogy.

1   Which he was initially calling *sidechain compression*, but I suggested a name change to avoid confusion with the sidechains of compressors. This technique was publicized by Mike Bevelle in the article **Compressors and Limiters**, *Studio Sound*, October 1977 (also reprinted June 1988). Engineers have been playing with parallel compression techniques for many years.

2   This was the principle of the Dolby A/SR systems, which used a direct signal path summed with a compressed one, *doing as little harm to the audio as possible*.

3   Voltage controlled amplifier. In a console such as a Solid State Logic, all the audio in a channel passes through a VCA. The gating and compression are accomplished by summing sidechain information and feeding it to the control voltage element of the VCA. It is trivial to add upward expansion functions to any VCA-type dynamics processor.

CHAPTER 12
Noise Reduction

I. Introduction

Anthropologist Benjamin Whorf observed that the Eskimos have numerous words for *snow*. Similarly, audio engineers discern a great number of categories of what is collectively called **noise**. Laypersons generally do not distinguish **distortion** from **noise** but we find it useful: **Distortion is a subset of the general category we call** *noise*: it is a kind of noise that is correlated with the signal. Distortion can be low level and act much like what we normally call **noise**, or it can be high level and quite obtrusive, lying on the peaks of the signal. Noise itself can be continuous or intermittent, random or semi-random, colored (containing identifiable frequency components), impulsive, crackly, clicky, ticky (primarily high frequency), or poppy (primarily low frequency). Every kind of bothersome noise requires its own dedicated technical cure, but the most powerful cure is just to ignore the noise! Often we engineers tend to forget that the ear has a built-in noise-reduction mechanism which gives us the ability to separate signal from noise, and hear information buried within the noise. Thus the key to good-sounding noise reduction is not to remove all the noise, but to accept a small improvement as a victory. Remember that louder signals mask the noise, and also remember that the general public does not zero in on the noise as a problem. They're paying attention to the music, and

{ "No single-ended noise reduction system is perfect; all noise reduction systems take away some degree of signal with the noise." }

so should the engineer! So before considering any noise reduction technique, we need to judge whether a noise is truly distracting.

The noise-reduction methods described in this chapter are all **single-ended** as opposed to **complementary**. The Dolby™ system is an example of a complementary, or two-step, noise-reduction system which applies one process during recording and an opposite process during playback. An important fact: no single-ended noise reduction system is perfect; all noise-reduction systems take away some degree of signal with the noise. Artifacts of overaggressive denoising include: comb-filtering or phasing noises, known semi-affectionately as *space monkeys*; and low level thumps, pops. Overly aggressive noise reduction can also remove the critical ambience and atmosphere from a recording.

> The difficulty lies in the fact that reverberation tends to decay to noise. However, much of the directional information and ambience we perceive is from reverberation. Therefore, remove the reverb with the noise, and — in effect — you remove the walls, floor and ceiling from the room.[*]

Sonic Solutions No Noise™ and **Cedar De-Noise** permit fine-tuning of the frequency response of the noise-reduction curve, and a skilled engineer will tailor that response curve for the best compromise between artifacts and perceived noise reduction. What distinguishes a good noise reduction job from a bad one? — *Good Taste*. The engineer must continually retain perspective, because the more noise removed, the more noise revealed (noise itself

masks other noise below it)! It's like peeling the layers of an onion. If you remove some crackle from the right channel, suddenly you may hear some tics which were not previously audible in the left. In all cases, careful comparison between the source and the processed product is necessary to ensure that the music has not been damaged. Ironically, the quieter the original recording, the more effective a noise reduction process can be. In other words, the more separated the original signal is from the noise, the more easily can the noise-reduction system diminish the noise without hurting the signal. So a real noisy recording probably cannot be fixed without creating artifacts.

II. Noise reduction techniques

Simple Filtering

A passage with obtrusive hiss-like noise which contains no high-frequency instruments can be treated with a simple high-frequency equalizer. For example, an electric piano solo introducing a song may be hissy, but that noise will be masked when the rest of the instruments enter. This is a candidate for a selective filter; say 1 to 4 dB dip circa 3-5 kHz (this is the range where the ear is most sensitive to hiss), active only during the piano introduction. However, even here the filter will affect harmonics of the piano, so we must make a judgment call.

P-pops are a type of signal-related noise, so they are a form of distortion, and since they are primarily low frequency, can be treated with a selective high-

---

[*] Gordon Reid of Cedar, in a conversation on the Mastering webboard.

[†] A myth from the restoration community suggested by Gordon Reid of Cedar. In truth, it's nearly impossible to derive intelligible information from a tape if the voices are barely intelligible or audible in the first place.

MYTH:

"I know that you can't hear anything but noise on this tape, but if you get rid of it all, you'll be able to hear my husband having sex with his lover."[†]

pass filter, typically 100 Hz, but sometimes as high as 400 Hz. If the filter is applied briefly, the result can be artifact-free (invisible to the ear). In my DAW, I capture a short section with the filter, then, using the crossfade editor, narrow the extent of the filter to the p-pop; with practice the technique can be extremely fast. It is also possible to edit out just the offending portion of a p-pop.

Narrow-Band Expansion

Compression techniques used in mixing and mastering (make-up gain, especially noticed during the release time) can bring up noise in original material such as tape hiss, preamp hiss, noisy guitar and synth amplifiers, all of which can either be perceived as problems or just "part of the sound." This is what makes our work so subjective. Since compression aggravated the noise, expanders are its cure. As little as 1 to 4 dB of reduction in a narrow band centered around 3-5 kHz can be very effective and if done right, invisible to the ear, performed with a multiband (downward) expander. Typically these units have 3 to 4 bands, but we will only use one. Start by finding a threshold, with initially a high expansion ratio, fast attack and release time. Zero in on a threshold that is just above the noise level. You'll hear ugly *chatter* and bouncing of the noise floor because the time constants are so fast. Now, reduce the ratio to very small, below 1:2, perhaps even 1:1.1, and slow the release until there is little or no perceived modulation of the noise floor. Too much expansion, and you will hear artifacts such as pumping or ambience reduction. The attack will usually have to be much faster than the release so that fast crescendos will not be

affected. Depending on the music, its dynamic characteristics and its original SNR, this subtle approach can yield artifact-free noise reduction. The other expander bands should be bypassed or ratios set to 1:1. A good expander will have look-ahead delay, which allows it to open before it's hit by the signal, thereby conserving transients. If the expander approach does not work, then we will have to apply more sophisticated, dedicated noise-reduction processors.

Complex Filtering

Tonal noise can be diminished by using narrow-band selective filtering at the critical frequency. **Sonic Solutions No-Noise**, developed by Dr. J. Andrew Moorer, has a complex filtering option that permits the insertion of many high-resolution narrow-band filters, suitable for removing hum and buzz (harmonics of the hum). Before inserting the filters, it's useful to do an FFT analysis of the noise floor to determine which harmonics are present so as to apply only the filters that are needed. In SADiE's 2496 or Artemis systems, there is enough DSP power to insert many narrow-band filters in real time, and I have a dehumming preset with about 25 filters set for a Q of 40 or higher. I've also found TC's **Backdrop**, developed by Dr. Gilbert Soulodre, to be very effective with tonal noise if I can find a sample of noise without signal. Systems like Backdrop, Cedar, and No-Noise must sample a brief piece of noise (even one second will do) in order to remove it without affecting the signal.[*] Which brings up the point that you should not tightly cut the beginnings or edit material which is being sent in for noise

---

[*] Cedar calls this the *noise fingerprint*.

reduction; the most likely candidate for a sample is a piece just before the downbeat.

Specialized Processors

GML Labs has a specialized noise-reduction unit for hiss and continuous noise. **Cedar** has just produced a new *miracle* process called **Retouch**, currently available only for **SADiE** DAWs. Retouch is able to remove impulsive noises that no previous system could handle, such as a baby crying, chair squeaks, even people talking in the middle of a take. It is very expensive, but there is no substitute when you need it.

Some manufacturers specialize in one kind of noise; some have separate (expensive) boxes to fix each of them. Each type of noise—scratch, crackle, hiss, buzz, rumble, thump, fitz, regular noise and irregular noise, high level and low level noise—needs its own dedicated correction algorithm. A decrackler is really a multiple-declicker, detecting and interpolating each moment of crackle, so it requires great DSP power. Sonic and Cedar have the most popular high-end noise-reduction systems, with interesting entries from Algorithmix, Audiocube, TC Electronic and Waves. Sonic's approach to continuous noise, such as hiss, or rumble, is to use 2048 individual contiguous filters, constituting a serious multiband expander. Artifacts are minimized since multiband processing avoids interaction between bands. Sequoia has an excellent FIR filter which allows you to visually and ergonomically pick each offending harmonic and reduce it. When the noise source is varying in frequency, as from analog tapes with varying speed, a special kind of **tracking filter** is required.

TC's **Backdrop** is based on psychoacoustics and noise-masking, and is very effective on continuous or tonal noise such as hum, buzz, hiss and rumble, with minimal artifacts when properly adjusted. You get what you pay for, and the critical ear can tell the quality difference between the most expensive and cheapest systems.

III. One Man's Meat Is Another Man's Poison

I once mastered a punk rock album where the opening of one tune had an obvious electrical tic on top of the bass player's note. I removed the tic and the note was restored to its beauty—I thought. But then I heard from the producer that he missed the tic and so I had to put it back. Thus proving that *beauty is in the ear of the behearer*, and many noises are considered to be part of the music. Get to know each musical form (especially punk rock) and in some cases think about leaving it dirty instead of clean!

IV. Manual Declicking, Dethumping, De-Distortioning, Depopping....

A good mastering system should have integrated manual denoising, which allows us to quickly and selectively clean up momentary noises. Declicking, dethumping, de-distortioning, depopping, and other techniques are critical mastering system features. The next figure, part A shows a *thunk* from an LP record. The left channel of this figure (top panel) has already been *dethunked*, as can be seen by the horizontal marker above the left channel waveform. When reproduced, the slight DC level shift that remains does not translate to an audible noise. The right channel contains a severe thunk manifested by an instantaneous upward, then downward DC level shift (which causes woofers to rattle). With Sonic Solutions' manual declicking, the correction process is as simple as marking the noise with the gates and selecting **D Type** from the menu.

D Type is a powerful interpolator which can stitch together "impossible" waveforms and even remove brief dropouts or holes with no audible effects.

In figure part B, the low frequency thunk and most of the DC discontinuity have been repaired, and the ramped DC level shift that remains (probably record warp) does not produce an audible noise.

LP records are not the only sources that need declicking. Something as simple as an obtrusive vocal "lip smack" can be cleanly and quickly excised, and brief overload distortion can also be cleaned up by the interpolation technique. Sonic's E-type decrackler can also selectively reduce sibilance. I use it instead of an overall sibilance controller when there are only a small number of offending s's in the recording. E-type can also reduce and sometimes eliminate the harsh sound quality of clipping and digital overs.

In the figure on the next page, on top, a severe click is marked manually by the gates, and on the bottom it has been removed. Note that Sonic Solutions' automatic vertical gain conveniently amplifies the display to the highest amplitude

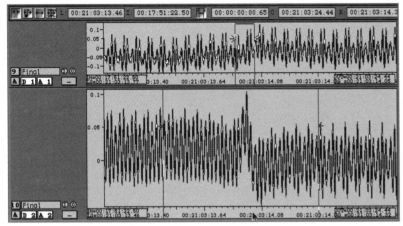

A: LP Thunk in the Right Channel (different panel heights reflect different visual magnifications, not different amplitudes). The left channel has already been denoised (red bar).

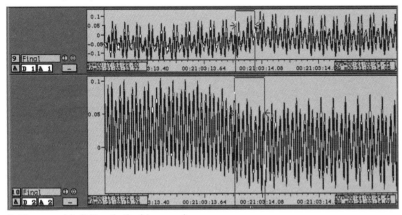

B: After manual declicking, the thunk is removed.

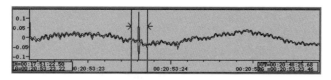

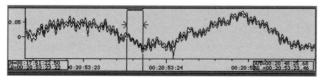

Click in the top panel has been removed in the bottom (marked by the red bar).

Manual declicking is extremely labor-intensive but very rewarding; it's like hiring a meticulous gardener to remove each weed in your garden by hand, instead of using harmful chemicals.

in the view, which is no longer the click after it has been removed.

Here's another remarkable before/after example (with a modern G4 computer, the repair takes about 3-5 seconds).

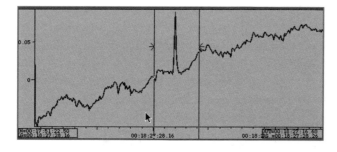

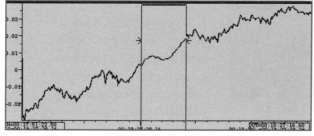

On top, a click is surrounded by the gates. At bottom, after choosing D-Type from the No-noise menu, the click is removed.

CHAPTER 13

Other Processing

I. Introduction

In this chapter we'll discuss important techniques such as how to determine proper polarity and inter-channel balance. In addition we'll introduce specialized processing including MS Equalization or MS compression... and the world of mastering processors including reverberation, ambience extraction, "replicators," exciters, etc.

II. The Balancing Act

First Check the Monitor Balance

Adjusting inter-channel balance seems like a simple procedure, but many people have misconceptions about how to achieve correct stereo balance. Before making any judgments of program channel balance, first verify that your stereo monitors themselves are balanced. Play a mono pink noise signal at equal level to both stereo speakers and confirm the pink noise image is tightly centered between the speakers at all frequencies of the pink noise. Ride the monitor level control up and down within the normal ranges and confirm that the image of the pink noise remains centered. If it's not tightly centered, then suspect the crossovers, drivers, level control, preamplifier channel balance or room acoustics. Chapter 14 covers the monitor calibration process in more detail.

Polarity is "direction," positive- or negative-going for an electrical signal, outward or inward for a transducer and the recommended standard is that positive voltage means positive pressure. If there's an audible "hole" between the left and right loudspeakers (especially obvious at low frequencies), then one loudspeaker is moving inward while another is moving outward, hence the two wavefronts are canceling acoustically to some degree. This is defined as **incorrect relative polarity**, caused by improper wiring. Many of us still use the antiquated phrase "the speakers are out of phase," but we really mean they're "out of polarity with each other" (because *phase* really means *time*). In a 2-channel reproduction system, incorrect relative polarity yields a hollow sound, imaging way to the sides and not in the middle, with reduced bass and lower midrange response. The solution is to search each balanced line and speaker connection for the pair of wires which are reversed.

Stereo Balance of the Program Material

Music feels much better when the balance is "locked in." When making stereo balance judgments on program material, I consider left-right channel balance errors of >0.2 dB to be significant, but try to keep balance errors to <0.1 dB. It is difficult to use meters to judge channel balance because at any moment in time, one channel will likely be higher than the other. I've seen songs where one channel's meter (peak or VU) is consistently a dB or so higher than the other, but the balance is exactly correct. This is because some high-frequency-dominant instruments project better than flat meters indicate; for example, with a mandolin on the right and viola on the left, proper balance will likely occur with the left meter reading higher, and it also depends on who's doing the lead part! If in doubt, change the balance 0.1 dB at a time until it sounds just right.

$\left\{ \textit{"Never Use The Meters to Make Channel Balance Judgments"} \right\}$

A stereo position indicator (see *Figure C16-01* in the Color Plates) may help, but most times it just tends to confirm what you've already heard. Judge balance by ear, and when in doubt, check with the producer, since the lead vocal is sometimes intentionally placed off-center. Other times, even if the lead vocal is supposed to be centered, this may not produce the best balance between two accompanying instruments located left and right, or you may feel that the instruments on one side are competing with the vocal's intelligibility. In that case you have to think like a mix engineer, so it pays to check the producer's intentions. Sometimes the producer will say, "oh, we didn't get that mix quite right, it's possible the violins on the left need to come up against the trumpets, use your judgment." But if it takes more than about a dB of balance adjustment to fix the problem, a remix may be in order or the sound image may end up lopsided, and it bears repeating—check with the producer.

Fixing Relative Polarity

The so-called *phase switches* on consoles do not change time, they invert the polarity. If two sources are 180° out of phase at all frequencies (or a large band of frequencies), then we conclude they are *out of polarity with each other*, and we must correct the polarity of one channel. If the correlation meter (see *Figure C16-01* in the Color Plates) shows a large phase difference approaching 180°, check for interchannel (relative) polarity errors by switching the monitor to mono and inverting one channel's polarity. The position that gives the most bass is the correct one. Sometimes this is the only method to verify the correct polarity when two spaced omnidirectional microphones were used, since there is a lot of random phase information in such a recording. When several mikes are mixed together, if only one pair is out of relative polarity, there's little or nothing we can do about it in the mastering. For example, if the percussion drops out in mono but the vocal remains fine, there's nothing you can do short of a remix.

Fixing Absolute polarity

By convention, absolute polarity is correct when the loudspeaker moves outward (toward the listener) with a positive-going pulse. First, check the absolute polarity of your reproduction system, with a polarity tester and polarity test signal. If you do not have a polarity tester, play a Telarc orchestral recording and confirm your woofers move outward on the attack of the big bass drum.

It is debatable whether the human hearing mechanism can detect absolute polarity. If both speakers are moving inward when they should be moving outward, can you hear the difference? Many listeners claim to be sensitive to absolute polarity reversals, but scientists have shown that this may only be due to a non-linearity in the loudspeaker driver or magnet structure. Nevertheless, I produced an absolute polarity test for Chesky Records, using a solo trumpet recorded in a natural space with a Blumlein microphone pair. When the polarity is incorrect, the trumpet appears (to most listeners) about a meter further back. This is evidence that incorrect absolute polarity can affect how we mix and master.

As a digital mastering engineer, I try to look for evidence in the DAW waveform that the polarity is correct. Most instruments produce waveforms with ambiguous polarity, but major bass drum attacks should be positive-going, and a solo trumpet on a held note produces a distinct, positive-going waveform. Sampled bass drum tracks have often been so mangled that you cannot tell the polarity from the waveform. Other than this direct evidence, all you can do is experiment with both polarities to see which sounds better. Of course, make sure **both** channels' polarity are changed.*

Fixing Phase shifts and Azimuth Error

Modern-day digital consoles also have controls to manipulate timing. A small timing error between two sources is a *phase error*, which can cause comb filtering especially if combined to one channel. If the two sources are 180° out of phase at only a few frequencies, then they are out of timing (phase shift), not out of polarity.

The procedure for correcting small interchannel phase shifts requires a keen and experienced ear. You must have a timing control calibrated in samples. Switch the monitor to mono, increase the delay on both channels equally, by about 5 samples. Then increase and decrease the relative timing of one channel a sample at a time. Use the timing control like the focus on a camera, with the goal being greatest high frequency response and minimum comb filtering at the center of focus.

---

* Reversing wires on pins 2/3 of an AES/EBU cable **does not affect the audio** in any way. Polarity reversal can be accomplished in the analog domain, or with a digital processor.

This procedure also can be used to align spot microphones with main mikes, (as described in Chapter 17) and it's how we adjust analog azimuth if there are no tones on the tape. Single-sample increments are very coarse at 44.1 kHz SR, which is why Cedar has invented the digital azimuth corrector, which has sub-sample timing increments, accurate to 1% of a sample.

DC Offset Removal

Sometimes poorly-calibrated A/D converters add a DC offset, where the centerline of the waveform at rest is not exactly 0 volts. Also, some poorly-implemented DSP processes add DC offset. When DC offset is excessive, headroom is reduced in the direction of the offset, in other words; raising gain would cause the audio to clip prematurely because the centerline is offset. But when using digital limiters, slight loss of headroom due to DC offset is not a problem. DC offset reveals itself on a digital meter as a static low level signal, but this could be noise, not DC; with DC offset, the waveform in the EDL during a quiet passage will appear offset from center. But the best way to determine if there is a problem is to repeatedly play and stop the material. If you hear a meaningful click or a pop when starting or stopping, the DC offset should be repaired. Prior to the advent of high-resolution digital equalizers, I preferred not to fix DC offset, but now the easiest solution is a very steep high-pass filter, below, say, 20 Hz.

{ *"A pitch corrector that sounds transparent and maintains the original timing—does not yet exist."* }

Pitch and Time Correction

It's impossible to fix the pitch of a vocalist when he's mixed with other instruments that are on pitch, so mastering engineers are not often called upon to correct pitch. However, when a soloist is playing *a capella*, we've been asked to make corrections. **The simplest and cleanest form of pitch correction is one where both the length (timing) and the pitch of the material are altered**, exactly like playing an analog tape recorder faster or slower. This is done by a sample rate conversion, and then reinserting the material of the "wrong" sample rate into the EDL—this technique can sound excellent if a good SRC is used. But sometimes we're called upon to change the speed of an entire song without changing the pitch, or the pitch without changing the speed, which are big challenges. I have never done it without creating an audible degradation in the sound; at worst the splicing in these algorithms yields a *gurgling* or *wavering* sound quality, and at best there is a fidelity reduction,[*] so we always prefer to use the simpler SRC method if permissible. As DSP has gotten more sophisticated, pitch and time correctors have become much better, and I have gotten away with using one for short periods; but I have not yet heard a transparent one and some degradation can be heard in a high-resolution environment.

---

[*] A popular song by Cher, "Believe," takes advantage of the weaknesses of such devices.

III. "Remixing" at the Mastering Session

Vocal Up and Vocal Down Mixes

The mixing session is often hectic and it's a good idea to hedge your bets by printing alternate mixes, e.g., "vocal up," and "vocal down" (by about ½ to ¾ dB). Later, in the pristine acoustics of the mastering environment we can choose the best mix, that which works best in the context of mastering processing.

Mastering from Multitrack Stems

A client brought a DAT with 10 songs. On one of the songs, the bass was not mixed loudly enough (this can happen to even the best producer). We were able to bring up the bass with a narrow-band equalizer that had little effect on the vocal, but when the producer took the ref home, he was dissatisfied. In his view the advantages of the increased bass were offset by the effect it had on the delicacy of the vocal. He asked if he could bring me a DAT of just the bass part so that it could be raised in mastering.

I asked for a DAT with a full mix reference on one channel for synchronization purposes, and the isolated bass on the other. I was able to load the DAT into my workstation, synchronize the isolated bass, and raise the bass instrument in the mastering environment, without affecting the vocal. It was an unequivocal success. This is an example of an unsynchronized *stem*, and since the bass is also present in the full mix, there is danger of phase cancellation between the full mix and the added bass track if they are not perfectly synchronized. I do not recommend this practice; instead, all stems should be sample-accurate synchronized, begin at the same

timestamp, and ideally, each stem should have unique elements.[*]

Another client doing the album of a pianist with orchestra brought a four-track Exabyte archive in Sonic Solutions format, with the piano isolated on two tracks. In the mastering we could adjust or equalize the solo piano separately.

When a stereo mix is done to multiple *stems*, there are typically six tracks (3 pairs), each with its own reverb: vocal, rhythm, and melody instruments. Mastering engineer Bob Olhsson has pointed out that surround mixing demands the stem approach, because clients certainly are not going to make multiple "vocal up" 6-channel surround mixes. Instead, mastering will become an extension of the mix environment. Producers will send 24-track tapes with stems divided into multiple 5.1 groups, such as vocals, bass, rhythm, etc., which if reproduced at unity gain, represent the mix as the producer put it down in the control room.

MS Mastering

Mastering engineers are always seeking ways of repairing or enhancing one element of a recording without detriment to any other. There are always tradeoffs, but judicious use of MS tools can be lifesavers, turning a good recording into a great one, or saving a so-so recording from the dust-heap. (Nothing can repair bad musicianship, and autotune doesn't work on mixed material).

A client had mixed in a bass-light room and his bass was very boomy, right up to about 180 Hz. At first the vocal came down slightly when I corrected

---

[*] Films are always mixed to stems, e.g dialog, music, effects.

the boomy bass, but through MS processing techniques, I was able to produce a perfectly-balanced master. MS stands for *Mid/Side*, or *Mono/Stereo*. In MS microphone technique, a cardioid, front-facing microphone is fed to the **M**, or mono channel, and a figure 8, side-facing microphone is fed to the **S**, or stereo channel. A simple decoder (just an audio mixer) combines these two channels to produce **L**(eft) and **R**(ight) outputs. Here's the decoder formula: M plus S equals L, M minus S equals R.[1] Here's how to decode in the mixer: feed M to fader 1, S to fader 2, pan both to the left. Feed M to fader 3, S to fader 4, invert the polarity of fader 4 ("minus S"), pan both to the right. Start with all faders at unity gain, and change the M/S ratio to taste. With more M in the mix, it becomes more monophonic (centered); with more S, the more wide-spread, diffuse, or vague the sound becomes. If you mute the M channel, you will hear a hole in the middle, containing largely the reverberation and the instruments at the extreme sides. Mute the S channel, and you will largely hear the vocalist; the sound collapses, missing richness and space. There's little separation between M and S channels, but enough to accomplish a lot of control on a simple 2-track. It's great for film work—the apparent distance and position of an actor can be changed by simple manipulation of two faders.

The MS technique doesn't have to be reserved to a miking technique. We can separate an ordinary stereo recording into its center and side elements, and then separately process those elements. I tell my clients I'm making three tracks from two. For example, let's take a stereo recording with a weak,

center-channel vocalist. First we feed it through our MS encoder, which separates the signal into M and S and we decrease the S level or increase the M level. Listening at the output of the MS decoder, presto, the vocal level comes up, as does the bass (usually) and every other centered instrument. In addition, the stereo width narrows, which often isn't desirable. But at least we raised the vocalist and saved the day! Similarly, I've used MS to fix the ratio between a center-located lead vocalist and side-located background singers, even varying the MS ratio between verse and chorus of the song. Some processors have built-in width controls; what they do is internally convert to MS format, adjust the M/S ratio, and then reconvert to LR format. The Waves S1 plug-in processor's width control is gain-compensated, so the apparent total level is held constant as the width is changed. You can accomplish the same thing by lowering the S as you raise the M, or vice-versa.

Automating the MS correction. When vocal (or center instrument level) has to be selectively tweaked, either the plug-in can be automated, or we can correct the problem directly in an EDL without using any processor. To raise the (centered) vocal, add a duplicate of the material in another stream, with the channels reversed. Add this in at as low a level as tolerable (typically −12 to −16 dB), for if taken to an extreme it will turn the entire material to monophonic. I may add a tetch of K-Stereo processing (described later) to compensate for any loss of ambience, width or sense of space, and lower the bass gain to reduce center-channel bass build-up. By contrast, in places where the center vocal sticks out too much, **subtract** a duplicate of the material in another stream, with the channels reversed. In other words, add in a reversed-polarity, reversed-channel duplicate of the source material. A crossfade into and out of the material in the extra stream is the automation that raises or lowers the level of the center-channel material. Another way to automate this process is to add an MS encode-decode plug-in to the mixer, and automate the panning between the M and S channels on the encode side.

MS EQ. We can accomplish a lot by manipulating the M and S signals with equalization. Let's take our stereo recording with weak centered vocalist, encode it into MS, and apply separate equalization to the M and S channels. Since the M channel has most of the vocal, we can raise the vocal slightly by raising (for example) the 250 Hz range, and perhaps also the presence range (5 kHz, for example) in just the M channel. This brings up the center vocal with little effect on the other instruments, and doesn't affect the stereo separation as much as if we had raised the M/S ratio of the entire spectrum.

The **Weiss EQ-1** has an optional MS encode/decode which can be placed around the equalizer section. Raising or lowering the EQ on one channel of the equalizer affects the stereo separation. Spread the cymbals without losing the focus of the snare, tighten the bass image without losing stereo separation of other instruments, and so on. The TC Electronic **Finalizer** 96K's spectral stereo imager is essentially an MS equalizer "on its side;" it's an MS width control divided into frequency bands. See **Finalizer** image below.

MS Compression. Consider a mix that sounds great, but the vocal is sometimes slightly buried when the instruments get loud. If we try compressing the overall mix, or even narrow band compression of the vocal frequency range, we might be disappointed that the compressor action ruins the great sound of the instruments. MS compression can help us isolate the compression to the center or M channel—by only compressing the M channel, we delicately bring up the center when signals get loud.[2] Or compress the M channel and expand the S, which helps control the vocalist and open up the band![3] Or, by doing multiband MS compression, we could keep the bass instrument from being affected by our vocal range compression. In other instances, we might achieve that special kick drum sound by compressing only the low frequencies of only the M channel. The possibilities are solely limited by our imaginations.

Patching Order of Processes

Sometimes it's better to compress before equalizing. For example, if the EQ is being used to enhance the level of some instrument (e.g. if we're

The TC Electronic Finalizer 96K is an all-in-one Mastering Processor.

looking for a punchy or thumpy bottom), a compressor after the EQ might undo the effect of the equalizer by pushing the strongest sound downward. 90% of the time my equalizer is patched before the compressor; as I make changes in the EQ, I alter the compressor's threshold to retain the same action. I almost always put sibilance controllers early in the chain, so they will operate with a constant threshold (sensitivity) regardless of how other devices are adjusted.

IV. An Eclectic Collection of Mastering Processors

Here is a brief (alphabetical) collection of processors used for mastering at major studios worldwide. Please do not draw conclusions about the inclusion or exclusion of a particular unit in this set; it represents items that either I have used or which have gained a strong reputation among other mastering engineers whose ears I trust. Some additional popular units are described in Chapter 16.

Plug-ins vs. Stand-Alone Processors

Currently, Sonic Solutions uses proprietary plug-in formats to preserve the highest sound quality, so we must feed an external program that can run plug-ins as an effects loop. Sadie V. 5 has a proprietary plug-in format but also accepts Direct-X. Ergonomically speaking, plug-ins are a mixed bag. It's much easier to operate a stand-alone box with real knobs than a plug-in with a mouse, but there are also stand-alone processors whose user interface leaves a lot to be desired. And some plug-ins feature a user interface which is so ergonomic that it's a lot easier to adjust the parameters of multiple channels simultaneously than with any standalone box. Sonically speaking, plug-ins have improved tremendously in the past few years, particular those Native Plug-ins employing 64-bit floating point architecture (see Chapter 16). At this point, the sound quality of a processor is up to its designer more than whether the process is a plug-in or an outboard box. However, pressure to reduce CPU demand often results in Plug-Ins with compromised sound quality.

Classic (and near-Classic) Analog and Digital Processors

The **Cranesong STC-8** (image below) is a high quality stereo analog compressor combined with a peak limiter, and is gaining a reputation amongst mastering engineers. The STC-8's compressor's attack and release times are optimized for mastering

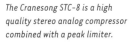

The Cranesong STC-8 is a high quality stereo analog compressor combined with a peak limiter.

purposes, and it is capable of both emulating vintage equipment and creating distinctive new sounds.

The **DBX Quantum II** is a powerful multi-function digital processor with up to 96 kHz operation. All DSP is calculated in 48-bit fixed-point notation, accurately dithered to 24 bits on its output for low-distortion sound. It has multiband and M/S options as well as parametric EQ, compression, expansion and limiting. One of the rare dynamics processors which include ratios below 1 (see Chapter 11), it's particularly valuable for *uncompression*. However, I have trouble adjusting to DBX's approach of naming release time in dB/sec; I just turn the knob and go by my ears. Since all the

characterized by a too-bright, edgy, fatiguing sound. I advise mix engineers to avoid using exciters on the mix bus until mastering in a more controlled acoustic environment (though moderate use of exciters on individual instruments can help a mix). However, the **Cranesong HEDD-192** (pictured below) is a digital processor that has almost no *digititis* and thus is in a class by itself. It uses natural distortion patterns derived from classic analog gear (see Chapter 16). Other digital exciters include the **SPL Machine Head** and **Steinberg Magneto**, which are digital processors, the latter being a plug-in. Analog exciters include the **Aphex** and **BBE**. A number of multifunction boxes contain exciter

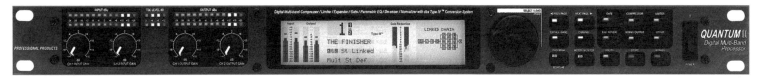

functions are crammed on one LCD screen with multiple menu levels, ergonomics can be daunting. This is the case with many such multi-function units; examine and test the menu structure before you buy—in the best units, critical functions will be no more than one or two menu levels below the top.[4]

Exciters

An exciter is a distortion generator. The use of Exciters can often lead to unmusical sonic results

modules, including the **TC Electronic Finalizer 96** and **Drawmer DC 2476** mastering processor, another multifunction processor.

The **Fairchild** tube limiter and **Pultec** equalizers have not been constructed since the 1960's, but have attained such legendary status for their *fat sound* that I am obliged to mention these unobtainables en passant. There may be some modern-day substitutes which do as well or perhaps better, with cleaner, quieter electronics. If you're looking for the

Top: The DBX Quantum II processor is a multi-function unit with up to 96 kHz operation.

Bottom: Cranesong HEDD-192 Analog Simulator.

Pultec or Fairchild sound or beyond, consider units from Cranesong, Manley, or Millennia.

Massenburg Equalizer Model GML-9500

George Massenburg is the design engineer for **GML** and the inventor of the very concept of parametric equalization. The model 9500 mastering equalizer (pictured above) is the mastering version of the popular 8200 analog parametric, which has been an industry standard and popular with mastering engineers for over 20 years. GML also manufacture an analog dynamic range controller and a digital noise reduction unit.

Digital Domain Model DD-2 K-Stereo Ambience Recovery Processor

K-Stereo. DSP permits us to accomplish tricks which were not possible in analog. I invented the **K-Stereo**™ and **K-Surround**™ processes to enhance the depth, ambience, space and definition in stereo mixes that otherwise would sound small. K-Stereo extracts existing ambience, giving the mastering engineer a handle on reverb returns after the mix has been made. It should be the first enhancement choice before trying a reverberator, because overall reverberation can muddy an existing mix, whereas K-Stereo selectively enhances elements in a mix which already contain ambience. For example, if a mix has a wet vocal that needs enhancement but also has a dry snare drum, K-Stereo will affect the vocal reverb but not the snare drum. It does this using a psychoacoustically-based process that's subject of a patent application. Digital Domain manufactures the **Model DD-2 K-Stereo Processor** (pictured at left); Z-Systems has licensed the K-Surround process in the model **Z-K6**, a 2 channel to 2-channel converter, and **Weiss** Engineering has licensed K-Stereo for a multifunction unit.

The L2 is the first hardware product produced by **Waves** and has become an obligatory mastering limiter (above top). This device helped spawn the narrow-minded philosophy "I can make anything louder than you can." However, an exceptional auto-release and 48-bit processing make the L2 the least damaging limiter I've encountered. Yes, this is a left-handed compliment, but the L2 can sound pure and transparent **at low gain-reduction settings**. It also contains Waves' IDR dither, which is among the better-sounding 16-bit dithers I have encountered, and an excellent 24-bit A/D converter.

I found the **Manley Massive Passive Equalizer** (pictured middle) to be remarkably transparent and quiet for a tube equalizer. It gains its name by employing a passive equalizer section followed by a quiet, high-gain tube amplifier. To my ears it has just the right amount of tube distortion yet retains clarity without being too "fat." It also has far more versatility than the apparent four bands-per-channel because the Q or shape control affects the shelving curve as well as the bell, giving the effect of a 7 or 8 band equalizer. It's well worth downloading the informative and humorous manual written by Manley's versatile Craig "Hutch" Hutchinson.

A mastering house should have a variety of compressors to choose from, since no two sound alike, even with similar attack and release settings. Several outstanding mastering engineers report that the Manley tube **Vari-Mu Compressor** (bottom image, previous page) can help provide desirable *punch* and *fatness* with modern rhythmic music and is a good replacement for the classic Fairchild, which also employed variable Mu techniques (Mu is tube shorthand for *gain*). Distortion can be varied from very low to *screaming* by changing the input/output gain ratio.

MaxxBass, a Plug-In from Waves.

MaxxBass. Mixing is a tough job. One problem we sometimes encounter is a bass instrument with inadequate definition or unclear notes. Obviously the best solution is to turn around and remix with better EQ or compression on the bass, but that's not always possible. **Waves**' plug-in called **MaxxBass** (pictured at left) is designed to help clarify the definition of the bass instrument with minimal effect on the rest of the mix. **It's a form of a dedicated exciter** and a very powerful process that's easy to overuse and dangerous to employ without high resolution monitoring.

This is not the fault of the processor, but a limitation of working on any mixed material, since it cannot distinguish the bass instrument from the toms or the bass drum and if overused, the result can be thin-sounding. Essentially the process works by low-pass filtering the source, synthesizing harmonics and then mixing them back into the full mix. Don't try this with a standard exciter, because another key to MaxxBass is that it retimes the harmonics with the main signal, which is not easy to accomplish using external boxes.

Another use of MaxxBass is to give an impression of bass response for small systems, by taking advantage of a psychoacoustic property of the ear that supplies missing fundamentals when the harmonics are only present. Watch an old movie on television and you may not notice that the dialogue has been sharply high-pass filtered below about 200 Hz. If using MaxxBass for this purpose, be aware that the sound is tailored for a particular small system and will not translate to every other. In fact, the tailored product can sound embarrassingly ugly if reproduced on a full-range system.

Millennia Media manufactures a *Twin Topology* line which can be either tube or solid state at the flip of a switch. The **NSEQ-2** equalizer (pictured)

Millennia Media NSEQ-2 Tube and Solid State Analog Equalizer.

used to align a monitor system or simply to send test tones to external devices.

Reverberation Processors—How Real Can You Get?

probably has the shortest internal signal path of any analog equalizer, with a single DC-coupled solid state or tube opamp performing the duties of input conditioning, equalization, and line driving. In common with many top-of-the-line analog processors, headroom is exceptional, clipping at +37 dBu (solid state) and in solid state mode it is as close to an analog *straight wire with equalization* as I have ever heard (see Chapter 16).

Measurement Devices and Interfaces

The Metric Halo Mobile I/O (pictured above right) is a portable high-resolution recording studio, and in conjunction with SpectraFoo, it serves as a multi-channel Firewire interface, portable jitter and spectrum analyser for digital and analog audio problems. Attached to a Titanium G4 Powerbook, it's a highly functional portable measurement and analysis system. The jitter and distortion analyses in this book were made with the MIO and SpectraFoo.

Another useful portable measurement and setup device is the Audio Toolbox by **Terrasonde**. Complete with measurement microphone, it can be

A small percentage of the work that comes in for mastering requires added reverberation. Some clients have purposely mixed dry because they did not have access to the quality of reverberation that we have at the mastering house; but the music must be of a nature that will not suffer if reverberation is added to every element. Most mastering requires a very natural-sounding reverberator, unless we're looking for a brief special effect. My requirements for a natural-sounding reverberator include excellent simulation of the early reflections that would be present in a real room (see Chapter 17); if soloing the early reflections, they should sound natural and be able to stand on their own. In 1994 I produced a unique audiophile test CD, **Chesky JD111**, containing a dry-versus-wet test that you can use to evaluate the sound of a reverberator. I placed a drum set on the stage at BMG studio A, in front of a single Blumlein microphone pair. The figure-8 microphone pattern has equal pickup front and rear, so it captures the reverb coming from the hall in stereophonic perspective. But first I closed the thick stage curtains, isolating the drums to the small stage area, and recorded a very dry-sounding one-minute drum solo (track 25 on the test CD). Then I opened the curtains, and recorded the solo once again with the identical mike, whose rear side picked up the reverb from the 60 x 40 foot, 2-story high diffuse-treated room (track 26). Compare the sound of the real room against any simulator.

Sibilance Controllers (De-Essers)

Sibilance (exaggerated `s' sounds) is a natural artifact of compressors as well as bright microphones and certain mouth and teeth shapes. A standard compressor exaggerates sibilance because the compressor doesn't correspond with the frequency response of the ear; the sibilant is in the ear's most sensitive frequency range, but typical s sounds fall **below** the compressor threshold. The solution is to employ a very fast, narrowband compressor working only in the sibilance region (anywhere from 2.5 kHz to as high as 9 kHz in some cases). A standard compressor can be adapted to a sibilance controller by equalizing the sidechain, or by using one band of a multiband compressor. Nearly every multi-function processor or plug-in manufacturer has a sibilance controller option, but it's not an easy process to get right. Listen for artifacts such as distortion or pumping, or ineffective reduction of the s's. I've found the best-sounding

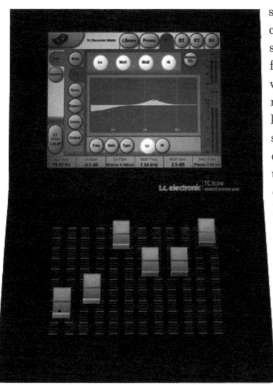

TC Electronic Icon remote. Visible on its screen the equalization capabilities of one of its four 96 kHz/48 bit 8-channel engines.

sibilance control in dedicated units such as the digital **Weiss DS1-MK2**, whose attack, release and filtering characteristics are idealized for processing premixed material with little or no artifacts. Several mastering engineers also recommend the analog **Maselec 2012** HF and peak limiter as an excellent de-esser.

Sintefex Convolution Processor

Convolution is a mathematical process which combines two functions as though one was run through the other function. A company called **Sintefex** uses convolution in its model **FX8000 Replicator** (pictured below), which some mastering engineers report can very effectively sample and duplicate the sound qualities of well-known compressors, limiters, equalizers and reverberation units. Too good to be true? As of this writing, I have yet to audition a unit.

The Sintefex FX8000. Does it really replicate? Many people think so.

The TC Electronic System 6000, TC's flagship multichannel product, is extremely easy to use (I figured it out without reading the owner's manual), has impeccable sound and is modularly upgradeable. The ICON remote (pictured at left) can control numerous 6000 mainframes at once. Four 8-channel 96 kHz/48-bit digital engines can perform artificial reverberation (among the best that I have heard), compression, expansion, limiting, de-essing, mixing, noise reduction, delay, special effects, monitor control and other processing. It would take an entire chapter to do justice to all the possibilities of this unit, for which third-party providers such as GML have written modules. In addition to digital processing, the frame contains high-quality A/D/A, whose approach to jitter reduction I've described in Chapter 19.

Weiss Engineering holds a special place in the hearts of old-time digital mastering engineers (if that's not a contradiction in terms), since they invented the first usable high-resolution digital processing system, still available as the modular 102 series. The Gambit line of rackmount processors is designed for superb ergonomics and sound quality. With a one-knob-per-function philosophy, the Gambit series feels just like an analog processor, with the added versatility of memory storage and MIDI remote control. I analyse the performance of the dynamics processor **DS1-MK2** and the linear phase **EQ1-LP** (pictured above right) in Chapter 16; the latter has become a favorite equalizer. Another useful device is the model **SFC-2** dual synchronous sample rate converter, which I often use to up- and down- sample (see Chapter 1).

Z Systems ZQ-2 is a 6-band stereo digital equalizer that sounds very clean and relatively *undigital* (pictured at right). I analyse its near textbook-perfect performance in Chapter 16.

Z Systems Z-link 96+ is an asynchronous sample rate converter (ASRC) employing the Analog Devices 1896 chip. We can use it to monitor CDs if the system's DAC/Master clock is not at 44.1 kHz, so as not to disturb the delicate lock between processors which are locked at a different rate.

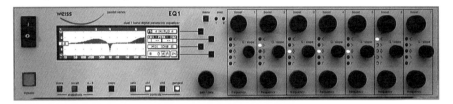

Above: Weiss DS1-MK2 Compressor/Limiter/Expander/De-Esser
Below: EQ1-LP/7-band linear-phase equalizer

Above: Z Systems ZQ-2 6-band stereo digital equalizer
Below: Z-Systems ZK-6 6-Channel K-Surround Processor

Z-Systems also manufacture digital surround processors including the aforementioned ZK-6 K-Surround processor (pictured previous page), which converts 2-channel material to 6-channel, a 5.1 compressor and equalizer as well as the ubiquitous digital routers described in Chapter 2.

1 The formally correct formulas are:

Encode:

$M = 0.5 * (L + R)$ which is 6 dB less than the mono sum. The encoder sums and attenuates by 6 dB.

$S = 0.5 * (L - R)$ which is 6 dB less than the mono difference. The encoder takes the difference and attenuates by 6 dB.

Decode:

$L = M + S$

$R = M - S$. Be aware that an MS encoder and decoder are identical except for the amplitude, and if you use a typical encoder to decode, you will have to raise the level by 6 dB.

2 Remember that a downward compressor brings sound down when it goes over the threshold, so the actual loudness increase of the compressor is accomplished by raising the gain makeup control. In the MS case, very slight compression, say 0.5 dB, may be all that is necessary to control that "lost" vocalist above the band.

3 If a unit which allows downward compression of M and upward expansion of S is not available, I may compress the M channel in one unit and then upwardly expand both channels in another; when properly adjusted, the net result is the same as if I had compressed the M channel and expanded the S.

4 The best way to take advantage of multifunction boxes is to load an existing preset, then bypass nearly all the unnecessary and often exaggerated settings that manufacturers habitually toss in, and save the preset as a blank slate. Apparently they can't sell a box to its intended market without presets, but the preset concept is foreign to the way in which mastering engineers work, especially a preset ludicrously named **Reggae, Rock and Roll** or **Smooth Jazz**. How can they give you a setting without having heard the recording you are working on?

"WE'LL FIX IT IN THE mastering."

—Anon

"
MAKING
GOOD SOUND
IS LIKE PREPARING
GOOD FOOD.
IF YOU OVERCOOK,
IT LOSES ITS TASTE.
"

— BOB KATZ

CHAPTER 14

How To Make Better Recordings in the 21st Century

PART ONE:
MONITOR
CALIBRATION

I. Introduction

Calibrated monitors are the critical tools of the 21st century audio engineer. Some engineers think (mistakenly) that the need for monitor calibration is only for making of 5.1 theatrical mixes. But we'll all make better recordings if we use calibrated stereo or surround monitors. A good-sounding monitor system does not come out of the box, it takes work and care. But after the work is done, there's nothing like the pleasure of hearing great-sounding music!

What is a Calibrated Monitor System?

A calibrated monitor system is one that is adjusted to a known standard gain and frequency response. The monitor gain control is repeatable and marked in decibels. *Repeatable* means that you can return the monitor to a particular gain at any time, and *calibrated* means that the standard decibel markings on the monitor scale mean the same thing to any engineer, whether in Calcutta, New York, or Hong Kong .... This will help us collaborate, to be more consistent in our work, and to produce mixes that will perform together when later assembled at the mastering house. As we shall see, the absolute value of the numbers also defines the sound quality of the mix that will result.

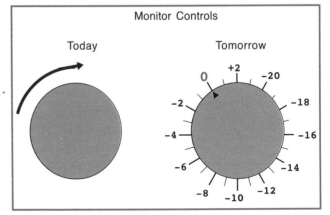

Tomorrow's monitor control will be marked in 1 dB steps, and the 0 dB position will be calibrated to the SMPTE RP 200 standard (to be explained).

II. Getting Rid of Slippery Language

21st Century audio will be integrated with television, home theater, computer audio, computer games, and music playback, often all coming from a central source. During the last century most of us worked in uncalibrated listening rooms, adjusting our recording levels as we pleased, and just turning the monitor knob until it sounded "loud enough."

Try this: Put your favorite high-end effects movie into the DVD player, and adjust the loudness for a big, enjoyable presentation. Next, put one of last year's hypercompressed pop-music CDs into the same player. Watch out when you hit PLAY, because the loudness will be overbearing and in danger of damaging components and your ears. No wonder the consumers are beginning to complain. We can no longer produce recordings in isolation without regard to monitor calibration, since the same consumer equipment that plays DVDs will also play compact discs, videos, MP3s, DVD-As and SACDs!

{ *"**Level** is often confused with **Gain**!"* }

This is why, in the 21st century, we need to learn how to adjust our monitor gain first to a known standard, and then make the recording fit to that gain. One obstacle is the slippery daily language that we use to describe audio.

So to avoid confusion, the first step is to pick words that mean the same thing to everyone. Here is a brief glossary of the language of levels:[*]

VOLUME... **usually associated with an audio level control, is an imprecise consumer term with no fixed definition.** The words more properly used in the art are **Intensity** and **Loudness**.

INTENSITY... (aka SPL, Level, Pressure) a measure of the amplitude or energy of the physical sound present in the atmosphere.

LOUDNESS... is used specifically and precisely for the **perceptual level created inside the listener's brain**. Psychoacousticians can create subjective experiments that measure loudness, and have found that loudness versus intensity is quite similar across a population of listeners. However, *loudness* is much more difficult to measure in a metering system, in fact, it's best presented as a series of numbers rather than as one overall "loudness." Because of the big difference between typical metering systems and our perception, two pieces of music that measure the same on an SPL or VU meter can have drastically different *loudness*, depending on many factors, including transient and frequency response, and the duration of the sound. Exposure time affects our perception; after a five minute rest, the music seems much louder, but then we get used to it again—good reason to keep a sound pressure level meter around to keep us from damaging our ears.

LEVEL... is a measure of intensity, but when used alone means absolutely nothing, because it can

---

[*] Thanks to Jim Johnston (in correspondence) for helping to clarify some of these definitions.

mean almost anything! To avoid confusion, always accompany *level* with another defining term, e.g. *voltage level, sound pressure level. Level* is very often confused with *Gain.* Engineers can have a whole conversation about "levels" and not even know what they're talking about, unless they clearly distinguish *gain* from *level.*

SOUND PRESSURE LEVEL *(SPL)...* is one of the units of intensity. SPL measurements can be repeatable if taken in the same fashion.* 74 dB SPL is the typical sound intensity of spoken word 12 inches away, which increases to 94 dB SPL at one inch distance. While we often see language like *95 dB SPL loud*, this usage is both inaccurate and ill-defined as *loud* refers to the user's perception, and *SPL* to the physical intensity.

Decibels are always expressed as a ratio

A **decibel** (dB) is always a relative quantity; it's always expressed as a ratio, compared to a *reference.* For example, what if every length had to be compared to one centimeter? You'd say, "this piece of string is ten times longer than one centimeter." It's the same thing with decibels, though sometimes the reference is implied. +10 dB means "10 dB more than my reference, which I defined as 0 dB." Decibels are logarithmic ratios, so if we mean "twice as large," we say "6 dB more" [20 * log (2) = 6].

DBU, DBM, DB SPL, DBFS... are expressions of decibels with defined references. I believe the term dBu was introduced in the 1960's by the Neve Corporation, and it means *decibels compared to a voltage reference of 0.775 volts. dBm* means *decibels*

compared to a power reference of one milliwatt. *dBFS* means *decibels compared to full scale PCM; that is,* 0 *dBFS represents the highest digital level we can encode.*

GAIN or **AMPLIFICATION...** is always a relative term expressed in plain decibels, the ratio of the amplifier's output level to the input. It is wrong to use an absolute level (e.g. dBu or dBm or dBv) with the term *gain.* It is sufficient to say that an amplifier has, for example, +27 dB gain, and a nominal output level of + 4 dBu when fed with a given level source, as in this figure.

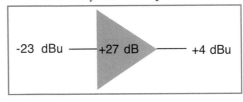

The meaning of Gain vs. Level. An amplifier with 27 dB gain is fed an input signal whose level is -23 dBu to yield an output level of +4 dBu. The decibels of gain should never need a suffix.

MONITOR GAIN VS. MONITOR LEVEL. Similarly, the sound pressure level from your monitor loudspeakers is often confused with the monitor gain. In fact, the term *monitor gain* is so slippery that I have started using a much more solid term that everyone seems to understand: **MONITOR POSITION.** For example, we say "the monitor control is at the 0 dB position."

AVERAGE VS. PEAK. As we learned in Chapter 5, the instantaneous peak level of a good recording can be as much as 20 dB greater than its average (long term) level. Generally, we measure average sound pressure level with a sound level meter; sometimes we look at the peak level. For monitor calibration, the SPL meter should use the *RMS* averaging method, as opposed to a simple average (mean); simple averaging can produce as much as 2 dB error. Unless otherwise specified, when we say *average* in

---

* SPL measurements must include the weighting curve used, e.g. A, or C, the speed of the meter (slow or fast), and method of spatial averaging (how many mikes were used and how they were placed).

this book, we are referring to the RMS-measured level as opposed to the peak level.

Crest Factor is the difference between the average level of a musical passage and its instantaneous peak level. For instance, if a fortissimo passage measures -20 dBFS on the averaging meter and the highest momentary peak is -3 dBFS on the peak meter, it has a crest factor of 17 dB.

III. Using A Calibrated Monitor System for Level and Quality Judgment

An experienced engineer can make a good mixdown just by listening and without looking at the meter. The key is understanding how to use the calibrated monitor control. In simple terms, the monitor level control is calibrated so that the 0 dB position produces 83 dB SPL with a pink noise calibration signal (to be explained). The recorded level of this *calibration signal* is set to -20 dBFS RMS (20 dB below full scale digital). What this means is that a comfortably loud average SPL has been set to 20 dB below the peak system level. Since the ear generally judges loudness by average level, and the most extreme crest factor anyone has measured for normal music is 20 dB, then our peak level will never overload!* Typical mixed material has crest factors from 10 to 18 dB, so this mixdown may reach peaks from -10 to -2 dBFS, more than adequate levels for 24-bit recording, as shown in Chapter 5.

What this means is that a high monitor position will permit us to produce music with high crest factor. Conversely, as you lower the monitor control position, you tend to raise the average recorded

---
* Assuming the mix engineer's ears have normal sensitivity to loud sounds. While no mix engineer works without glancing at the peak meter, you get my point.

level to produce the same loudness to the ear. In the 20th century, we approached this from the opposite

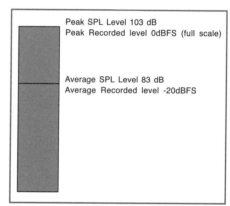

Peak SPL Level 103 dB
Peak Recorded level 0dBFS (full scale)

Average SPL Level 83 dB
Average Recorded level -20dBFS

When monitor gain is calibrated so average SPL is 83 dB at -20 dBFS, and you then mix by the loudness of the monitor, then the music will never overload and you will never have to look at a record level meter!

way; as we raised the average recorded level, we were forced to turn down the monitor to keep our ears from overloading!

Monitoring by the numbers

Judging Loudness. If we become familiar with how various known recordings reproduce on our calibrated system, and the monitor position we use to reproduce those recordings, then we can judge the absolute loudness of any master in the making just by noting the monitor position, without having to compare it with other known recordings.

Judging Sound Quality. As the average level increases and approaches the peak level, more compression and peak limiting will be required to keep the medium from overloading. As we described in Chapter 10, some amount of *compression* can enhance a recording, but extreme compression is self-defeating, it lowers the crest factor and dilutes the clarity, impact, spaciousness, and liveliness of the presentation. It's ironic that mastering engineers are being asked to do some damage to recordings in the name of loudness. Of course, the point where damage occurs is subjective and depends a lot on the music and the message, but we all agree there is such a thing as too much.

Work to a predetermined and fixed monitor gain. In the 21st century of mastering, we should work to a predetermined and fixed monitor gain; if the music becomes too loud, turn down the amount of processing or the output of the processors rather than turn down the monitor! We should use the measured *position* of the monitor control as a guide to the sound quality we are probably going to produce. In other words, if we find the monitor control drifting down too far, our recording is also probably deteriorating. 0 dB position is typically necessary to reproduce audiophile classical and acoustic jazz recordings that have used no compression or limiting. I've found that -6 dB position (corresponding with a crest factor of about 14 dB) is the lowest monitor gain that still produces a high-quality musical product with typical pop music, and most of the pop music recorded in the last century until about 1993 sounds "just right" at the -6 dB position. Slowly but surely, as we are forced to turn the monitor below -6 dB to keep a comfortable loudness, the sound quality is reduced. By working hard, I can make masters geared for -7 or -8 dB monitor position that still sound pretty good.[*] But some current hypercompressed pop CDs exceed this loudness by as much as 6 more decibels!

---

[*] Some monitors are marked in "SPL," which designers think is very sophisticated. However, it's very misleading. **This is a classic case of confusing gain with level.** The 83 marker is meaningless after calibration.

Monitor gain for mixing versus mastering. Mixing and mastering should be collaborative processes. I recommend that you be conservative with average levels during mixing, so as not to deteriorate the recording, for we cannot restore quality that has been lost. When mixing pop music, set your monitor position from 0 dB to no lower than -6 dB to make a recording that falls in line with the vast majority and still has good clean transients; it will help you produce a recording with life and acceptable dynamic range for home and car listening. You will still be able to be creative with compression and other effects—a fixed monitor gain is liberating, not limiting. When such a well-made recording arrives for mastering, we have much more freedom; we will raise the apparent loudness if we can do so while preserving or enhancing the recording's virtues, but the clarity and beauty of the recording will not have been ruined prior to arrival at the mastering house.

> { *A fixed monitor gain is liberating, not limiting.* }

Different Size Rooms. Note that room volume and number of loudspeakers affect the apparent loudness of a system. The more loudspeakers, the louder the system for the same monitor control position. I determined these recommended monitor control positions in a large stereo mastering room with loudspeakers 9 feet from the listener. In an extra large theatre, as much as 2 dB additional gain may be needed, whereas in a small remote truck with loudspeakers a couple of feet from the listener, as much as 2 dB less gain may be necessary. Set your standards accordingly.

IV. Setting Up and Calibrating the System

Summary of Essential Tools

Now that we know the benefits of having a calibrated monitor, let's see what tools we need to construct a good-sounding, calibrated monitor system.

- A great room, whose dimensions, wall construction and layout have minimal obstructions/reflections between the loudspeakers and the listener, with low noise and good isolation from the outside world.
- For surround sound, five matched "satellite" loudspeakers and amplifiers with flat frequency response (preferably good down to 60 Hz), high headroom, each capable of producing at least 103 dB SPL before clipping. To repeat the adage from Chapter 6, high headroom monitors are necessary to make proper sound judgments: if our monitors are compressing, we cannot judge how much compression to use in the recording.
- One (preferably two) subwoofers, capable of extending the low frequency response of all the satellites down to about 25 Hz, and producing at least 113 dB SPL at low frequencies before clipping.
- A low distortion monitor matrix with versatile and flexible bass management, capable of repeatable, calibrated monitor gains, and of down mixing and comparing sources from 7.2 through mono. With

this, we can confidently produce recordings that can be interchanged with the rest of the world, and sound wonderful on systems large and small.
- A monitor selector to feed the matrix, with both digital and analog inputs.
- Measurement/calibration equipment:
  Preferable: A calibrated 1/3 octave real time analyzer (RTA) and microphone(s), with multiple memories, selectable response speed, and ability to integrate several microphone locations (spatial averaging).
  Alternate (less accurate): A high quality sound level meter with calibrated microphone, selectable filters and response speed.
  Test Signals: If using a sound-level meter, then you need RMS-calibrated sources of **filtered pink noise**. If using a 1/3 octave RTA, then you can use ordinary wide-band RMS-calibrated pink noise.
- **And let's not forget the most critical ingredient: Knowledge. The services of a trained acoustician may be needed on first-time setup,** to perform anechoic and early-reflection analysis of the room and loudspeakers, interpret the causes of measured frequency response errors, their audible significance, and suggest acoustically-based cures.

Placing the Main Loudspeakers

The ideal reproduction system should have no obstacles in the path between all the loudspeakers and your ears. This certainly turns most recording consoles and outboard racks into serious problems and is the reason why my rack gear is in the back corner, and my listening couch is placed in **front** of the computer and DAW. This forces me to go behind the ideal listening position when doing heavy editing, but all critical listening and remote control of transports and processors can be accomplished from the couch where there is little or no acoustical interference between loudspeakers and ear.

The Rope (Clothesline) Procedure

Tom Holman[*] describes how two pieces of string can be used to set up your monitors at the proper distances and angles to conform with the ITU 775[†] recommendation, illustrated below.

Here's a step-by-step embellished recipe. All speakers are equidistant from the center of an imaginary circle, with the center front being 0°, front left and right speakers at +/- 30°, and the surround speakers at +/- 110° (ITU accepts surrounds between 100° & 120°). Start with a long piece of rope or clothesline (which doesn't stretch so easily) a little longer than 3 times the length of the proposed distance to one loudspeaker. Tie one end to a mike stand located at the center of the circle (the prime listener). Run the rope to the approximate proposed position of the right front speaker, and

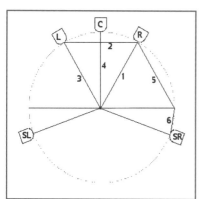

The ITU 775 recommendation for 5 channel loudspeaker placement.

put a piece of black tape on the string to mark the radius of the circle (see **1**). Then fold the long rope at the tape and add two more pieces of tape to mark three identical length sections. This radius is our "standard length," and equals 60° of angle when it runs between two points of the circle.

Spread the marked rope to create an equilateral triangle (see **1, 2, 3**), and now mark the floor at the points for the left front and right front speakers. Cut the rope at the first tape to leave a radius that can swing from the central mike stand. To find the center speaker location, fold a standard length of the remaining rope in half and mark its midpoint. Use that rope to find the midline between the LF and RF speaker and temporarily mark the floor there. Then cross the radius rope over this centerline and mark the position for the center speaker at the end of the radius rope (see **4**).

How to find 110° without a protractor? Use a standard length rope reaching from RF (see **5**) and temporarily mark the spot where it meets the radius rope. This is at 30°+60°= 90°. Now divide a standard length rope in thirds (see **6**), run it from the 90° spot and mark where this 1/3 distance meets the radius rope. This is 90°+20° = 110°, for SR. Do a mirror image of this procedure to find SL, and you're done!

Physically place the subwoofers just in front of, and slightly outside the centerlines of the satellites. Later you may "tweak" the position of the subwoofers for the flattest response at the listening position and best integration with the satellites.

* Holman, Tomlinson [2000] *5.1 Surround Sound: Up and Running*, Focal Press.

† International Telecommunication Union, specification ITU-R BS.775-1

Connecting and calibrating the system levels

The 5.1 monitor system has six outputs, which should be connected to the inputs of the corresponding loudspeaker/amplifiers. I'm going to be describing a system using true stereo subwoofers. One way to connect such a system takes advantage of a subwoofer with two inputs (which most of them have), as illustrated below. You will be using some of the bass management built into the sub and some built into the monitor matrix.

You will choose the low-pass setting on the subwoofer which produces the most seamless "splice" to the satellites; ideally as low as 40 Hz, but some systems need as high as 80 Hz. This depends on the low frequency response of the satellites.[1] Start with the frequency recommended by the manufacturer and later you can tweak according to your room response measurements, as I will explain. Set the woofer polarity to normal and the initial phase setting to 0 degrees (if the woofer has a continuous phase control). The phase control on the subwoofer lines up the apparent distance of the sub with that of the satellites. Leave the woofer phase at 0° if your monitor matrix has delay compensation—if the sub is closer than the satellites, add time delay to the sub based on 1 ms = 1 foot. Later this can be fine-tuned, preferably using time-delay spectrometry, or the real-time analyzer. If your room geometry does not permit the surrounds to be the same distance from the ear as the front speakers, then you can delay the appropriate sets of speakers to match.

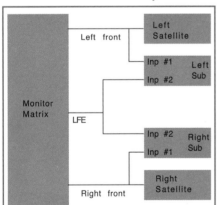

Connecting a monitor matrix with stereo subwoofers. By using the dual inputs of each sub, we can still have a mono LFE signal (the .1 channel) and stereo bass from the front main speakers.

Now let's check the integrity of each connection. **Turn the monitor gain control down all the way!** Feed a calibrated, **uncorrelated***, **5-channel** pink noise source at a level of -20 dBFS RMS into all digital inputs of the system, advance the monitor gain and the trim adjustment on each loudspeaker just a small amount to verify it's operating. Then, solo each output in turn and verify it's getting to the correct speaker.

SMPTE RP 200 Level Calibration

Now we'll be producing some loud test signals, so we suggest putting on earplugs. Place a calibrated measurement microphone pointing directly upwards, at ear height at the central listening position. Connect this to your 1/3 octave RTA. Set the RTA to an averaging time between about 3 and 10 seconds, and wait at least that long before taking any reading. **Turn the loudspeaker trim controls down all the way!** Set the master monitor level to the 0 dB (reference) position. Now, solo ONLY the Left loudspeaker. Slowly turn up the left trim gain until the midband energy (particularly in the 1 kHz band) reads 68 dB SPL (68.2 dB for perfectionists).[2] If all the individual bands were flat at 68 dB SPL, they would sum mathematically to 83 dB SPL, which is the SMPTE RP 200 standard. Inspect the RTA for a general smooth shape with peaks and dips ideally less than plus or minus 3 dB. If any band has a significant peak or dip, it's time to consult an acoustician! Generally I prefer to solve frequency anomalies with acoustic solutions first rather than equalization. Don't be concerned at this time about

---

*Uncorrelated means there is random, or no continuous relationship between channels. Correlated means there is some relationship. If the same, mono source is fed to all channels, then they are 100% correlated.

the absolute flatness of the high end, which will be rolled off.

Repeat this procedure for each of the 5 main loudspeakers, sending pink noise one channel at a time. If 68 dB is not an easy value to "read" with your RTA, then you may, for example, raise the pink noise to -18 dBFS RMS, which should result in 70 dB SPL per 1/3 octave band and (if all bands were equal) would sum to 85 dB SPL broadband. Remember, it's far more accurate to use the midband level measured with a 1/3 octave analyzer than a wideband SPL measurement, due to variations in microphone off-axis response, low frequency room resonances, filter tolerances, and so on. The alternative is to use a sound level meter with a band-limited 500 Hz to 2 kHz signal calibrated to -20 dBFS RMS, to read 83 dB SPL. If only full-range pink noise is available and an RTA is not available, an alternative method (though less accurate, with as much as 2-3 dB possible error) is to use a wideband SPL meter set to C weighting, slow response,

Note that the theatrical standard adjusts the surrounds each to 3 dB below the fronts, but for home music production, all five loudspeakers should have the same gain.

Total Sound Level

The subwoofers have not yet been calibrated and are turned down all the way. Five uncorrelated sources should sum approximately 7 dB higher than an individual channel. Release the solo button and verify that all five main speakers are operating, and the SPL in the midband rises about 7 dB (+/- 1 dB).

If not, then one or more of your cables may be wired out of polarity, speaker distances or level calibration could be off, or a component is defective.

Phantom Center Check

Now let's check the **phantom center** produced by an in-phase mono signal when listening at the central position. This confirms the front main speakers are in polarity and there are no acoustic anomalies. Turn the pink noise off and turn the monitor control to about −10. Change the pink noise source to mono, that is, the same signal to all channels. Solo both left and right front loudspeakers. Now remove your earplugs, turn on the **mono** pink noise and verify the phantom center appears as a fairly narrow virtual image at the physical location of the center loudspeaker. You might tweak the angles (toe-in) of the speakers until the phantom image is narrow in the critical midband. If the image is off-center, recheck the left/right gains and speaker distances. Try tweaking one channel's trim up or down slightly to recenter the image, then return to the previous section and recheck the measured left/right gains to verify they match acoustically within +/- 0.1 dB in the 1 kHz band. Loudspeakers must be well-matched to produce an excellent phantom center.

Now compare the sound of the phantom center with that of the center speaker itself, by alternating between soloing the center or the two sides. The center speaker should sound a little brighter, but the position of the pink noise should not change if you are sitting in the center and all speakers are equidistant from the listener.

Bass Management

Integrating a subwoofer or pair of subwoofers to extend the response of a stereo system is an art and a science. Extending that idea to 5.1 is serious science, with its own set of compromises. We're going to start by creating and verifying an exceptional full-range 2-channel system, then extending it to 5.1. Since we are using stereo subwoofers, it is logical to set the bass level on a per-speaker basis, but the two subs couple with each other and the distances between them and from the walls affect the total bass response. It's not an easy affair, and you should approach it systematically.

Objective Subwoofer Measurement: Put your earplugs back on and send uncorrelated pink noise at -20 dBFS RMS to the LF system: left satellite and sub. Turn up the left subwoofer's trim gain until the RTA shows the low end is in the same ballpark as the rest of the frequencies. You may see amplitude anomalies near the splice point, indicating some parameters are not yet optimized. Then check the polarity of the sub; the position that produces the most bass is the correct one; if the result is ambiguous, temporarily set the sub's cutoff frequency as high as possible and recheck the polarity. The next part is the most time-consuming, where art and science really combine, for the ideal splice will happen only when the low-pass frequency, high-pass frequency, subwoofer amplitude, time delay and phase are just right. Take your time, "focusing" each parameter until the flattest response is obtained at the splice point. If you must compromise, remember, the ear finds peaks more objectionable than dips. Now take a spatial average of the response over a few listening positions around the sweet spot, and continue working until you're satisfied the left sub is integrated according to the RTA.

You may have to move the subwoofer around to produce the flattest extreme low end; the closer the sub is to walls or corners, the higher the amplitude of the deep low bass. If you move the sub, then you will have to readjust its time delay.

Next, if your room is symmetrical, it makes sense to try placing the right subwoofer as a mirror image to the left. Though occasionally, this is not a good idea if the subs both end up at the peak or null of a standing wave (expert acousticians apply here). Repeat the above process with the right loudspeaker system. Now send a mono pink noise source to all channels and solo both the left and right system (including the sub), turning the master monitor down until the 1 kHz band reads 68 dB, and see if the bass response with both channels operating is still within tolerance. Don't be surprised to see a heavier bass response than with the individual channel reading. If it rises, even as little as a dB, consider spreading the subs further apart to reduce their coupling, but then again, if they approach the walls, the low bass will go up from wall proximity. This interaction is at different low frequencies, so hopefully you will find a position with the least compromise.

Subjective Assessment, Stereo First

We have not yet set the bass management for the center speaker or the satellites, but now is a good time to check out the sound of the full-range stereo pair with bass management. It would be nice to discover a definitive piece of music that confirms your subwoofers are now perfectly integrated with the rest of your system. Since a subwoofer is not supposed to be a "boom machine" for most music, it really should be conspicuous by its absence rather than its presence. And that's the first way to listen. Listen to music with the subwoofers on and off. They should not feel "lumpy," they should simply add a sense of weight to the extreme low end. If the crossover frequency is 60 Hz or below, then you may hardly notice a difference except for the solidity of the sound. That's the way it should be!

Finding the right recording to evaluate bass is difficult because recordings of bass are all over the map. It could take days to check your subs by using a variety of recordings. An excellent way to evaluate a full range system is with a recording of a string bass whose level is very naturally-recorded. I have been using one of my own stereo recordings as a bass test record: my recording of Rebecca Pigeon, "Spanish Harlem" on Chesky JD115

This song, in the key of G, uses the classic I, IV, V progression. Here are the frequencies of the fundamental notes of this bass melody:

49	62		73		
		65		82	98
			73	93	110

If the system has proper bass response, the bass should sound natural; notes should not stick out too far or be recessed. Start with the subs turned off and verify the lowest note(s) are a little weak. Then turn the subs on and verify they restore the lowest notes without adding any anomalies. Verify that the addition of the subs does not move the instrument forward in the soundstage (an indication the bass level is set too high) or become vague in its placement (an indication the subwoofers are too far apart). It's that simple. Then, take a break and enjoy Rebecca's performance for its natural acoustic reproduction of voice, string and percussion instruments, and the acoustic depth of a good recording hall. If you get this sound quality, then you are off to a good start with an excellent 2-channel stereo system.

Bass Management for Center and Surrounds

Our next job is to smoothly extend the low frequency response of the center and surround loudspeakers. Once again insert uncorrelated, calibrated level pink noise, with the master monitor to 0 dB position. Solo the center loudspeaker, and set the bass management to feed the low frequencies of the center speaker to the subwoofer(s). Adjust the highpass frequency of the center loudspeaker to the same frequency used for the left and right (if the center speaker is the same model as the sides). Then tweak the bass management level trim of the center (the amount of energy from center redirected to the subwoofer) until the total bass response is as flat as possible with the RTA. Determining a correct bass level from the two surrounds is a bit more complicated, since they are electrically summed into a single mono bass (unless the bass management is sophisticated enough to redirect the left surround's bass to the left sub and vice versa). Soloing each surround in turn, adjust the bass-management trim from each one for flattest response, then check the bass response from both surrounds at once with both uncorrelated and mono pink noise. Favor the response with mono pink noise since we are assuming that in typical music recording the bass will be in phase in both surrounds.

LFE Gain Setting

The LFE, or .1 channel is an auxiliary channel designed to increase the headroom of the bass channels. This is because when extra bass is desired below about 50 Hz, the ear (which is insensitive to bass) could require digital levels as much as 10 dB hotter than full scale digital! In a properly-designed 5.1 system, this headroom is taken care of in the design of the subwoofer. If in doubt, check with the manufacturer. To meet the RP 200 standard, the individual RTA bands **for the LFE channel only** should read 10 dB higher than the 1 kHz band. That is, 78 dB SPL if the 1 kHz band is at 68 with -20 dBFS RMS pink noise. **Solo** the LFE output and adjust the level of the LFE channel trim until the 50 or 63 Hz band reads 78 dB.

This completes the monitor calibration. Now you're on the same page as the most advanced 21st century mastering engineers. To speak the same language, tell all your fellow engineers: "My monitor system is calibrated with 0 dB reference SMPTE RP 200." Now sit back and enjoy your calibrated multichannel reproduction system!

V. Taking it Beyond: Monitor Equalization?

My philosophy is to avoid monitor equalization unless absolutely necessary. I believe that we should do everything possible to fix room-induced problems acoustically, and to relocate subwoofers and/or satellites if necessary for more linear response. Equalization, if performed, should be done by a skilled and experienced acoustician who understands the trade offs of electrically equalizing the direct response when a room anomaly is the root cause. When EQing, remember that the ear responds to the direct and room sound differently than an RTA. Finally, consider the tradeoff of additional noise and distortion if an equalizer is added to a system.

---

1   If the satellites are good down to 40 Hz, so much the better, because the stereo imaging will probably be more coherent with a lower crossover frequency. However, when mastering for Dolby Digital, it is important to make a test listen with a mono crossover at 100 Hz to be compatible with consumer bass management systems. Many authorities recommend a 4th order (24 dB per octave) low pass on the woofer and a 2nd order (12 dB per octave) high pass on the satellites.

2   Holman shows an individual band SPL of 70 dB SPL, but note that this was taken with a pink noise signal of −18 dBFS. If the source noise is higher, then we must expect a higher output SPL. Measurements will be much more repeatable from room to room when you measure the 1 kHz band, as described in the text. So, determine the level to use when measuring the 1 kHz band by subtracting 14.8 (which all but perfectionists round to 15 dB) from the official broadband SPL. For example, if the source of pink noise is at -20 dBFS RMS broadband, the broadband SPL would be 83 dBC, and set the monitor gain until the 1 kHz band reads 68.2 dB. If the source of pink noise is at -18 dBFS RMS broadband, then the broadbadn SPL would be 85 dBC, and the 1 kHz band 70 dB (70.2). This is partly explained in a footnote to the SMPTE RP200 specification.

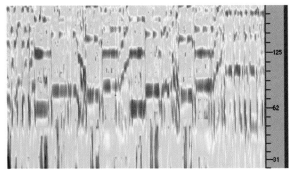

Figure C8-01: *SpectraFoo™ spectragram of the bass frequencies of several measures from a rock piece . Read it like an orchestra score, time runs from left to right. Red represents the highest levels. Note the bass runs in the 62-125 Hz fundamental range are paralleled by second and third harmonics.*

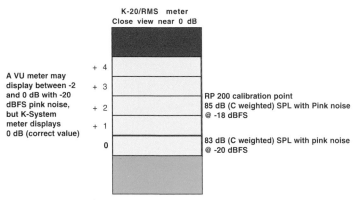

Figure C15-02: *A K-20/RMS meter in close detail, with the calibration points.*

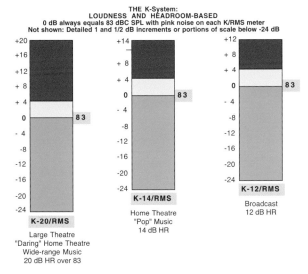

Figure C15-01: *The three K-System meter scales are named K-20, K-14, and K-12. I've also nicknamed them the papa, mama, and baby meters. The K-20 meter is intended for wide dynamic range material, e.g., large theatre mixes, "daring home theatre" mixes, audiophile music, classical (symphonic) music, "audiophile" pop music mixed in 5.1 surround, and so on. The K-14 meter is for the vast majority of moderately-compressed high-fidelity productions intended for home listening (e.g. some home theatre, pop, folk, and rock music). And the K-12 meter is for productions to be dedicated for broadcast.*

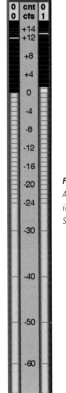

Figure C15-03:
A K-14/RMS Meter as implemented in Spectrafoo

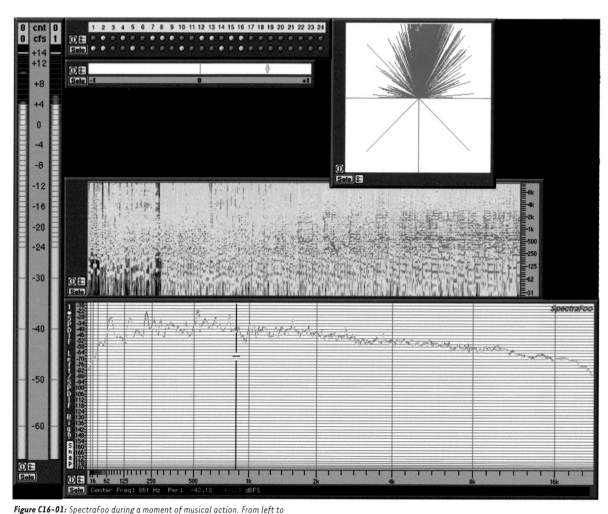

Figure C16-01: SpectraFoo during a moment of musical action. From left to
right at top: K-14 Meter, bitscope, and stereo position indicator. Directly
below the bitscope is a phase/correlation meter. In the middle of the screen is
a Spectragram, quiet section at left part, then the song begins. At top right is
a stereo position indicator, and at the bottom, the Spectragram, left channel
in green, right channel in red.

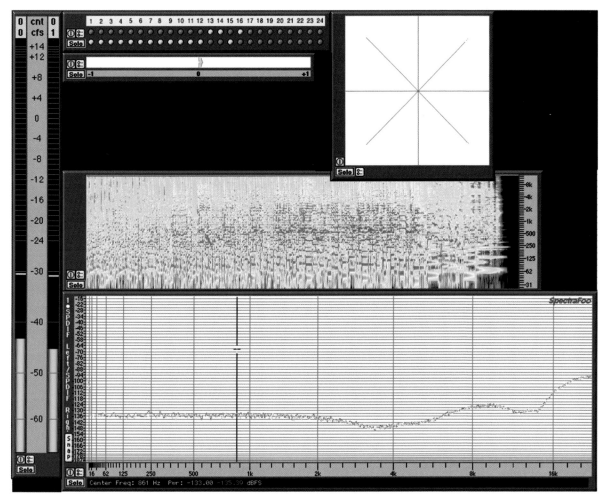

Figure C16-02: *SpectraFoo during a pause in the music. Only the bottom four bits are toggling on the bitscope, and the characteristic curve of POW-R dither type 3 is revealed on the Spectragram. The last notes of the music "fading to black" can be seen at the right of the timeline on the Spectragraph.*

Figure C16-03: Comparing 16, 20, and 24 bit flat-dithered noise floors (red, orange, green traces, respectively).

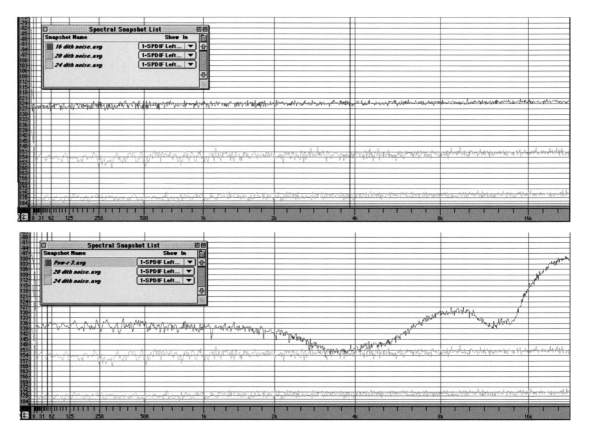

Figure C16-04: POW-R type 3 at 16-bit (red trace) noise floor, with 20-bit flat dither (orange) and 24-bit flat dither (green) for reference.

Figure C16-05: Distortion and noise performance of Millennia Media NSEQ-2 analog equalizer in tube mode (red), 20-bit random noise floor for reference (blue), 24-bit noise floor (green), and Z-Systems ZQ-2 digital equalizer (yellow).

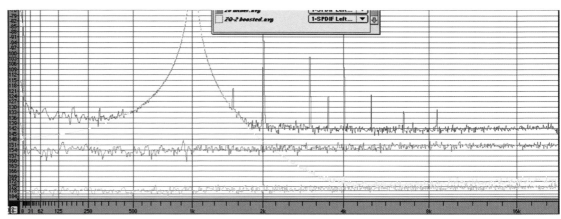

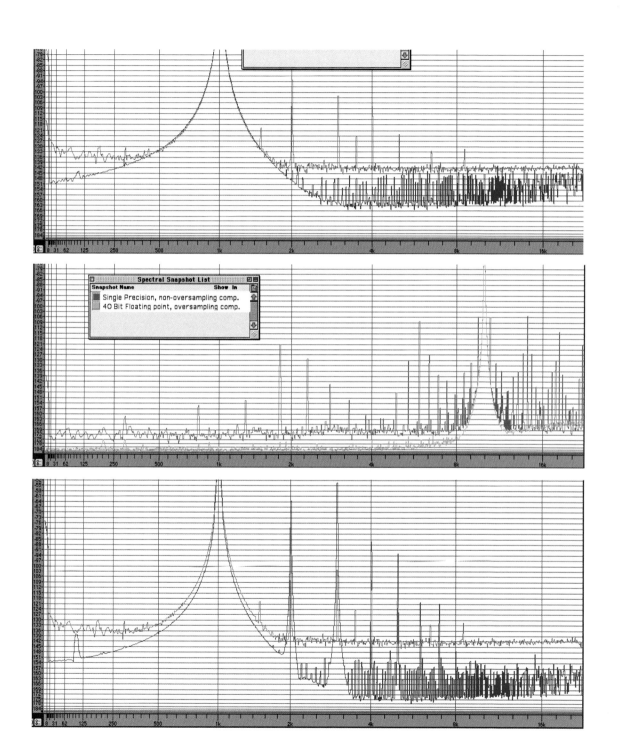

Figure C16-06: *Distortion and noise performance of analog Millennia Media NSEQ-2 (red trace), versus Digital Z Systems set to truncate at 20 bits, no dither (blue trace).*

Figure C16-07: *Comparing two digital compressors, both into 5 dB of compression with a 10 kHz signal. Red trace: Single Precision, non-oversampling. Green: 40-bit floating point, double-sampling and dithered to 24-bit fixed level.*

Figure C16-08: *Comparing Cranesong HEDD-192 digital analog simulator (blue trace) to NSEQ (red).*

Color Plates

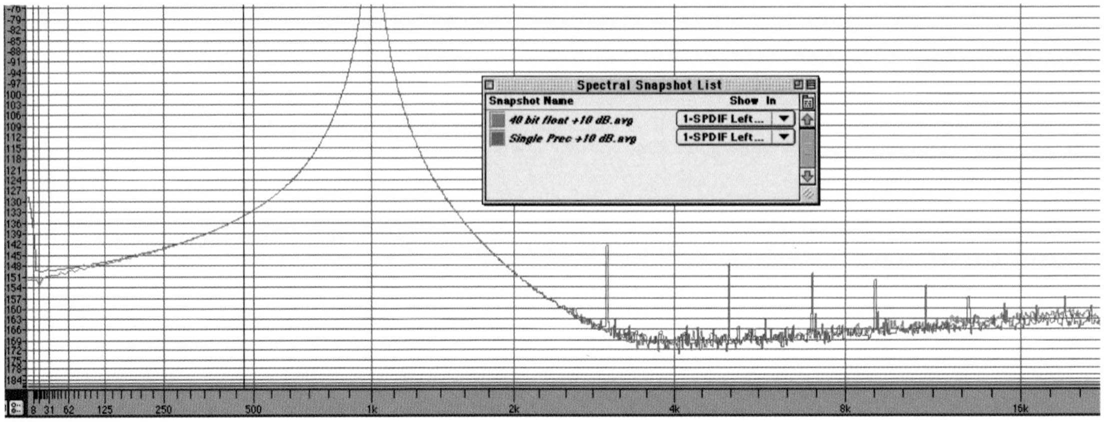

Figure C16-09: *A simple 10 dB boost applied in two different types of processors. In red, a single-precision processor, whose distortion is the result of truncation of all products below the 24th bit. And in blue, the output of a 40-bit floating point processor which dithers its output to 24 bits.*

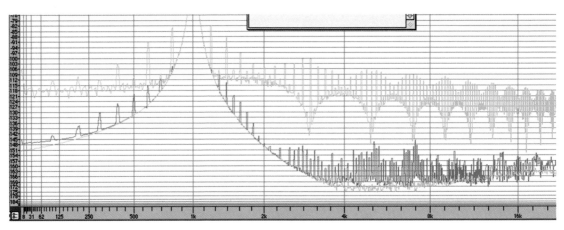

Figure C16-10: *Compares two excellent-sounding digital dynamics processors, the oversampling Weiss DS1-MK2 (green trace), which uses 40-bit floating point calculations, and the standard-sampling Waves L2 (red), which uses 48-bit fixed point. The switchable safety limiter of the Weiss, which is not oversampled, is shown in orange.*

Figure C19-01:

Jitter testing:

16-bit J-Test signal (blue trace) overlayed with the Noise floor of UltraAnalog A/D converter (red trace) which together define the limits of resolution of my jitter test system.

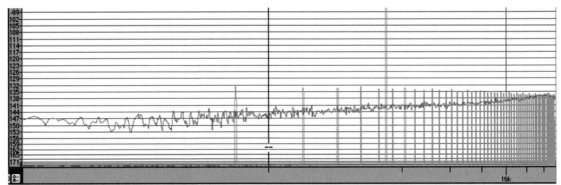

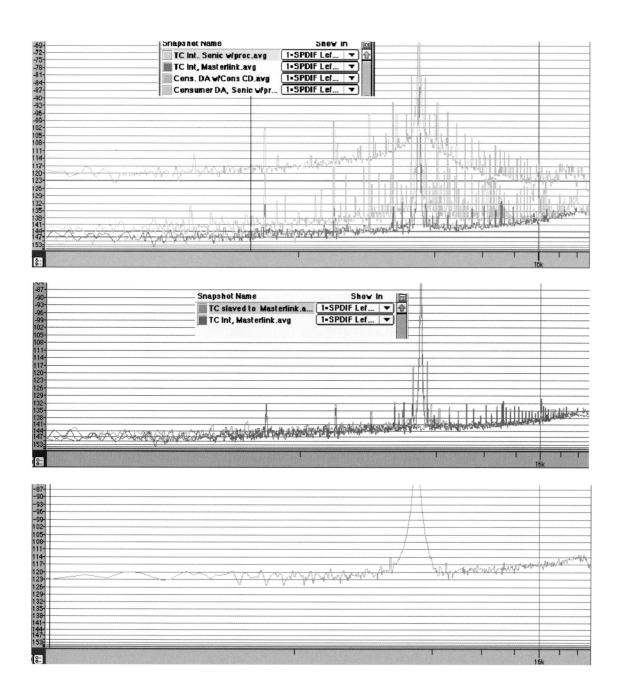

Figure C19-02:

Jitter measurements with J-Test signal:

Orange Trace: TC DAC jitter on internal sync, fed from Sonic Solutions.

Red: TC DAC jitter on internal sync, fed from Masterlink.

Blue: Consumer DAC fed from consumer CD Player.

Green: Consumer DAC fed from Sonic Solutions.

Figure C19-03:

Jitter measurements, demonstrationg how different clocking methods may produce different sound with the same source transport.

Masterlink transport feeding J-Test Signal to TC D/A.

Blue: TC D/A slaved to Masterlink transport via AES/EBU.

Red: TC D/A on internal sync.

Figure C19-04:

Jitter Measurements:

J-Test signal feeding Weiss DAC on AES/EBU sync

View from the bridge. *Digital Domain's Mastering studio. Visible in front of the listening couch are: Rolling rack with Weiss EQ-1 LP Equalizer, Weiss DS1-MK2 dynamics processor, and Digital Domain DD-2 K-Stereo Processor; One pair of Dorrough meters; Reference 3A (satellite) loudspeakers on sand-filled stands plus Genesis Servo-controlled subwoofers.*

Color Plates   *184*

CHAPTER 15
How To Make Better Recordings in the 21st Century

PART TWO: THE K-SYSTEM, AN INTEGRATED APPROACH TO METERING, MONITORING, AND LEVELING PRACTICES

I. History: The VU Meter

On May 1, 1999, the VU meter celebrated its 60th birthday. 60 years—but still widely misunderstood and misused. The VU meter has a carefully-specified time-dependent response to program material that I call *averaging* to simplify discussion, but really means the particular VU meter response. This instrument was intended to help program producers create consistent loudness amongst program elements, but as it was a poor indicator of recording overloads, the meter's designers depended on the 10 dB or greater *headroom* over 0 VU of the analog media then in use.

Summary of VU Inconsistencies and Errors

In general, the meter's ballistics, scale, and frequency response all contribute to an inaccurate indicator. The meter approximates momentary loudness changes in program material, but reports that moment-to-moment level differences are greater than the ear actually perceives.

Ballistics: The meter's ballistics were designed to "look good" with spoken word. Its 300 ms integration time does give it a syllabic response, but does not make it accurate. One time constant cannot sum up the complex multiple time constants that make up the loudness perception of the human listener. Skilled users soon learned that an occasional short "burst" from 0 to +3 VU would probably not cause distortion, and usually was meaningless with regard to loudness change.

Scale: In 1939, logarithmic amplifiers were large and cumbersome to construct, and it was

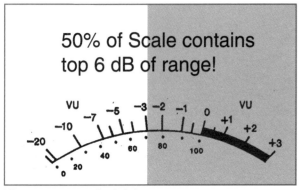

50% of Scale contains top 6 dB of range!

VU meter operators are often fooled into treating the top and bottom halves of the scale with equal weight, but the top half has only 6 dB of the total dynamic range.

desirable to use a simple passive circuit. The result is a meter where every decibel of change is not given equal merit. The top 50% of the physical scale is devoted to only the top 6 dB of dynamic range, and, as illustrated, the meter's useable dynamic range is only about 13 dB. Not realizing this fundamental fact, inexperienced and experienced operators alike tend to push audio levels and/or compress them to stay within this visible range. The extreme needle movements make it difficult to distinguish compressed from uncompressed material. Soft material may hardly move the meter, but be well within the acceptable limits for the medium and the intended listening environment.[5]

Frequency response: The meter's relatively flat frequency response results in meter deflections that are far greater than the perceived loudness change, since the ear's response is non-linear with respect to frequency. Frequency distribution and average level both affect loudness. For instance, when mastering reggae music, which has a very heavy bass content, the VU meter may bounce several dB in response to the bass rhythm, but perceived loudness change is probably less than a dB.

Lack of adherence to standards: In current use, there are large numbers of improperly-terminated mechanical VU meters and inexpensively-constructed indicators which are labeled "VU." I've seen fights break out amongst program producers reading different "VU" instruments. A true VU meter is a rather expensive device and it can't be called *VU* unless it meets the standard.

Over the past 60 years, psychoacousticians have learned how to measure loudness much better than a VU. Despite all these facts, *the VU meter is a very primitive loudness meter*. In addition, digital technology lets us correct the non-linear scale, its dynamic range, ballistics, and frequency response.

II. The Magic of *83* with Film Mixes

Unlike music CDs, films are consistent from one to another, because the monitoring gain has been standardized, as we learned in Chapter 14. In 1983, as workshops chairman of the AES Convention, I invited Tomlinson Holman of Lucasfilm to demonstrate the sound techniques used in creating the Star Wars films. Dolby systems engineers labored for two days to calibrate the reproduction system in New York's flagship Ziegfeld theatre. Over 1000 convention attendees filled the theatre center section. At the end of the demonstration, Tom asked for a show of hands. "How many of you thought the sound was too loud?" About four hands were raised. "How many thought it was too soft?" No hands. "How many thought it was just right?" At least 996 audio engineers raised their hands.

The choice of 83 dB SPL has stood the test of time, as it permits wide dynamic range recordings with little or no perceived system noise when recording to magnetic film or high-resolution digital. 83 dB also lands on the most effective point on the Fletcher-Munson equal loudness curve, which is where the ear's frequency response is most linear. When digital technology reached the large theatre, the SMPTE attached the SPL calibration to a point 20 dB below full scale digital instead of 0 VU.[*] When we converted to digital technology, the VU meter was rapidly replaced by the peak program meter, which didn't faze the film world, but definitely caused the music industry to suffer, as we shall see.

III. United We Stand At Home

As we saw in Chapter 14, with the integration of media into a single system, it is in the direct interest of music producers to think holistically and unite with video and film producers for a more consistent consumer audio presentation. New program producers with little experience in audio production are coming into the audio field from the computer, software and computer games arena. We are entering an era where the learning curve is high, recording engineer's experience is low, and the monitors they use to make program judgments are less than ideal. It is our responsibility to educate new engineers on how to make loudness and quality judgments. A plethora of peak-only meters on every computer, DAT machine and digital console do not provide information on program loudness. Engineers must learn that the sole purpose of the peak meter is to protect the medium and that something more like average level affects the program's loudness.

***

* See Appendix 9 for discussion on how "85" became "83".

Current-day leveling problems: The Loudness Race

The loudness race is not new; in the days of vinyl, mastering engineers competed to produce the loudest LP. But what is new is the fantastic magnitude of the problem: due to the nature of the digital medium, there is no longer the physical limit which was previously imposed by analog mechano-electrical systems and magnetic analog recording. Without that limit it is possible to produce CDs whose average level is almost the same as the peak level, an incredible 20 dB above the old average levels! Powerful digital compressors and limiters enable mastering engineers to produce a distorted signal for which there is no precedent in over 100 years of recording.[1] So, as we converted to digital technology, the result became chaos, yielding unprecedented differences in loudness between recordings.

On the next page is a waveform taken from a digital audio workstation, showing three different styles of music recording. The time scale is about 10 minutes total, and the vertical scale is linear, +/- 1 at full digital level, 0.5 amplitude is 6 dB below full scale. The "density" of the waveform gives a rough approximation of the music's dynamic range and crest factor. On the left side is a piece of heavily compressed pseudo "elevator music" I constructed for a demonstration at the 107th AES Convention. In the middle is a four-minute song from a popular compact disc produced in 1999. On the right is a four-minute popular rock and roll recording made in 1990 that's quite dynamic-sounding for rock and roll of that period. The perceived loudness difference between the 1990 and 1999 CDs is

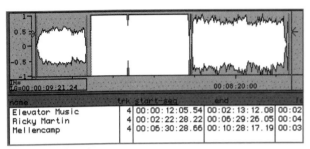

On the left, moderately compressed "Elevator Music." In the Middle, a "top of the pops" selection from the year 1999. At right, a rock and roll record from 1990. Vertical and horizontal scales are identical.

greater than 6 dB, though both peak to full scale! Auditioning the 1999 CD, one mastering engineer remarked, "this CD is a light switch?! The music starts, all the meter lights come on, and it stays there the whole time." To say nothing about the distortion. Are we really in the business of making square waves? Why has the average sound quality of popular music CDs gone downhill since the introduction of the digital medium, and what can we do to fix the problem?[2]

The psychoacoustic problem is that when two **identical** programs are presented at slightly differing loudness, the louder of the two often appears "better," **but only in short term listening**. This explains why CD loudness levels have been creeping up until sound quality is so bad that everyone can perceive it (illustrated below). And why there is a remarkable (and unnacceptable) 15 dB difference in average level among pop CDs! Remember that the loudness "race" has always been an artificial one, since the consumer adjusts their

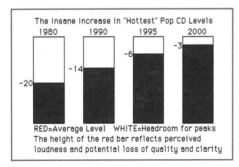

Is this what will happen to the next generation carrier? (e.g. DVD-A, SACD). It will, if we don't take steps now to stop it.

volume control according to each record anyway. This uncontrolled situation is an obstacle to creating quality program material in the 21st century. What good is a 24-bit/96 kHz digital audio system if the programs we create only have 1 bit dynamic range?

There are, of course, specific places where heavy compression is needed: background listening, parties, bar and jukebox playback, car stereos, headphone-wearing joggers, the loudspeakers at the record stores, headphone auditioning at the record store *kiosk*, and so on. In each of these cases, it should be possible to either produce a custom-compressed CD just for the purpose, or to install a compressor in the jukebox, CD changer, or reproduction system. Certainly this is a lot less damaging than compromising recorded music for all listeners. What we wish for is a low-fidelity replacement for the analog cassette. Ironically, the compact disc has become its own worst enemy, for it cannot be different things to different needs.[3] I dream of a perfect world where all the MP3 singles are heavily compressed and all the CD albums undamaged.

IV. The relationship between SPL and 0 VU

Around 1994 I installed a pair of Dorrough meters, in order to view the average and peak level simultaneously on the same scale. These meters use a scale with 0 "average" (a quasi-VU characteristic I'll call **AVG**) placed at 14 dB below full digital scale, and full scale marked as +14 dB. Music mastering engineers often use this scale, since a typical stereo

1/2" 30 IPS analog tape has approximately 14 dB headroom above 0 VU.

The next step is to examine a simple relationship between the 0 **AVG** level and the sound pressure level. For many pop productions, our calibrated monitor control sits at -6 dB (which yields 77 dB SPL with -20 dBFS RMS pink noise).

Since on the meter, -20 dBFS reads -6 AVG, then 6 dB higher, or 0 AVG must be 83 dB SPL. This means we're really running average SPLs similar to the theatre standard (our sound quality is not as

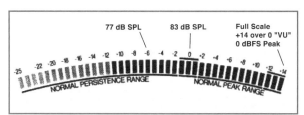

The Dorrough Meter. With the monitor control's position set to 6 dB below the film reference, 77 dB SPL lands at −20 dBFS, or −6 AVG on the meter. Not by coincidence, this corresponds with 83 dB SPL at the meter's 0 AVG point, revealing the obvious correlation between a mastering engineer's meter ZERO and 83 dB SPL.

clear as that of the theatre, and our loudness is probably slightly lower because some high-frequency transients have been clipped by 6 dB of compression). Our "pop studio" headroom is only 14 dB above 83 instead of 20. The absolute loudness[*] of our pop presentation is nominally 6 dB louder than a film in the theatre, necessitating turning down the monitor gain by 6 dB.

Running a sound pressure level meter during the mastering session confirms that the ear likes 0 AVG to end up circa 83 dB (~86 dB with both loudspeakers operating) on forte passages, even in this compressed structure. If the monitor gain is further reduced by 2 dB the mastering engineer judges the loudness to be lower, and he raises average recorded level—and the AVG meter goes up by 2 dB. It's a linear relationship.[†] **This leads us to the logical conclusion that we can produce programs with different amounts of dynamic range by designing a loudness meter with a sliding scale, where the moveable 0 point is tied to the same monitor SPL. Regardless of the scale, production personnel would tend to place music near the 0 point on forte passages.**

V. The K-System Proposal

This leads us to my K-System proposal, a metering and monitoring standard that integrates the best concepts of the past with current psychoacoustic knowledge in order to avoid the chaos of the last 20 years. It also develops a common *language of levels*, so that engineers can properly communicate.

In the 20th Century we concentrated on the *medium*. In the 21st Century, we should concentrate on the *message*. We should avoid meters which have 0 dB at the top—this discourages operators from understanding where the message really is. Instead, we move to a metering system where 0 dB is a **reference loudness**, which also determines the monitor gain. In use, programs which exceed 0 dB

---

[*] **ABSOLUTE LOUDNESS:** A term I use when comparing the apparent loudness of different sources **without moving the monitor control.**

[†] Linear until things get so squashed that the increasingly compressed sound is not equally louder for the same measured increase in the flat meter's average level.

> *"The K-system is not just a meter scale, it is an integrated system tied to monitoring gain."*

give some indication of the amount of processing (compression) which must have been used. There are three different K-System meter scales, with 0 dB at either 20, 14, or 12 dB below full scale, for typical headroom and SNR requirements. The dual-characteristic meter has a bar representing the average level and a moving line or dot above the bar

representing the most recent highest instantaneous (1 sample) peak level.

Several accepted methods of measuring loudness exist, of varying accuracy (e.g., ISO 532, LEQ, Fletcher-Harvey-Munson, Zwicker and others, some unpublished). The extendable K-system accepts all these and future methods, plus providing a "flat" version with RMS characteristic that resembles the classic VU meter.

Note that full scale digital peak level is **always** at the top of each K-System meter, it does not change. Only the average level calibration slides, the 83 dB SPL point slides relative to the maximum peak level. Using the term K-(N) defines simultaneously the meter's 0 dB point and the monitoring gain, making this **the first integrated metering and monitoring system.** *

Simplified Explanation

Many mastering engineers have recognized that the peak meter is inadequate for judging loudness, so they use a traditional analog VU meter. But because of the wide range of average levels on current pop CDs, they use a variable VU meter attenuator to prevent the VU from pinning or reading out of range. Think of the K-System as a **coordinated attenuator for both the averaging meter and the monitor gain**. The principle is that as we attenuate the average meter while going from K-20 to K-14 we must also turn **down** the monitor gain, to arrive at the same loudness to the ear. If the monitor gain were **not attenuated**, then K-14 material reaching 0 dB average on its scale would

[K-System Meter.
For a color image,
please see the
Color Plates
section, ***Figure***
C15-01.]

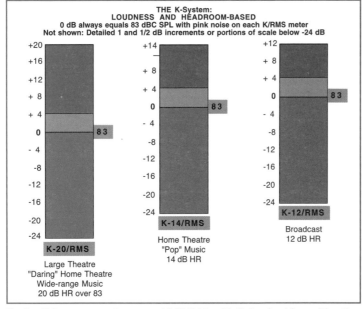

THE K-System:
LOUDNESS AND HEADROOM-BASED
0 dB always equals 83 dBC SPL with pink noise on each K/RMS meter
Not shown: Detailed 1 and 1/2 dB increments or portions of scale below -24 dB

K-20/RMS
Large Theatre
"Daring" Home Theatre
Wide-range Music
20 dB HR over 83

K-14/RMS
Home Theatre
"Pop" Music
14 dB HR

K-12/RMS
Broadcast
12 dB HR

The three K-System meter scales are named K-20, K-14, and K-12. I've also nicknamed them the papa, mama, and baby meters. The K-20 meter is intended for wide dynamic range material, e.g., large theatre mixes, "daring home theatre" mixes, audiophile music, classical (symphonic) music, "audiophile" pop music mixed in 5.1 surround, and so on. The K-14 meter is for the vast majority of moderately-compressed high-fidelity productions intended for home listening (e.g. some home theatre, pop, folk, and rock music). And the K-12 meter is for productions to be dedicated for broadcast.

---

* I invented these K-(N) terms because it was getting very awkward to describe the crest factor or loudness of music in a simple but useful way.

sound 6 dB **louder** than K-20 material going to 0 dB average on its scale.

Peak and Average calibrated to same decibel value with sine wave

The peak and average scales are calibrated as per AES-17, so that peak and average sections are referenced to the same decibel value with a sine wave signal. In other words, +20 dB RMS with sine wave reads the same as + 20 dB peak, and this parity will be true only with a sine wave. Analog voltage level is not specified in the K-system, only SPL and digital values. There is no conflict with -18 dBFS analog reference points commonly used in Europe.

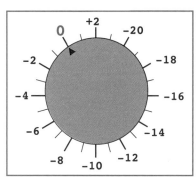

For medium-size control rooms, typical monitor gain (control position) will be 0 dB with the K-20 meter, -6 dB with the K-14 meter, and –8 dB with the K-12 Meter. 0 dB monitor gain is the calibration point that corresponds with the RP200 standard (see Chapter 14).

VI. Production Techniques with the K-System

To use the system, first choose one of the three meters based on the intended application. Wide dynamic range material probably requires K-20 and medium range material K-14. Then, calibrate the monitor gain to RP200 as in Chapter 14. 0 dB always represents the same calibrated(83 dBC) SPL on all three scales, unifying production practices worldwide. If console and workstation designers standardize on the K-System it will make it easier for engineers to move programs from studio to studio. Sound quality will improve by uniting the steps of pre-production (recording and mixing), post-production (mastering) and metadata (authoring) with a common "level" language. By anchoring operations to a consistent monitor reference, operators will produce more consistent output, and everyone will recognize what the meter means.

If making an audiophile recording, then use K–20; if making "typical" pop or rock music, or audio for video, then probably choose K-14. It will be hard for current pop mastering engineers to convert to K-14 or even K-12 in some cases, because much of today's damaged pop music is significantly hotter than even K-12—but we must find a way to back off from the loudness race. Ideally, K-12 should be reserved strictly for audio to be dedicated to broadcast; broadcast recording engineers may choose K-14 if they feel it fits their program material. Pop engineers are encouraged to use K-20 when the music has useful dynamic range. The two prime scales, K-20 and K-14 will create a cluster near two different monitor gain positions. People who listen to both classical and popular music are already used to moving their monitor gains about 6 dB (sometimes 8 to 12 dB with the hottest pop CDs). It will become a joy to find that only two monitor positions satisfy most production chores. With care, producers can reduce program differences even further by ignoring the meter for the most part, and working solely with the calibrated monitor.

Using the Meter's Red (Fortissimo) Zone. This 88-90 dB+ region is used in films for explosions and special effects. In music recording, naturally-recorded (uncompressed) large symphonic ensembles and big bands reach +3 to +4 dB on the average scale on the loudest (*fortissimo*) passages. Rock and electric pop music take advantage of this *loud zone*, since climaxes, loud choruses and occasional peak moments sound incorrect if they only reach 0 dB (*forte*) on any K-system meter. Use the *fortissimo* range *occasionally*, otherwise it is musically incorrect (and ear-damaging). If engineers find themselves using the red zone all the time, then either the monitor gain is not properly calibrated, the music is extremely unusual (e.g. *heavy metal*), or the engineer needs more monitor gain to correlate with his or her personal sensitivities. Otherwise the recording will end up overcompressed, with squashed transients, and its loudness quotient out of line with K-System guidelines.

Equal Loudness Contours. Mastering engineers are more inclined to work with a constant monitor gain. But music mixing engineers often work at a higher SPL, and vary their monitor gain to check the mix at different SPLs. I recommend that mix engineers calibrate your monitor attenuators so you can easily return to the recommended standard for the majority of the mix. Otherwise it is likely the mix will not translate to other venues, since the equal-loudness contours indicate a program will be bass-shy when reproduced at a lower (normal) level.

Tracking/Mixing/Mastering. The K-System will probably not be needed for multitracking—a simple peak meter is sufficient. For highest sound quality, use K-20 while mixing and save K-14 for the calibrated mastering suite. If mixing to analog tape, K-14 may prove more appropriate. K-20 doesn't prevent the mix engineer from using compressors during mixing, but I hope that engineers will return to using compression as an esthetic device instead of trying to win the loudness race.

Using K-20 during mix encourages a clean-sounding mix that's advantageous to the mastering engineer. At that point, the producer and mastering engineer should discuss whether the program should be converted to K-14, or remain at K-20. *The K-System can become the lingua franca of interchange within the industry, avoiding the current problem where different mix engineers work on parts of an album to different standards of loudness and compression.*

When the K-System is not available. Current-day analog mixing consoles equipped with VUs are far less of a problem than digital models with only peak meters. Calibrate the mixdown A/D gain to -20 dBFS at 0 VU (sine wave), and mix normally with the analog console and VUs. However, mixing consoles should be retrofitted with calibrated monitor attenuators so the mix engineer can repeatably return to the same monitor setting.

Adapting large theatre material to home use may require a change of monitor gain and meter scale. Producers may choose to compress the original 6-channel master, or better, remix the entire program from the multitrack stems (submixes). With care, most of the virtues and impact of the original production can be maintained in the home. Even audiophiles will find a well-mastered K-14 program to be enjoyable and dynamic. We should try to fit this reduced-range mix on the DVD with the wide-range theatre mix.

Multichannel to Stereo Reductions. The current legacy of loud pop CDs creates a dilemma because DVD players can also play CDs. Producers should try to create the 5.1 mix of a project at K-20. If possible, the stereo version should also be mixed and mastered at K-20. While a K-20 CD will not be as loud (absolute loudness) as many current pop CDs, it will probably be more dynamic and enjoyable, and importantly there will not be a serious loudness jump compared to K-20 DVDs in the same player. If the producer insists on a hotter CD, try to make it no louder than K-14, so there will be no more than a 6 dB loudness difference between the DVD and the audio CD. Tell the producer that the vast majority of great-sounding pop CDs have been made at K-14 and the CD will be consistent with the lot, even if it isn't as hot as the current hypercompressed fashion. The hypercompressed CD is the one that's out of line, not the K-14.

Full scale peaks and SNR. As we've discussed (Chapters 5 and 14) it is not necessary to peak a 24-bit recording to full scale. Another good reason is that a program's signal-to-noise ratio is determined by its **actual loudness**, the position of the listener's monitor level control determines the perceived loudness of the system noise. If two similar music programs reach 0 on the K-system's average meter, even if one peaks to full scale and the other does not, both programs will have similar perceived SNR. Use the averaging meter and your ears as you normally would, and with K-20, even if the peaks don't hit the top, the mixdown is considered normal and ideal for mastering.

Multipurpose Control Rooms. With the K–System, multipurpose production facilities will be able to work with wide-dynamic range productions (music, videos/films) one day, and mix pop music the next. A simultaneous meter scale and monitor gain change accomplishes the job. Operators should be trained to change the monitor gain according to the K-standard.

In Color Plate **Figure C15-02** is a picture of the K-20/RMS meter in close detail, with the calibration points. Individuals who wish to use a different monitor gain should log it on the tape (file) box, and try to use this point consistently. Even with slight deviations from the recommended practice, the music world will be far more consistent than the current chaos. Everyone should know the monitor gain they like to use.

In Color Plate **Figure C15-03** is a picture of an actual K-14/RMS Meter in operation at the Digital Domain studio, as implemented by Metric Halo labs in the program SpectraFoo™ for the Macintosh computer. SpectraFoo versions 3f17 and above include full K-System support and a calibrated RMS pink noise generator. On the PC, Pinguin has implemented meters that conform exactly with the K-System. The Dorrough and DK meters nearly meet K-System guidelines but be sure to use an external RMS meter for calibration since they use a different type of averaging. In practice with program material, the difference between RMS and other meter averaging methods is imperceptible. I hope soon a company will implement the K-System with a truer loudness characteristic.

Audio Cassette Duplication. Cassette duplication has been practiced more as an art than a science, but it should be possible to do better. The K-System may finally put us all on the same page, ironically just in time for the cassette's obsolescence. It's been difficult for mastering engineers to communicate with cassette duplicators, finding a reference level we all can understand. The cassette tape most commonly used cannot tolerate average levels greater than +3 over 185 nW/m (especially at low frequencies) and high frequency peaks greater than about +5-6 are bound to be distorted and/or attenuated. Displaying crest factor makes it easy to identify potential problems; also an engineer can apply cassette high-frequency preemphasis to the meter. An engineer can make a good cassette master by using a "predistortion" filter with gentle high-frequency compression and equalization. Use K-14 or K-20, and put test tone at the K-System reference 0 on the digital master. Peaks must not reach full scale or the cassette will distort. Apparent loudness will be less than the K-standard, but this is a special case.

Classical music. The dilemma is that string quartets and Renaissance music, among other forms, have low crest factors as well as low natural loudness. Consequently, the string quartet will sound (unnaturally) much louder than the symphony if both are peaked to full scale digital. For example, dedicated classical producers have avoided mastering their harpsichord recordings to full scale, or they sound unnaturally loud at standard monitor gains. It's hard to get out of the habit of peaking our recordings to the highest permissible level. I strongly feel it is much better for the consumer to have a consistent monitor gain than to peak every recording to full scale digital. Attentive listeners prefer auditioning at or near the natural sound pressure of the original classical ensemble.[4]

Classical engineers should mix by the calibrated monitor, and use the average section of the K-meter only as a guide. It's best to fix the monitor at the 0 dB position and always use the K-20 meter even if the peak level does not reach full scale. There will be less monitoring chaos and more satisfied listeners. However, some classical producers are concerned about loss of resolution in the 16-bit medium and may wish to peak all recordings to full scale. I hope you will all reconsider this thought when 24-bit media reach the consumer. Until then chaos will remain in the classical field, and perhaps only metadata will sort out the classical music situation at the listener's end.

Narrow Dynamic Range Pop Music. We can avoid a new loudness race and consequent quality reduction if we unite behind the K-System before we start fresh with high-resolution audio media such as DVD-A and SACD. Similar to the above classical music example, pop music with a crest factor much less than 14 dB should not be mastered to peak to full scale, as it will sound too loud.

Recommended:

1) Author with metadata to benefit consumers using equipment that supports metadata

2) If possible, master such discs at K-14 or even K-20.

3) Legacy music, remasters from often overcompressed CD material should be reexamined for its loudness character. If possible, reduce the gain during remastering so the average level falls within K-14 guidelines. Even better, remaster the music from unprocessed mixes to undo some of the unnecessary damage incurred by the loudness race. Some mastering engineers already have made archives without severe processing.

Multichannel

There's good news for audio quality: 5.1 surround sound. Current 5.1 mixes of popular music sound open, clear, beautiful, yet also impacting. Six speakers provide much more headroom and sound output than two, so if you work by the monitor gain, the channel meter levels will tend to run a bit lower. What became clear while watching the K-20 meter is that the best engineers are using the peak capability of the 5.1 system strictly for headroom, the way it should be. System hiss is not evident at 0 dB monitor position with long-wordlength recording, good D/A converters, modern preamps and power amplifiers.

Labeling The Boxes

Since the K-System is extendable to future methods of measuring loudness, program producers should mark their tape boxes or digital files with an indication which K-meter and monitor calibration was used. For example, *K-14/RMS*, or *K-20/Zwicker*. I hope that these labels will someday become as common as listings of nanowebers per meter and test tones for analog tapes.

VII. Metadata and the K-System

Metadata is *data within data*, that is, control data embedded in the digital audio stream. Dolby Digital, MPEG2, AAC, and hopefully MLP will take advantage of metadata control words (defined below); note that standard PCM, as used in the Compact Disc, has no provision for metadata, and to the best of my knowledge, neither does SACD. Pre-production with the K-System will speed up the authoring of metadata for broadcast and digital media. Music producers must become familiar with how metadata affects the listening experience.

Metadata Control Words

Dialnorm, *dialogue normalization*, also known as *volume normalization*, is used in digital television and radio as "ecumenical gain-riding." Program level is controlled at the decoder, producing a consistent average loudness from program to program; with the amount of attenuation individually calculated for each program and carried as a command on the metadata word. At each program change, the receiver decodes the dialnorm control word and attenuates the level by the calculated amount, resulting in the "table radio in the kitchen" effect. In a somewhat unnatural manner, like the radio, average levels of sports broadcasts, rock and roll, newscasts, commercials, quiet dramas, soap operas, and classical music all end up at the loudness of dialogue, a rather strange effect, but no different loudness-wise than standard radio today. The listener can turn his receiver up and experience the intended loudness—without the noise modulation and squashing of current analog broadcast techniques. Or, he can choose to turn off the dialnorm on some receivers, and hear a loudness variance from program to program.

Dialnorm is a simple gain change, without compression, and maintains the crest factor and dynamic range of the studio mix. For example, in variety shows, the music group will sound pleasingly louder than the presenter. Sports crowds will be excitingly loud, and the announcer will no longer "step on" the effects, because the bus compressor will be banished from the broadcast chain.

Mixlev. *Dialnorm* does not reproduce the dynamic range of real life from program to program. This is where the optional control word *mixlev* (mix level) enters the picture. The *dialnorm* control word is designed for casual listeners, and *mixlev* for audiophiles or producers. Very simply, *mixlev* sets the listener's monitor gain to reproduce the SPL used by the original music producer. If the K–system was used to produce the program, then K–14 material will require a 6 dB reduction in monitor gain compared to K-20, and so on. Attentive listeners using *mixlev* will no longer have to adjust monitor gains for different music types.

The use of *dialnorm* and *mixlev* can be extended to other encoded media, such as DVD-A. Proper application of metadata and the K-System for pre-

production practice—will result in a far more enjoyable and musical experience than we had at the end of the 20th century of audio.

In Summary

The designers of the compact disc never anticipated that an all-digital recording system would yield an alarming loudness race and seriously distorted music, worse than ever took place in the days of the LP. I propose a new system with a common language, integrating monitoring and loudness metering to produce more consistent masters, and move audio practice into the 21st century. Teach everyone how—the Rosetta stone is in this chapter.

---

1 Ironically, current-day compression practices (especially in pop music) are far more aggressive than necessary, even stronger than our approach to the noisier analog medium of the past! CDs can and should be produced to the same audio quality standard as the DVD, but I'd be satisfied with the leveling practices that made good LPs.

2 I see an interesting analogy of the loudness race and the migration of pitch since the 16th century. Music seems to be racing to be just a little more sharp than the previous generation, so that an A played on an instrument tuned to previous standards is now the G or G# of today, so it ultimately turns into a problem of transposition. Unfortunately, audio systems cannot accomodate an infinite loudness rise. We must voluntarily "transpose" back, or go deaf.

3 This is what the DVD and DVD-A proclaim to be, a single audio medium for all needs, because the table radio or the car can contain built-in compression, following the metadata coefficients laid down by the program producer. Let's meet again in 20 years and see if that promise has been met.

4 The late Gabe Wiener produced classical recordings noting in the liner notes the SPL of a short passage. He encouraged listeners to adjust their monitor gains to reproduce the "natural" SPL which arrived at the microphone. I used to second-guess Wiener by first adjusting monitor gain by ear, and then checking against Wiener's number. Each time, I found my monitor gain was within 1 dB of Wiener's recommendation. Thus demonstrating that the natural SPL is desirable for attentive, foreground listeners.

5 One of my first lessons in the inaccuracy of the VU meter was in 1972, when I heard William Pierce, voice of the Boston Symphony, clearly and distinctly in the noisy control room at Channel 24, yet he hardly moved the needle. The trained operator must use his ears and learn how to interpret this instrument.

CHAPTER 16

Analog and Digital Processing

The mastering engineer must recognize when a recording is so good that the interests of the client are best served simply by leaving it alone. And there are recordings for which so little work is needed that the gains due to processing would not warrant the losses due to the same processing! For although equipment is getting better, there is no such thing as a transparent audio processor. This chapter is about how we measure and interpret performance, as there is an interaction between objective degradation and subjective improvement. Let's take a journey into the twilight zone between the objective and the subjective.

I. The Ironies of Perception vs. Measurement

Although we'll be using test measurements, we must remember that each single measurement only provides a small part of the picture. An audio processor is like an object inside a house with no doors, only a number of small windows that you can peer into. By looking at the object through each window's unique angle we can find out more, and add up the clues, but we can never be totally sure of what we are seeing, and must always leave open the possibility that there may be some aspect we cannot see, some mystery as to why this equalizer sounds "good" and this other one sounds "bad."

For example, here are a couple of "objective" measurements that just don't add up!

What Makes it Sound Bright?

I've discovered a digital filter that measures "dull" but sounds bright! The TC Electronic System 6000 lets the user choose between different low-pass filters for the A/D and D/A converters. Some of the filters roll off significantly above 16 kHz (at 44.1 kHz sampling), so you'd think they would sound dull. But instead, to my ears, the 16 kHz filters called *Natural* and *Linear* sound more *open* and *clear* than the particular 20 kHz filter called *Vintage*. However, there are other converters whose filters extend to 20 kHz and which sound even more *open* than the TC's Linear filter. So measured bandwidth cannot tell the whole psychoacoustic story. We look into the audible effects of filtering in Chapter 18.

The Fallacy of Typical Weighting Curves

We have equipment in our studio whose noise floor measures as low as −120 dBFS to as high as −50 dBFS (after A/D conversion). However, much of this equipment is perceptually quiet: if I have to put my ear up to the loudspeaker to hear the hiss, then I consider it insignificant. Interestingly, the weighting methods[1] by which converter manufacturers commonly measure noise bear little relationship to human perception. One particular converter whose A-weighted noise floor is −108 dBFS sounds significantly quieter than another converter whose A-weighted noise floor is −115 dBFS! The reason is that the often-cited, A-weighted curve does not adequately consider the ear's greater sensitivity in critical bands. It turns out that the converter which measures better (A-Weighted) produces significantly more energy circa 3 kHz, where the ear is most sensitive, and the A-weighting filter does not take into account the significance of this critical band. To be psychoacoustically accurate, noise measurement standards should adopt a curve closer to the measured noise floor of the human ear, such as the 9th order curve used by some of the best-sounding dithers (see Chapter 4). This curve is called "F" weighting.[2]

There are many other areas in which traditional measurements do not correlate with what our ears tell us, particularly in the evaluation of low bit rate coding systems. These systems measure quite well with standard techniques, but once the ear has been trained to hear their errors, we can easily identify artifacts we've never heard before with analog technology: described by some as *chirping,* or *space monkeys.* Let's see if we can objectively find out why some analog and digital processors sound better than others. Just remember that measurements look at an object through a few narrow windows, and there may be a different, or better, explanation for sound quality than what I've come up with.

II. Measurement Tools We Can Use While Mastering

FFT Measurements

FFT stands for *Fast Fourier Transform*. To really learn how to interpret (and not misinterpret) an FFT requires a college-level engineering course, and although I cannot claim to be such an expert, I have learned just enough to be dangerous! High-resolution FFT analysers, such as SpectraFoo™, are very reasonably priced, thanks to the exponential increase in CPU power and they provide an essential *early warning system*, a protection from the

{ *"Never turn your back on digital."*
—Bob Ludwig. }

vicissitudes (bugs) of digital audio. *Never turn your back on digital*, says Bob Ludwig, or as I say, *you're only one mouse click away from disaster!* It's a whole new world based on software designed by fallible human beings.

FFT for Music

Figure C16-01 in the Color Plate section shows SpectraFoo in action during a CD mastering session.

At the middle top is a bitscope, currently showing 16 (and only 16) active bits, an indication that the dither generator is probably doing its job. This bitscope can reveal if some digital device is malfunctioning, since one of the symptoms of a disfunctional processor is to toggle unwanted bits, or hold some bits steady when there is no signal. Bitscopes can also show if there are any unwanted truncations caused by defective or misused processors. However, the bitscope is only one of the small windows we can look through; it can easily miss problems, or seem to indicate problems which require further interpretation. For example, some equalizers produce idle noise when the music goes to silence. This can be perfectly normal, but will show up on the bitscope as activity. Toggling the equalizer in and out while observing the bitscope will ascertain if that is the source of the problem or some other anomaly in the signal chain.

At top right is a stereo position indicator, which is frozen at a moment when the information is slightly right-heavy. At left is a meter that conforms to the K-14 standard (see Chapter 15). The meter shows the hottest moment of a rather hot R&B piece (which I would have preferred to reduce, but the client desired it this hot!). For the record, this material was monitored at -8 dB, which really makes it K-12 material. Just below the bitscope is a correlation indicator, revealing that the material is significantly monophonic. I prefer a correlation indicator to an oscilloscope; meter deflections closer to the center of the scale indicate less correlation from channel to channel and likely a larger or more spacious stereo image. However, I always use my ears to confirm the image is not too "vague" and perform a mono (folddown) test to make sure the sound is mono-compatible.

At mid-screen is the spectragram, showing spectral intensity over time. This can be useful to identify the frequencies of problem notes, or simply to entertain visitors! At bottom is the spectragraph, whose general rolloff shape gives a vague idea of the program's timbre (though most times I disregard the spectral displays, since the eye candy of the visual display distracts our aural senses).

Figure C16-02 in the Color Plates shows SpectraFoo during a pause in the music, with only the bottom four bits toggling, confirming that the dither is working correctly, since dithers which use heavy noise-shaping exercise several bits. Note that the bitscope shows four bits toggling (since dither is random, in this snapshot, bit 15 is at zero) and that the spectragraph shows the curve of the dither noise, which can be identified by its shape as POW-R type 3 or a similar 9th order curve. Using this analyzer, you can often determine the type of dither used by the mastering engineer on recorded CDs.

The level meters had not decayed fully when this shot was taken. The correlation meter fluctuates very slightly near the meter's center, showing that the dither is uncorrelated between channels (random phase). I always glance at this display at the beginning and end of the program, to make sure no bugs or patching errors have crept in. I carry a SpectraFoo umbrella even if it's not raining!

II. Measurement Tools to Analyze your Equipment

Let's sort out what happens beneath the knobs. As in geometry, the shortest distance between two points is a straight line, so too in audio — both digital and analog — the cleanest signal path contains the fewest components. The converter used to be the most degrading piece in the studio, but although they have greatly improved in recent years, we should still avoid extra conversion whenever possible. For analog tapes, it's best to do all the analog processing on the way to the first and only A/D conversion. But these days mixes are often on digital tape, and as there are a lot of desirable analog processors which the mastering engineer may prefer because they sound more *organic* than their digital equivalents, the tonal benefits of analog processing might outweigh the transparency losses of an extra conversion.* The best defense is a good offense, and it is possible to reliably measure signal below the noise with an FFT analyzer. An FFT can confirm if a digital processor is not truly bypassed when it says *bypass*, which can be pretty deleterious (see Chapter 4). Jitter (see Chapter 19) is irrelevant to FFT analysers, which strictly look at data.

Even though the analyzer can only examine 24 bits (the limitation of the AES/EBU interface), it can measure distortion 40 dB below the 24-bit noise floor! This is because Spectrafoo is a 64-bit floating point system. So we can compare the distortion of processors which truncate at the 24th bit versus others which use 48 bits or so internally and then dither up to 24 bits. Whether we can hear these differences is a different question. Psychoacoustician J. Robert Stuart has demonstrated that we can hear a 24-bit truncation in an 18-bit system. The ear's dynamic range is approximately 20 bits (120 dB), but this varies with frequency. At certain frequencies we can even hear below 0 dB SPL!

How Many Bits is Enough?

In color plate *Figure C16-03*, we compare 16, 20, and 24-bit flat-dithered noise.[3] The levels of all the "bins" add up, so at 16 bits, the curve which looks like it rides at approximately -124 dBFS (level of individual bins) totals to an RMS level of about -91.2 dBFS RMS, the theoretical limit of a properly-dithered 16-bit system. But discrete signals at some frequencies can be heard as low as -115 dBFS in a properly-dithered 16-bit system, below which they are buried in the noise. Psychoacoustically, for the vast majority of popular and classical music, 16 bits properly done are just enough to do the job right. But as soon as we post-produce, copy, process and change gain, we accumulate noise and need professional headroom, or perhaps we should call it *footroom*† since the top, at 0 dBFS, is a constant.

Psychoacousticians studying the limits of the human ear have determined that 20-bits is enough

---

* And losses can be minimized using upsampling (see Chapter 1).

† This is a made-up word, not an official term!

for good A/D and D/A performance. Anything more is just gravy, and it's very rare to find a "24-bit" converter with better than 18-20-bit noise level. For processing, however we need the additional *footroom*, better than 24 bits, because the frequency-content of digital distortion is far more annoying to the ear than analog distortions which are much louder. **This is because distortion created during digital processing yields harmonic components which beat against the sample rate, producing dissonant inharmonic beat or intermodulation products.** For purist processing, we may need as much as 48 to 72 bits, especially for extreme gain changes, complex filtering, compression, or to avoid cumulative distortion when cascading processes. It's a myth that there's no generation loss in digital processing; **little by little, bit by precious bit, sound suffers with every DSP operation.**

Figure C16-04 in the color plates shows the noise floor of a popular dither called POW-R type 3 at 16-bit (red trace). For reference, we show the noise of flat 20-bit dither (orange), and 24-bit dither (green). POW-R's shape is designed to maximize performance by keeping the noise at or near the ear's low-level sensitivity at various frequencies. POW-R dither reaches 20-bit performance in the critical upper midrange (circa 3.5 kHz) where the ear is most sensitive. Thus, much of the low level ambience and reverberation that would have been masked is revealed, even with 16-bit reproduction. This performance can only be achieved by recording at a longer wordlength to begin with, as noise accumulates and the SNR gets slightly worse when

you add final dither to the processed source.

Analog versus Digital Processing

Cheap versus Good...Is It Really Accurate?

Many people have argued that the reason we notice harshness in some digital recordings is that digital audio recording is more *accurate* than analog. Their claim is that the *accuracy* of digital recording reveals the harshness in our sources, since digital recording doesn't compress (mellow out) high frequencies as does low speed (15 IPS) analog tape. *Accuracy*, they say, is why we have regressed to tube and vintage microphones. But I say this is only a half-truth, since most of these arguments come from individuals who have not been exposed to the sound of good digital recording equipment, which is not only accurate, but can even be *warm and pretty*. Cheap digital equipment is subject to edgy sounding distortion which can be caused by sharp filters, low sample rates, poor conversion technology, low resolution (short wordlength), poor analog stages, jitter, improper dither, clock leakage in analog stages due to bad circuit board design and many others, such as placing sensitive A/D and D/A converters inside the same chassis with motors and spinning heads. It takes a superior power supply and shielding design to make an integrated digital tape recorder that sounds good; compare the sound of an inexpensive modular digital multitrack (MDM) with the Nagra Digital recorder—4 very expensive tracks versus 8 cheap ones.

When it comes to processing, numeric precision is also expensive, even though it's all software. Numeric imprecision in digital consoles

MYTH:
It's a digital processor, so there's no generation loss.

MYTH:

It's a Digital Console. It must be better than my old analog model!

produces problems somewhat like noise in analog consoles, but there is an important difference: noise in analog consoles gradually and gently obscures ambience and low-level material and usually does not add distortion at low levels. However, numeric imprecision in digital consoles causes quantization errors (which increase at low levels) destroying the body and purity of an entire mix, creating edgy, colder, sound, which audiophiles call **digititis**. Since digital consoles do not make sound warmer, depending on the quality of their digital processing—and the number of passes through that circuitry—it might be better to mix through a high-quality analog console.

Even though good digital equipment is getting cheaper at an exponential rate, it is still expensive to produce excellence in digital recordings. That's why analog tape and analog mixing remain very much alive at this point in the 21st century.

Two Fine Equalizers, One Analog, One Digital

In my opinion, much inexpensive tube equipment is overly warm, noisy, unclear and undefined, and the common use of "fuzzy" analog equipment to cover up the problems of inexpensive digital equipment is a band-aid, not a cure for the loss of resolution. Not many people have been exposed to recent audiophile-quality tube equipment, and only the best-designed tube equipment has quiet, clear sound, tight (defined

> *"Audio processing is the art of balancing subjective enhancement against objective degradation."*
> —Bob Olhsson.

bass), is transparent and dimensional, yet still warm. Audiophiles feel a well-designed tube circuit can be more linear and resolving[4] than a low-cost solid state circuit. I certainly feel I hear more through some amplifiers than others. Modern-day tube designers often make innovative use of low-noise regulated power supplies on filaments and cathodes, a practice which was impractical in the 50's.

Figure C16-05 in the Color Plates section shows the low distortion and noise performance of a well-designed, popular state-of-the art analog tube equalizer, the **Millennia NSEQ-2** (red trace). For reference, 20- and 24-bit noise are shown in blue and green, respectively. Notice that the tube noise of the NSEQ is about 10 dB greater than 20-bit, making it a *virtual 18-bit analog equalizer*. However, this performance is dependent on the analog gain structure used. If you drive the equalizer harder, its noise floor will be lower compared to maximum signal, and distortion may or may not be a problem. Since the Millennia's clipping level is around +37 dBu, it may be perfectly legitimate to drive it with nominal levels of +10 dBu or even higher, provided the source equipment doesn't overload! Yet even with nominal levels of 0 dBu as was used for this graph, this tube equalizer is extremely quiet. Its noise is inaudible at any reasonable monitor gain unless you put your ears up to the speaker, demonstrating that noise-floor is probably the least of our worries. 1/2" 30 IPS 2-track analog tape has even higher noise, but no one complains about it for popular music.

For this FFT, we set up a D/A converter, feeding the NSEQ and then an A/D and the FFT. A digitally-generated 1 kHz -6 dBFS 24-bit dithered sine wave feeds the D/A. We adjust converter gain so 0 dBFS is +18 dBu, and boost the equalizer about 6 dB, till just below A/D clipping. The equalizer is coasting at this level, since it's around 19 dB below its clip level! If you are looking for extreme "tubey" effects, you can drive the equalizer even harder, and also realize a greater SNR, provided the converters can handle the hotter level, certainly the equalizer can.

Notice that the equalizer's distortion is dominated by second, third, and fourth harmonics, which tend to *sweeten* sound. For comparison, in yellow is the performance of the superb **Z-Systems** digital equalizer, dithered to 24 bits, boosting 1 kHz 5.8 dB with a Q of 0.7. Its harmonic distortion performance is textbook-perfect (no visible harmonics on the FFT). Some engineers use the word "dry" to describe the sound of a component that has little or no distortion. Looking through other "windows" we find that harmonics are far from the only sonic differences between these pieces of gear. Tubes, power supplies and transformers can *loosen* the bass, which can sometimes be desirable; the digital equalizer retains the tightness of the bass;[*] the digital and analog equalizer's curves are also different, though the ZQ-2 does a nice job of simulating the shapes of gentle

analog filters. Equalizer curve shape and phase shift probably make up other areas of delicate sonic difference between models of equalizers.

The premium price of both the ZQ-2 and the NSEQ reinforce my point that high-quality analog or digital recording is expensive. At the time of this writing, it will be a number of years before there's enough power in a typical computer plug-in to come up to the quality of the best outboard processors.

"Nasty" Digital Processors

Truncation distortion can be fairly "nasty." For example, in *Figure C16-06* of the Color Plates section, we compare the analog Millennia NSEQ (orange trace) versus the digital Z Systems set to truncate at 20 bits, no dither (black trace).

Don't try this at home! I think there are better ways to add *grunge* than turning off the dither. Much of the ambience, space, and warmth of the original source have been truncated, lost forever, converted to low level grunge (severe inharmonic distortion and noise). Even a small amount of non-harmonic distortion can be bothersome. Which sounds better, an analog processor with a smooth but higher noise floor, plus second and third harmonic distortion, or an undithered digital processor with a lower average noise floor plus inharmonic distortion?

Poorly-implemented digital compressors produce severe inharmonic distortion, which is without integer relationship to the fundamental. *Figure C16-07* in the Color Plates compares two digital compressors, both into 5 dB of compression with a 10 kHz signal.

---

[*]   Since digital equalizers don't soften the bass like some tube units, you may wish to "loosen" the bass with compression or some other tool.

In orange is a single-precision, non-over-sampling compressor, and in black a double-sampling compressor implemented in 40-bit floating point. Note the single-precision compressor produces many non-harmonic aliases of the 10 kHz signal, especially in the critical midband. Nasty-sounding first-generation compressors are still common in low-cost digital consoles and DAW plugins. It takes a lot of processing power to double-sample. I'm convinced that the proliferation and misuse of cheap digital processing has degraded the sound quality of much recently-recorded music.

The Magic of Analog?

Static distortion measurements don't explain every reason why some compressors sound *excellent* and others hurt your ears. There are analog processors which are so *magical* that though they are not transparent, they add an interesting and exciting sonic character to music, or to put it another way, *their subjective cure is better than their objective disease.* Analog tape recording is a perfect example of this type of process; measured objectively it's noisy and distorted, but subjectively it can kick ass! If psychoacoustic research had been a bit more advanced on the audible effects of masking distortion and noise, then perhaps we may not have pursued this expensive search for 144 dB extremes. For example, the noise floor of the Sony-Philips DSD system is not particularly special (about 120 dB in the audible band), but it sounds excellent, indicating that low-noise must not be our only goal. We may even conclude that part of the good sound is due to masking; maybe -120 dB is just enough to cover the ugly parts of the distortion of even some of our best analog and digital gear. In addition, noise-free recording media can be very *sterile-sounding* because all the nits and cracks and distortions caused by the musicians and their amplifiers are completely revealed by the quiet media. So, sometimes, adding extra noise can be more beneficial to the music than working noise-free. Perhaps one of the many reasons why analog tape sounds more *musical* to many people...noise can be very euphonic. We should certainly experiment with noise-masking and make our decisions on what is best for the music. *[Please see sidebar, **Clarity or Fuzz**.]*

I think that many classic analog compressors' warm, fat yet clear sound signatures come from a unique combination of attack and release character-istics, which may be emulated in a digital processor. There are some plug-ins which emulate classical analog compressors but to my ears they do not come up to the job; I think they will get better over time when the cost of DSP goes down. Currently, plug-in designers are forced to minimize the DSP load of their processors or users complain they can't fit a plug-in on every channel strip (as if this is desirable). Certainly the Weiss digital compressor does not sound *digital*, so we know that it can be done with programming skill and expensive DSP.

An Analog Simulator-Pick your flavor of grunge

Figure C16-08 in the Color Plates compares the NSEQ to the Cranesong HEDD-192, a digital *analog simulator* of excellent sound quality.

The Cranesong (blue trace) has been adjusted to produce a remarkably similar harmonic structure to the NSEQ. For this graph, its levels have been purposely set to produce more distortion than the Millennia was producing. Amazingly, the ear thinks it's hearing an excellent analog processor without any imaging or resolution loss. But the low-level grunge at the bottom of the picture looks mighty suspicious; looking through this "window" you might think the Cranesong was truncating important information. But two important factors ameliorate: First, the Cranesong's grunge is about 12 dB lower than that of a truncated device and thus is likely masked by the noise and the euphonic harmonics. Secondly, the HEDD has a unique summing internal architecture that does not alter, truncate or recalculate the original source signal. The Cranesong clones the original source and sends that to its output, while mixing in the calculated distortion, thereby largely preserving the ambience and space of the original. The low level distortion in the figure is part of the additive distortion signal and not a result of recalculations to the source. In other words, only the distortion is distorted! We took this measurement first at 44.1 kHz; at 88.2 and 96 k. As you can see in the two figures on the next page, at 96 k the low level grunge is virtually gone, and the Cranesong's distortion is even cleaner, if that's not a contradiction in terms!

Cooking Better Sound—Naturally

There are certain analog consoles whose character is highly prized because they add spice, dimension and even punch to a mix. One name that comes to mind is API, which to my ears has an excellent combination of desirable linearities (like headroom and bandwidth) and nonlinearities. I think the subtle "grit" in their discrete opamps

could even be slight intermodulation distortion, which does just the right thing for rock and roll yet is subtle enough for jazz and classical depending on how you drive the stages (a matter of taste). I think the transformers add some punch or fattening via saturation and 2nd and 3rd harmonic distortion as well as some upper harmonics and a touch of phase shift (which could add some *dimensionality*).

Our role as mastering engineer is like that of the master chef who knows just how much and what kind of spice is useful to add *pizzazz* without overcooking or spoiling the flavor. By the middle of our careers we have collected a sizable analog and digital spice rack! The Cranesong can mimic three types of naturally-occurring analog distortion, called **Triode**, **Pentode** and **Tape.** The **triode** control adds a pinch of **salt**, pure second harmonic, which, being the octave, is quite subtle, almost inaudible with some music. It can *clear up* the low end by adding some definition to a bass, but it can also thin out the sound too much. The **pentode** is extremely versatile; it provides both

Clarity or Fuzz, which is best?

There's nothing wrong with using fuzz if it produces the right esthetic result. With high-resolution digital recording, tube equipment can add a nice flavor.

Or, it can be used as a useful cover-up, a *fuzzy band-aid*. A client once told me, "Bob, your mastering is so much clearer than the mix, I'm starting to hear all the mistakes!" Yes, high-resolution processing revealed more and more of the source, but this came at a price, all the warts were revealed. I solved the problem by fuzzing up the sound slightly with some delicate *tape style* harmonics.

For if the performance is not the absolute best, or the mix is not wonderful, or the sound is just better when it's not perfectly clear—then *fatness, masking fuzz, or analog distortion magic* may be just the right approach for the music. In mastering I usually prefer to accomplish this by first passing the signal through the highest resolution electronics, which add little or no distortion, and then add a touch of the fuzzy sauce with a selectively *fuzzy* component or a noisy dither. This approach is methodical, controllable, and reversible.

Clearly, artful use of noise can mask and therefore ameliorate some low-level distortions. Ironically, digital recording's super low-noise may be its greatest enemy.

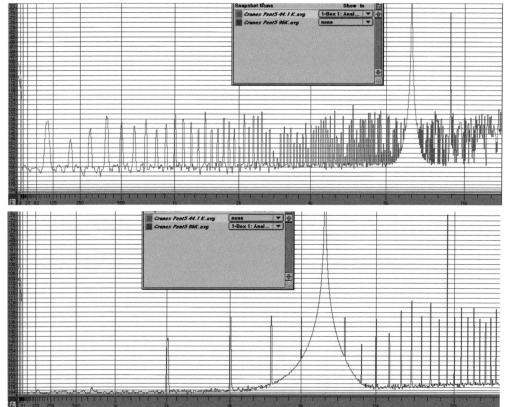

Comparing Cranesong HEDD 192 in Pentode mode at two different sample rates with a 10 kHz −15 dBFS test tone.

At top, 44.1 kHz SR, at bottom, 96 kHz. Note the different frequency scales since the higher sample rate displays harmonic frequencies of the audio signal up to 48 kHz.

which when mixed in, can sweeten the pentode pepper, yielding flavors from red to yellow, green or Jalapeño! The celebrated third harmonic (an octave plus a fifth) sweetens and fattens the sound, much like analog tape. **Tape** also produces the fat sound of analog tape, which helps to "glue" a mix together. **Tape** can help digitally-mixed sources that may be well-recorded but miss some of that "rock and roll fatness." The control produces largely second and third harmonic distortion, but as it's advanced, some additional higher harmonics, emulating analog tape performance. Too much sugar gives slow, muddy molasses, a rarely desirable quantity, but available if you need it. But just a light amount can act as a sweet-sounding bandaid to ameliorate truncated or edgy recordings. Regardless, space and depth have been permanently lost if there was truncation prior to the use of the Cranesong.* No one is sure why, but critical listeners have observed that adding delicate amounts of harmonic distortion in just the right proportion appear to enhance the depth and clarity in a recording. The trick is to know the exact amount.[5]

salt and *pepper*. At lower levels it adds third and fifth harmonics, which are dangerously seductive, producing a unique presence boost and brightness with little grunge or digititis, especially at 96 kHz SR (pictured). At higher levels, additional odd harmonics add grit and some fatness, like an overdriven pentode tube—a Marshal amplifier in a 1 U rack-mount box! Past the fifth, subtle amounts of seventh and ninth harmonics add a sometimes desirable "edge."

The Cranesong's **tape** control is the **sugar,**

Single Precision, Double Precision, or Floating Point?

First-generation digital processors gave digital processing a bad name. But single precision 24-bit processors are going the way of the Dodo, at least in respectable audio equipment. All things being equal

---

* Though Digital Domain's K-Stereo process does a pretty good job of restoring that lost ambience.

(and they never are) 32-bit floating point processors are generally regarded as inferior-sounding to 48-bit (double-precision fixed), and 40-bit float. Some newer floating-point devices, such as the software program **ChannelStrip** by Metric Halo, work in 64-bit and have impressively low measured distortion. However, one designer, Z-Systems, has produced a 32-bit floating point digital equalizer using proprietary distortion-reducing techniques that sounds very good and measures as well as some other equalizers using longer wordlengths. Ultimately the skill of the designer determines how nice the device sounds. The mathematics involved are not trivial, and the designer's choice of filter coefficients can make as much difference as his choice of wordlength.

Figure C16-09 in the Color Plates shows that with a single precision processor, even a simple gain boost can ruin your digital day. A dithered 24-bit 1 kHz tone at -11 dBFS is passed through two types of processors, each boosting gain by 10 dB. The distortion of the single precision processor (red trace) is the result of truncation of products below the 24th bit. Nevertheless, the highest distortion product, at -142 dBFS, is extremely low. I believe the *sound* of a single 24-bit truncation may not be audible, but cumulative truncation adds enough inharmonic distortion to become annoying to the sensitive ear. In blue we compare the perfectly clean output of a 40-bit floating point processor which dithers its output to 24 bits. I measured similar performance with a 48-bit (double precision) processor and 32-bit floating point processor, which both dither to 24 bits.

Double Sampling?

The most advanced digital equalizers and dynamics processors use double sampling technology, which means that the internal sampling rate is doubled to reduce aliasing distortion. High-quality linear phase filters are used in the internal sample rate converters. I'm not certain this has audible meaning for equalizers,[6] but dynamics processors benefit because non-linear processing generates severe aliases of the sampling rate, and the higher the sample rate, the less aliasing.

Figure C16-10 in the Color Plates compares two excellent-sounding digital dynamics processors, the oversampling Weiss DS1-MK2, which uses 40-bit floating point calculations, and the standard-sampling Waves L2, which uses 48-bit fixed point.

To compare apples to apples, both processors are limiting by 3 dB, with the Waves in red, and the Weiss in green, set to 1000:1 ratio. Note the oversampling processor exhibits considerably lower quantization distortion. However, the switchable safety limiter of the Weiss, which is not oversampled, produces considerable alias distortion even at 1 dB limiting (orange trace). At 88.2 kHz and above (not shown), the Weiss safety limiter and the Waves perform measurably better, and double sampling may not be needed. Thus there is considerable advantage of doing all our processing at higher rates, which moves the distortion products into the inaudible spectrum above 20 kHz. Then, sample rate convert to 44.1 kHz during the last step, which filters out most of the high-frequency by-products.

Despite the measured differences, the "window" we've chosen, (steady-state sinewave performance) probably has little to do with the perceived performance of these two excellent-sounding limiters. Because steady state measurements have little or no relationship to audible performance of limiters. I believe the key to the ear's reaction is the *duration* of the limiting action. In typical use, limiters go into gain reduction for a very short time. At limiting ratios of 1000:1, with instantaneous attack, and fast release, these processors produce only momentary distortion, shorter than the human ear's sensitivity to distortion (about 6 ms according to some authorities). But if a user overpushes a limiter so that it is working on the RMS levels of the material as well as the peaks, then its sinewave-measured distortion becomes audibly significant.

Compressors, however, are different animals, and double sampling is critical for them, because a compressor may be into gain reduction for a good percentage of the time. I feel that double-sampling contributes to the Weiss's *robust* and *warm* sound when used as a compressor. While Heavy Metal recordings employ considerable distortion for effect, classically they employ analog processors for this purpose to avoid the inharmonic aliases of typical digital processors.

Better Measurement Methods?

It should be clear by now that we can easily measure simple phenomena that are probably too subtle to hear (such as single tone harmonic distortion near the 24 bit level). But we can hear (perceive) very complex phenomena that are difficult to describe with measurements (such as the sound quality of one equalizer versus another). What we will need to better describe such complex audible phenomena are *psychoacoustically-based* measurement instruments that have not yet been invented. Current research and development of coded audio such as MP3 (that benefits from the ear's masking) could lead to better noise and distortion analysers that can discriminate between distortion we can and cannot hear.

The Bonger—A Listening Test

Since current steady-state sine-wave measurements are misleading when measuring nonlinear processors like compressors, a more effective measurement method is by listening: using the **gonger** aka **bonger**, originally developed by the BBC's Chris Travis and available on a test CD from Checkpoint Audio (see Appendix 10). This test is a pure sine wave that modulates through various amplitudes, in order to exercise and reveal any amplitude non-linearities in the signal path. Just play the bonger through the device under test and listen to the output for noise modulation, buzz or distortion.

Identity Testing—Bit Transparency

Any workstation that cannot make a perfect clone should be junked. The simplest test is the identity test, or bit-transparency test. Set a digital equalizer to flat and unity gain, then test to see if it passes signal identical to its input. Some people scoff at this test, since analog equipment almost never produces identical output. But the test is important, since digital equipment can produce egregious distortion as we have seen. The bit scope can aid in null testing: it is quite likely that a device is bit-transparent if you selectively put in 16 bits, then 20, then 24, and get out the same as you put in. You can also watch a 16 or 20-bit source expand to 24-bits when the gain changes, during crossfades, or if any equalizer is changed from the 0 dB position. A neutral console path is a good indication of data integrity in a DAW. After the bitscope, your next defense is to perform some basic tests, for linearity, for distortion with the FFT, and finally, test for perfect clones (perfect digital copies). The **null test** confirms bit-for bit identity: Play two files at the same time, inverting the polarity of one and mixing the two together. There must be zero output or the two files are not identical. Since designers are fallible human beings, you should carry out basic tests on your DAW for each software revision.

Choose Your Weapon

So, which to use, analog or digital processing? A few years ago, I didn't like the sound of cumulative digital processing. I could tolerate a couple of the best-designed single-precision units in series. After that, it was back to analog.

If processing digitally, be aware of the weaknesses of the equipment. Until manufacturers adopt more powerful processors, and processing power catches up, limit the number of passes through any digital system. Each pass will sound a little bit colder even using 24 bit storage. **A mix made through a current-day digital console may or may not sound better than one made through a high-quality analog console**, depending on several factors: the number of passes or bounces that have been made, the number of tracks which are mixed, the quality of the converters which were used, the outboard equipment, and the internal mixing and equalization algorithms in the digital console. While no console equalizer currently has the power of a $6000 Weiss, economically it's a lot simpler to replicate a good equalization algorithm for 144 channels than performing the equivalent in analog hardware, so there is hope for the digital console's future, when silicon will be cheaper.

And there's no turning back; 24-bit recording and high sample rates are taking over, and they sound better, so for mastering we can

{ *"The Source Quality Rule: Always start out with the highest resolution source and maintain that resolution for as long as possible into the processing."* }

choose from the best of several worlds, and we make our choices by balancing the benefits and the losses:

- (some) very transparent, low-noise, pure-sounding digital gear
- (some) good-sounding, reasonably-transparent, low-noise analog gear that we can use to add a little sugar, salt, pepper, or spice, or simply to prevent the sound from getting colder
- a digital processor that simulates analog distortion or warmth.

Why Is Good DSP So Expensive?

Intellectual property is the most nebulous thing to a consumer. It's easy to see why a two-ton

Mercedes Benz costs so much, but the amount of intellectual work that has gone into a one-gram IC is not so obvious. It can take five man-years to produce good audio software, created by individuals with ten or more years of schooling or experience. Similarly, when the doctor takes ten minutes to examine you, prescribes a 10-cent pill and then presents you with a $100 invoice, remember you're paying for all that knowledge and experience. This doesn't mean I'm against socialized medicine, I just want to re-emphasize the reasons why intellectual property and good DSP are so expensive.

The Source-Quality Rule

An important corollary of this discussion is the **source-quality rule:** *Source recordings and masters should have higher resolution than the eventual release medium.* **Always start out with the highest resolution source and maintain that resolution for as long as possible into the processing.** When mastering, one consequence of this rule is to reduce the number of generations and copies, and if possible, go back one or more generations when a new process must be added or applied.

This rule even applies when you're making an MP3 or other data-reduced final result. Consider a lossy medium like the (rapidly obsolescing) analog cassette. Dub to cassette from a high quality source, like a CD, and it sounds much better than a copy from an inferior source, like the FM radio, by avoiding cumulative bandwidth losses, as wider bandwidth sounds better. In other words, the higher the audio quality you begin with, the better the final product, whether it's an audiophile CD, a multi-media CD-ROM, MP3, or a talking Barbie doll. It may seem funny, but you'll never go wrong starting at 96 kHz/24 bit if the product is to end up on 44.1 k/16 bit CD. Sample rate conversion should be the penultimate process, followed by dithering.

In Summary

Mastering engineers do not have to think about the meaning of life every time they perform their magic; many engineers simply plug in their processors, listen, and make music sound better. But I also like to consider just why things sound better, because it helps me avoid problems that are not obvious at first listen, and also dream up innovative solutions. I hope that this chapter has inspired you to dream up some innovations of your own!

---

1   See the Appendix for references on noise filters. Ironically, all the standard noise-weighting filters should be revised, because they have no relationship with human perception of very quiet devices such as A/D and D/A converters.

2   And even then, the F-curve is an approximation, since the ear's perception of noise is much more than just a frequency response curve, as Jim Johnston explains: Noise should be measured separately in each critical band and compared to the ear's threshold for that critical band.

3   Most of the SpectraFoo™ screenshots were taken at an FFT resolution of 32K points (32000 "bins") with about 4 second average time and Hanning weighting. The actual amplitude of details on an FFT depends on its resolution, so FFTs are only directly comparable if the same methods are used.

4   The term *resolving*, when applied to the sound of tube circuits, is itself an unquantifiable audiophile subjective term. It's fair to say that audiophile negative reactions to some ugly-sounding solid-state circuits use inexact terms such as *resolution* and *transparency*, which may be proved to be simply distribution of harmonics or differences in frequency response. And maybe not!

5   For the curious, K-Stereo and K-Surround do **not** use harmonic distortion to enhance depth. They use other psychoacoustic principles.

6   Although the makers of the double-sampling Weiss Equalizer, GML plugin, and the Audiocube feel that double sampling is important for equalizers. Some engineers like the sound of high frequency curves that extend beyond 20 kHz, even if that is later cut off when the sample rate is halved at the output of the equalizer. And Jim Johnston (in correspondence) states that when a digital filter has response extending to half the sampling rate, it can produce some really odd and unexpected frequency responses, indicating that double sampling is important for such type of equalizers.

CHAPTER 17

How to Achieve Depth and Dimension in Recording, Mixing and Mastering

I. Introduction

I placed this acoustics lesson in the middle of a book on mastering because the creation of wonderful audio masters requires that some basic acoustic principles be understood. As we enter the era of surround recording and reproduction, many mix engineers are repeating their mistakes from two-channel work—panpotting mono instruments to discrete locations, and then adding multiple layers of uncorrelated stereophonic reverb "wash" in a vain and misguided attempt to create space and depth. It's important to learn how to manipulate the surprising depth available from 2-channel canvas before moving on to multi-channel surround.

It amazes me how few engineers know how to fully use good ol' fashioned 2-channel stereo. I've been making "naturalistic" 2-channel recordings for many years taking advantage of room acoustics, but it is also possible to use artificial means to simulate depth, and there are many engineers working in the pop field who know how to do so. Learn to discern the audible difference between simple pan-potted mono, and recordings which simulate or utilize the reflections from nearby walls to create a real sense of depth. Without such knowledge, your recordings will tend to produce a vague, undefined image; the musical instruments will be obscured and unclear.

Techniques here include using the Haas[1] effect, particularly when implemented binaurally, use of delays and alteration of phase, more naturalistic reverberators, and understanding how to unmask via placement. Also be aware that well-engineered

2-channel recordings have encoded ambience information which can be extracted to multichannel, and it pays to learn about these techniques.

Depth Perception in Real Rooms

Early Reflections versus Reverberation

At first thought, it may seem that depth in a recording can be achieved simply by increasing the proportion of reverberant to direct sound. But the artificial simulation of depth is a much more complex process. Our binaural hearing apparatus is largely responsible for the perception of depth and space, decoding the various early reflections from nearby walls that support and strengthen the sound of musical instruments and voices. First, we must define the terms *early reflections* and *reverberation*. Early reflections consist of the part of the room sound within approximately the first 50-100 milliseconds. There is a great deal of correlation between the direct sound and the early reflections; you can think of the early reflections as being *attached* to the direct sound. In a large and diffuse room, after about 100 milliseconds, enough wall bounces have occurred to make it impossible to hear discrete bounces; this is the onset of random (uncorrelated) reverberation, which we can say is *detached* from the direct sound. That's why it is the early reflections, even more than the reverberation, which largely affect our perception of the depth of the sound, giving it shape and dimension. The ear's decoding ability is such that a few simple well-placed echos actually solidify and clarify the location of the direct sound; this is why a simple, dead, panpotted mono source (without early reflections) is so hard to locate precisely.

Masking Principle/Haas Effect

Recording engineers were concerned with achieving depth even in the days of monophonic sound. In those days, many halls for orchestral recording were deader than those of today. Why do monophonic recording and dead rooms seem to go well together? The answer is involved in two principles that work hand in hand: 1) The masking principle and 2) The Haas effect.

The Masking Principle and Mono versus Stereo Recordings

The masking principle says that a louder sound will tend to cover (mask) a softer sound, especially if the two sounds lie in the same frequency range. If these two sounds happen to be the direct sound from a musical instrument and the reverberation from that same instrument, then the initial reverberation can appear to be covered by the direct sound. When the direct sound ceases, the reverberant hangover is finally perceived. This is why in mixing, we often add a small delay between the direct sound and the reverberation, it helps the ears to separate one from the other, reducing the masking.

In concert halls, our two ears sense reverberation as coming diffusely from all around us, and the direct sound as having a distinct single location. Thus, when music is perceived binaurally, there is less masking because the direct and reverberant sound come from different directions. However, in monophonic recording, the reverberation is reproduced from the same source speaker as the direct sound, and so we may perceive the room as deader than it really is, because the two sounds overlap directionally. Furthermore, if we choose a recording hall that is very live, then the reverberation will tend to intrude on our perception of the direct sound, since in monaural, both will be reproduced from the same location-the single speaker.

This is one explanation for the incompatibility of many stereophonic recordings with monophonic reproduction. The larger amount of reverberation tolerable in stereo becomes less acceptable in mono due to the physical overlap. As we extend our recording techniques to 2-channel (and multichannel) we can overcome masking problems by spreading artificial reverberation spatially away from the direct source, achieving both a clear (intelligible) and warm recording at the same time. One of the first tricks that mix engineers learn is to put reverberation in the opposite channel from the source. This helps unmask the sound, but can produce an unnatural effect.[2] As we get more sophisticated, we discover that instead of hard-panning the source and its mono echo or reverb return, using multiple delays or stereophonic early reflections can yield a far more cohesive, natural effect. The presence of the stereophonically-spread early reflections also serves to clarify the location of the dry source. In a sophisticated stereo mix, engineers take advantage of variations on these themes to produce variety and space in the recording.

The Haas Effect

The Haas effect can help overcome masking. In general, Haas says that echoes occurring within approximately 40 milliseconds of the direct sound become fused with the direct sound. We say that the echo becomes "one" with the direct sound, and only a loudness enhancement occurs; this is what happens in a real room with the earliest wall and floor reflections. Since the velocity of sound is approximately one foot per millisecond, 40 milliseconds corresponds to a wall that's 20 feet distant (assuming a flat wall perpendicular to the angle of the direct sound).

A very important corollary to the Haas effect says that fusion (and loudness enhancement) will occur even if the closely-timed echo comes from a different direction than the original source. However, the brain will continue to recognize (binaurally) the location of the original sound as the proper direction of the source. The Haas effect allows nearby echoes (greater than about 10 ms. and less than about 40 ms. delay) to enhance and reinforce an original sound without confusing its directionality. The maximum definition of the source's directionality will occur using the longest delay possible that is not perceived as a discrete echo.

The Magic Surround

We can take advantage of the Haas effect to naturally and effectively convert an existing 2-channel recording to a 4-channel or surround medium. When remixing, place a discrete delay in the surround speakers to enhance and extract the original ambience from a previously recorded source! No artificial reverberator is needed if there is sufficient reverberation in the original source. Here's how it works:

Because of the Haas effect, when the delay and source are correlated (e.g., a snare drum hit) the ear fuses them, and so still perceives the direct sound as coming from the front speakers. But this does not apply to ambience because it is uncorrelated—the ear does not recognize the delay as a repeat, and thus ambience will be spread, diffused between the location of the original sound and the location of the delay (in the surround speakers). Thus, the Haas effect only works for correlated material; uncorrelated material (such as natural reverberation) is extracted, enhanced, and spread directionally. Dolby laboratories calls this effect *the magic surround*, for they discovered that natural reverberation was extracted to the rear speakers when a delay was applied to them. Dolby also uses an L-minus-R matrix to further enhance the separation. The wider the bandwidth of the surround system and the more diffuse its character, the more effective the psychoacoustic extraction of ambience to the surround speakers.

Haas In Mixing

There's more to Haas than this simple explanation. To become proficient in using Haas in mixing, you can study the original papers which discuss the various fusion effects at different delay and amplitude ratios. During mixing, remember the 1 foot per millisecond relationship, and see what happens with carefully-placed and leveled delays in the 12 to 40 millisecond range. You will discover that they can enhance an instrument's clarity and position all due to psychoacoustics: the ear's own decoding power.[3] In fact, Haas delays are far more effective than equalization at repairing the sound of

a drumset which was recorded in a dead room, for example. Furthermore, multiplying the delays until they simulate the complex early reflections of real rooms can greatly improve our stereo mixing technique. More than a few delays is beyond our ability to do on a simple mixing board, and for early reflections we must use computerized simulations found in devices such as the TC Electronic, EMT, and certain models of Sony reverbs. The latest algorithm from TC, currently only available in the System 6000, is quite astounding.

Haas In Mastering

We often receive recordings for mastering which lack depth, spatiality and clarity because the mix engineer did not mix the early reflections or reverberation well enough or loudly enough. But since the mix has already been made, adding artificial reverberation can muddy the sound. This is why an ambience extraction technique should be employed instead. My K-Stereo processor, model DD-2, can enhance the depth of existing stereo mixes by extracting and spatially-spreading their inherent ambience.

Haas' Relationship To Natural Environments

In a good stereo recording, the early correlated room reflections are captured with their correct placement; they support the original sound, help us locate the sound source as to distance and do not interfere with left-right orientation. The later uncorrelated reflections, which we call reverberation, naturally contribute to the perception of distance, but because they are uncorrelated with the original source the reverberation does not help us locate the original source in space. If the recording engineer uses stereophonic miking techniques and a more lively room instead, capturing early reflections on two tracks of the multitrack, the remixing engineer will need less artificial reverberation and what little he adds can be done convincingly.

Using Frequency Response to Simulate Depth

Another contributor to the sense of distance in a natural acoustic environment is the absorption qualities of air. As the distance from a sound source increases, the apparent high frequency response is reduced. This provides another tool which the recording engineer can use to simulate distance, as our ears have been trained to associate distance with high-frequency rolloff. An interesting experiment is to alter a treble control while playing back a good orchestral recording. Notice how the apparent front-to-back depth of the orchestra changes considerably as you manipulate the high frequencies.

Recording Techniques in Natural Rooms to Achieve Front-To-Back Depth

Balancing the Orchestra with only a few micophones (minimalist). A musical group is shown in a hall cross section (see diagram at right). Various microphone positions are indicated by letters A-F.

Microphones **A** are located very close to the front of the orchestra. As a result, the ratio of **A's** distance from the back compared to the front is very large. Consequently, the front of the orchestra will be much louder in comparison to the rear, and the amount of early reflections reaching the

microphone from the rear will be far greater than from the front. Front-to-back balance will be exaggerated. However, there is much to be said in favor of mike position **A**, since the conductor usually stands there, and he purposely places the softer instruments (strings) in the front, and the louder (brass and percussion) in the back, somewhat compensating for the level discrepancy due to location. Also, the radiation characteristics of the horns of trumpets and trombones help them to overcome distance. These instruments frequently sound closer than other instruments located at the same physical distance because the focus of the horn increases direct to reflected ratio. Notice that orchestral brass often seem much closer than the percussion, though they are placed at similar distances. You should take these factors into account when arranging an ensemble for recording. Clearly, we perceive depth by the larger proportion of reflected to direct sound for the back instruments.

The farther back we move in the hall, the smaller the ratio of back-to-front distance, and the front instruments have less advantage over the rear.

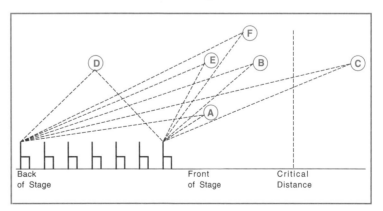

At position **B**, the brass and percussion are only two times the distance from the mikes as the strings. This (according to theory) makes the back of the orchestra 6 dB down compared to the front, but much less than 6 dB in a reverberant hall, because level changes less with distance.

For example, in position **C**, the microphones are beyond the critical distance—the point where direct and reverberant sound are equal. If the front of the orchestra seems too loud at **B**, position **C** will not solve the problem; it will have similar front-back balance but be more buried in reverberation.

Using Microphone Height To Control Depth And Reverberation

Changing the microphone's height allows us to alter the front-to-back perspective independently of reverberation. Position **D** has no front-to-back depth, since the mikes are directly over the center of the orchestra. Position **E** is the same distance from the orchestra as **A**, but being much higher, the relative back-to-front ratio is much less. At **E** we may find the ideal depth perspective and a good level balance between the front and rear instruments. If even less front-to-back depth is desired, then **F** may be the solution, although with more overall reverberation and at a greater distance. Or we can try a position higher than **E**, with less reverb than **F**.

Directivity Of Musical Instruments

Frequently, the higher up we move the mike, the more high frequencies it will capture, especially from the strings. This is because the high frequencies of many instruments (particularly violins and violas)

radiate upward as well as forward. The high frequency factor adds more complexity to the problem, since it has been noted that treble response affects the apparent distance of a source. Note that when the mike moves past the critical distance in the hall, we may not hear significant changes in high frequency response when height is changed.

The recording engineer should be aware of how all the above factors affect the depth picture so he can make an intelligent decision on the mike position to try next. The difference between a B+ recording and an A+ recording can be a matter of inches.

Beyond Minimalist Recording

The engineer/producer often desires additional warmth, ambience, or distance after finding the mike position that achieves the perfect instrumental balance. In this case, moving the mikes back into the reverberant field cannot be the solution. Another call for increased ambience is when the hall is a bit dry. In either case, trucking the entire ensemble to another hall may be tempting, but is not always the most practical solution.

The minimalist approach is to change the microphone pattern(s) to less directional (e.g., omni or figure-8). But this can get complex, as each pattern demands its own spacing and angle. Simplistically speaking, with a constant distance, changing the microphone pattern affects direct to reverberant ratio.

Perhaps the easiest solution is to add ambience mikes. If you know the principles of acoustic phase cancellation, adding more mikes is theoretically a sin. However, acoustic phase cancellation does not occur when the extra mikes are placed purely in the reverberant field, for the reverberant field is uncorrelated with the direct sound. The problem, of course, is knowing when the mikes are deep enough in the reverberant field. Proper application of the **3 to 1 rule**[4] will minimize acoustic phase cancellation. So will careful listening. The ambience mikes should be back far enough in the hall, and the hall must be sufficiently reverberant so that when these mikes are mixed into the program, no deterioration in the direct frequency response is heard, just an added warmth and increased reverberation. Sometimes halls are so dry that there is distinct, correlated sound even at the back, and ambience mikes would cause a comb filter effect.

Assuming the added ambience consists of uncorrelated reverberation, then in principle an artificial reverberation chamber should accomplish similar results to those obtained with ambience microphones. In practice, however, this has to be a qualified yes, by assuming not only that the artificial reverberation chamber has a true stereophonic response and is consonant with the sound of the original recording hall, but also that the main microphones have picked up sufficient early reflections for the depth effect to be convincing. Artificial reverberation alone, being uncorrelated, will not help the imaging or produce a focused depth picture.

What happens to the depth and distance picture of the orchestra as the ambience is added? In general, the front-to-back depth of the orchestra remains the same or increases minimally, but the apparent overall distance will increase as more reverberation is mixed in. The change in depth may not be linear for the whole orchestra since the instruments with more dominant high frequencies may seem to remain closer even with added reverberation.

The Influence of Hall Characteristics on Recorded Front-To-Back Depth

In general, given a fixed microphone distance, the more reverberant the hall, the farther back the rear of the orchestra will seem. In one problem hall the reverberation is much greater in the upper bass frequency region, particularly around 150 to 300 Hz. A string quartet usually places the cello in the back. Since that instrument is very rich in the upper bass region, in this problem hall the cello always sounds farther away from the mikes than the second violin, which is located at his right. Strangely enough, a concert-goer in this hall does not notice the extra sonic distance because his strong visual sense locates the cello easily and does not allow him to notice an incongruity. When she closes her eyes, however, the astute listener notices that, yes, the cello sounds farther back than it looks!

It is therefore rather difficult to get a proper depth picture with a pair of microphones in this problem hall. Depth seems to increase almost exponentially when low frequency instruments are placed only a few feet away. It is especially difficult to record a piano quintet in this hall because the low end of the piano excites the room and seems hard to locate spatially. The problem is aggravated when the piano is on half-stick, cutting down the high frequency definition of the instrument.

The miking solution I choose for this problem is a compromise; close mike the piano, and mix this with a panning position identical to the piano's virtual image arriving from the main mike pair. I can only add a small portion of this close mike before the apparent level of the piano is taken above the balance a listener would hear in the hall. The close mike helps solidify the image and locate the piano. It gives the listener a little more direct sound on which to focus.

Can minimalist techniques work in a dead studio? Not very well. My observations are that simple miking has no advantage over multiple miking in a dead room. I once recorded a horn overdub in a dead room, with six tracks of close mikes and two for a more distant stereo pair. In this dead room there were no significant differences between the sound of the minimalist pair, and the six multiple mono close-up mikes! (The close mikes were, of course, carefully equalized, leveled and panned from left to right.) This was a surprising discovery, and it reinforces the importance of good hall acoustics and especially early reflections on a musical sound. In other words, when there are no significant early reflections, you might as well choose multiple miking, with its attendant post-production balance advantages.

Miking Techniques and the Depth Picture

Coincident Microphones. The various simple miking techniques reveal depth to greater or lesser degree. Microphone patterns which have out of phase lobes (e.g., hypercardioid and figure-8) can produce an uncanny holographic quality when used in properly angled pairs. Even tightly-spaced (coincident) figure-8s can give as much of a depth picture as spaced omnis. But coincident miking reduces time ambiguity between left and right channels, and sometimes we seek that very ambiguity. Thus, there is no single ideal minimalist technique for good depth, and you should become familiar with changes in depth produced by changing mike spacing, patterns, and angles. For example, with any given mike pattern, the farther apart the microphones of a pair, the wider the stereo image of the ensemble. Instruments near the sides tend to pull more left or right. Center instruments tend to get wider and more diffuse in their image picture, harder to locate or focus spatially.

The technical reasons for this are tied in to the Haas effect for delays of under approximately 5 ms. vs. significantly longer delays. With very short delays between two spatially located sources, the image location becomes ambiguous. A listener can experiment with this effect by mistuning the azimuth on an analog two-track machine and playing a mono tape over a well-focused stereo speaker system. When the azimuth is correct, the center image will be tight and defined. When the azimuth is mistuned, the center image will get wider and acoustically out of focus. Similar problems can (and do) occur with the mike-to-mike time delays always present in spaced-pair techniques.

Spaced microphones. I have found that when spaced mike pairs are used, the depth picture also appears to increase, especially in the center. For example, the front line of a chorus will no longer seem straight. Instead, it appears to be on an arc bowing away from the listener in the middle. If soloists are placed at the left and right sides of this chorus instead of in the middle, a rather pleasant and workable artificial depth effect will occur. Therefore, do not rule out the use of spaced-pair techniques. Adding a third omnidirectional mike in the center of two other omnis can stabilize the center image, and proportionally reduces center depth.

Multiple Miking. I have described how multiple close mikes destroy the depth picture; in general I stand behind that statement. But soloists do exist in orchestras, and for many reasons, they are not always positioned in front of the group. When looking for a natural depth picture, try to move the soloists closer instead of adding additional mikes, which can cause acoustic phase cancellation. But when the soloist cannot be moved, plays too softly, or when hall acoustics make him sound too far back, then one or more *spot mikes* must be added. When the close solo mikes are a properly placed stereo pair and the hall is not too dead, the depth image will seem more natural than one obtained with a single solo mike.

To avoid problems, apply the 3 to 1 rule. Also, listen closely for frequency response problems when the close mike is mixed in. As noted, the live hall is more forgiving. The close mike (not surprisingly) will appear to bring the solo instrument closer to the listener. If this practice is not overdone, the effect is not a problem as long as musical balance is maintained, and the close mike levels are not changed during the performance.

We've all heard recordings made with this disconcerting practice. Trumpets on roller skates?

Delay Mixing. At first thought, adding a delay to the close mike seems attractive. While this delay will synchronize the direct sound of that instrument with the direct sound of that instrument arriving at the front mikes, the single delay line cannot effectively simulate the other delays of the multiple early room reflections surrounding the soloist. The multiple early reflections arrive at the distant mikes and contribute to direction and depth. They do not arrive at the close mike with significant amplitude compared to the direct sound entering the close mike. Therefore, while delay mixing may help, it is not a panacea. To adjust the delay of the solo mike(s) properly, start with a delay calculated by the relative distance between the solo mike and the main mike, then focus the delay up and down in 1 ms. increments until the sound is most coherent and focused and the soloist sounds clearest.

Influence Of The Control Room Environment On Perceived Depth

At this point, many engineers may say, "I've never noticed depth in my control room!" The widespread practice of placing near-field monitors on the meter bridges of consoles kills almost all sense of depth. Comb-filtering, speaker diffraction and sympathetic vibrations from nearby surfaces destroy the perception of delicate time and spatial cues. The recent advent of smaller virtual control surfaces has helped reduce the size of consoles, but seek advice from an expert acoustician if you want to appreciate or manipulate depth in your recordings.

Examples To Check Out

Standard multitrack music recording techniques make it difficult for engineers to achieve depth in their recordings. Mixdown tricks with reverb and delay may help, but good engineers realize that the best trick is no trick: learn how to use stereo pairs in a good acoustic. Here are some examples of audiophile recordings I've made that purposely take advantage of depth and space, both foreground and background, on Chesky Records. Sara K. *Hobo*, Chesky JD155. Check out the percussion on track 3, "Brick House." Johnny Frigo, *Debut of a Legend*, Chesky JD119. Check out the sound of the drums and the sax on track 9, "I Love Paris." Ana Caram, *The Other Side of Jobim*, Chesky JD73. Check out the percussion, cello and sax on "Correnteza." Carlos Heredia, *Gypsy Flamenco*, Chesky WO126. Play it loud! And listen to track 1 for the sound of the background singers and handclaps. Phil Woods, *Astor and Elis*, Chesky JD146, for the natural-sounding combination of intimacy and depth of the jazz ensemble.

Technological Impediments to Capturing Recorded Depth

Depth is the first thing to suffer when technology is incorrectly applied. Here is a summary of some of the technical practices that when misused, or accumulated, can contribute to a boringly flat, depthless recorded picture:

- Multitrack and multimike techniques
- Small/dead recording studios or large rooms with poor acoustics/missing early reflections
- low resolution recording media
- amplitude compression
- improper use of dithering, cumulative digital processing, and low-resolution digital processing (e.g., using single-precision as opposed to double or higher-precision)

In Summary: When recording, mixing and mastering—use the highest resolution technology, best miking techniques, and room acoustics. Process dead tracks with Haas delays and early reflections, and specialized ambience recovery tools. Then you'll resurrect the missing depth in your recordings.

---

1 Haas, Helmut (1951), *Acustica*. The original article is in German. Various English-speaking authors have written their interpretations of Haas, which you can find in any decent textbook on audio recording techniques.

2 Even if unnatural, it can be interesting, nevertheless. Listen to 1960's-70's era rock recordings from the Beatles, Beach Boys, Lovin' Spoonful, The Supremes, Tommy James and the Shondells, and many more, where mono instruments or vocals are panned to one side, and often their reverb return completely to the other side.

3 When adding Haas delays, listen closely in mono, because improper delay ratios can cause comb filtering in mono. A small degradation in mono may be tolerable if the improvement is significant in stereo. Early reflections, due to their more complex nature, are more compatible with mono folddowns than simple Haas delays.

4 Burroughs, Lou (1974), *Microphones: Design and Application*, Sagamore Publishing Company. (Out of Print). Burroughs quantified the effects of acoustic phase cancellation (comb filtering, interference) with real microphones and real rooms, and devised this rule: The distance between microphones should be three times the distance between each microphone and the source of the sound to which it is being applied. This is particularly important to avoid comb-filtering when both microphones are feeding a single channel; when the microphones are feeding different channels (e.g. stereo), the degradation will be much less noticeable in stereo but still be a problem in mono.

CHAPTER 18

High Sample Rates: Is This Where It's At?

I. Introduction

Now that we've cured the wordlength blues—it's time to tackle the sample rate issue. Whatever the eventual real benefits for the professional and the consumer, the current relentless drive for higher sample rates is certainly very lucrative for the hardware manufacturers. Clearly, engineers who must regularly replace their expensive high-resolution processors to keep up with the Joneses will spend big dollars.

I've been working with higher sample rates for several years,[*] but after some experiments that I will relate below, I have concluded that most hardware design engineers are having trouble seeing the forest for the trees. I think that a fresh look at how A/Ds and D/As are designed may reduce the need for extreme sample rates!

A great number of engineers think that the reason higher sample rate recordings sound better is because they permit reproduction of extreme high frequencies. They point out the *open, warm, extended* sound of these recordings as evidence for this contention.[1] However, most objective evidence shows that higher bandwidth is not the reason for the superior reproduction; remember that **the additional frequencies that are recordable by higher sample rates are inaudible.** But if we can't hear these frequencies, then why are we inventing expensive processors and wasting so much bandwidth and hard disc space? And how can 50-year-old ears detect differences between 44.1 kHz and 96 kHz and even 192 kHz sample rates, even though most of us can't hear much above 15 kHz?

---

[*] I was the recording engineer for the world's first 96 kHz/24 bit audio-only DVD.

I believe the answer lies in the design of digital **low-pass filters**, which are part of the requirements of digital audio. Digital filters are used in **oversampling A/D and D/A converters** and in **sample rate converters**. Digital filters employ complex mathematics, which is expensive to implement and so, cheaper filters have to include greater quality tradeoffs, such as lowered calculation resolution, ripple in the passband, or potential for aliasing.

One type of filter has a sharp cutoff; the consequences of sharp filtering include time-smearing of the audio, possible short (millisecond) echos which are caused by amplitude response ripples in the passband frequency response (20 Hz–20 kHz), even ripples as small as 0.1 dB. Moving the filter cutoff frequency to 48 kHz (for 96 kHz SR) relaxes the filtering requirement and makes it easier to engineer filters with less ripple in the passband and less phase shift near the upper frequency limit.

> *"The filters in a typical compact disc player or in the converter chips used in most of today's gear are mathematically compromised."*

Oversampling

One of the biggest improvements in digital audio technology came in the late 80's, with the popularization of oversampling technology by DBX's Bob Adams, in a high-quality, 128x oversampling 18-bit oversampling A/D. An **oversampling A/D** converter has a front end which typically operates at 64 or 128 times the base sample rate and produces 1-bit to 5-bit words in delta-sigma format,[2] depending on the model. In other words, for 44.1 kHz operation, the input of a 128X converter actually operates at 5.6448 MHz! Oversampling takes the converter's noise, spreads it around a wider frequency spectrum, and shapes it, moving much of the noise above the audible frequency range. In addition, when it is digitally downsampled to the base rate at the output of the converter, some of the higher frequency noise is filtered out, to yield as much as 120 dB or even better signal-to-noise ratio within a 20 kHz bandwidth.

The downsampling is accomplished with a digital circuit called a **decimator,** which is a form of divider or sample rate converter, and which must contain a filter at half the sample rate to eliminate aliases, requiring a 22.05 kHz cutoff at a 44.1 kHz SR. This filter must be designed without compromise or it will affect the sound. Some manufacturers concentrate on transient response, others on phase response, ripple, linearity, or freedom from aliasing. But all of these characteristics are important, and getting it right is expensive—precision construction requires more math, and math requires labor and parts (size of the integrated circuit die). **Thus, the filters in a typical compact disc player or in the converter chips used in most of today's gear are mathematically compromised.**

On the D/A (output) side, at low sample rates, sharp anti-imaging filters are required to retain frequency response to 20 kHz. It is impractical (probably impossible) to build a sharp analog filter with the required characteristics, so instead an **oversampling or upsampling** digital filter multiplies the base sample rate up 2x to 8x or more, moving artifacts and distortion above the audible band. The higher sample rate permits using a gentle, uncompromised analog filter. But the typical digital filters used in the inexpensive chips have poor performance. To minimize the effect of these concessions, the most progressive high-end D/A manufacturers add an additional upsampling filter of their own design, in front of the DAC chip. The additional filter reduces the error contribution of the chip's own filter, in essence because the internal DAC's filter does not have to work as hard. Internally, these advanced DACs are always operating at 88.2 or 96 kHz regardless of the incoming rate. At the double sampling rates, the supplementary filter is disabled. The supplementary filter would be unnecessary if the manufacturers of the converter chips used higher quality filters in the first place.

An Upsampling Experience

Audiophiles, and some professionals, have been experimenting with digital upsampling boxes which are placed in front of D/A converters. In some cases they report greatly improved sound. Although the improvement may be real, in my opinion they can be attributed to the various digital filter combinations, not to bandwidth or frequency response or (especially) the sample rate itself. Remember that all original 44.1 kHz SR recordings are already filtered, so they cannot contain information above about 20 kHz. An upsampler cannot "manufacture" any new frequency information that wasn't there in the first place.

I've compared the sound of upsamplers versus DACs working alone. Sometimes I hear an improvement, sometimes a degradation, sometimes the sound quality is the same either way. Sometimes the sound gets brighter despite a ruler-flat frequency response, which can probably be attributed to some form of phase or intermodulation distortion in the digital filter. **Sonic differences have come down to mathematics in this new digital audio world**.

The Ultimate Listening Test: Is It The Filtering or the Bandwidth?[†]

In December 1996, I performed a listening test, with the collaboration of members of the Pro Audio maillist. The idea was to develop a test that would eliminate all variables except bandwidth, with a constant sample rate, filter design, DAC, and constant jitter. The question we wanted to answer was this: Does high sample rate audio sound better because of increased bandwidth, or because of less-intrusive filtering?

The test we devised was to create a filtering program that takes a 96 kHz recording, and compare the effect on it of two different bandwidth filters. The volunteer design team consisted of Ernst Parth (filter code), Matthew

> *"The issues of the audibility of bandwidth and the audibility of artifacts caused by limiting bandwidth must be treated separately. Blurring these issues can only lead to endless arguments."* —Bob Olhsson[*]

---

[*] From the Mastering Engineer's Webboard.

[†] I previously published some of this information in *Audiomedia* Magazine; we publish the full story in this book.

Xavier Mora (shell), Rusty Scott (filter design), and Bob Katz (coordinator and beta tester). We created a digital audio filtering program with two impeccably-designed filters which are mathematically identical, except that one cuts off at 20 kHz and the other at 40 kHz. The filters are double-precision dithered, FIR linear phase, 255-tap, with >110db stopband attenuation, and <.01 dB passband ripple.

MYTH:

Upsampling makes audio sound better by creating more points between the samples, so the waveform will be less jagged.

After the filter program was designed, I took a 96 kHz SR orchestral recording, filtered it and brought it back into a Sonic Solutions DAW for the comparison. I expected to hear radical differences between the 20 kHz and 40 kHz filtered material. But I could not! Next, I compared the 20 kHz filtered against "no filter" (of course, the material has already passed through two steep 48 kHz filters in the A/D/A). Again, I could not hear a difference! The intention was to listen double-blind; but even sighted, 10 additional listeners who took part in the tests (one at a time) heard no difference between the 20 kHz digital filter and no filter. And if no one can hear a difference sighted, why proceed to a blind test?

I tried different types of musical material, including a close-miked recording I made of castanets (which have considerable ultrasonic information), but there was still no audible difference. I then created a test which put 20 kHz filtered material into one channel of my Stax electrostatic headphones, and the time-aligned wide-bandwidth material into the other channel. I was not able to detect any image shift, image widening or narrowing—there was always a perfect

mono center at all frequencies in the headphones! This must be a pretty darn good filter!

As a last resort, I went back to the list and asked maillist participant Robert Bristow Johnston to design a special "dirty" filter with 0.5 dB ripple in the passband. Finally, with the dirty filter, I was able to hear a difference… this dirty filter added a boxy quality that resembles the sound of some of the cheaper 44.1 k CD players we all know.

This 1996 test seems to show that a "perfect 20 kHz filter" can be designed, but at what cost? Also note that as this test was conducted in the context of a 96 kHz sample rate, the artifacts of two other 48 kHz steep filters already in use may have obscured or masked the effect of the filter under test. Since I conducted my test, several others have tried this filtering program, and most have reached the same conclusion: the filter is inaudible. One maillist participant, Eelco Grimm, a Netherlands-based writer and engineer, performed the test and reported that there were no audible differences using the Sonic Solutions system, yet he and a colleague were able to pick out differences between filtered and non-filtered blind using an Augan workstation. He did not compare the sound of the 20 kHz versus 40 kHz filters, so we are not sure if he's hearing the filter or the bandwidth, but I believe he was hearing the filter, which must not be ideally-designed. I believe the reason he did not hear the differences on the Sonic system is perhaps its jitter was high enough to mask the other differences, which must be very subtle indeed!

Regardless of whether Eelco's group did reliably hear the bandwidth differences, it should be clear by now that the so-called "dramatic" differences people hear between sample rate systems are not likely to be due to bandwidth, but probably to the filter design itself. Ironically, it was necessary to make a high sample rate recording in order to prove that high sample rates may not be necessary.

As I mentioned, 44.1 kHz reproduction has improved considerably in recent DACs employing add-on high-quality upsampling filters. The next figure illustrates Weiss's THD measurement of their SFC, showing that its filter has textbook-perfect distortion and noise performance.

Why can't more manufacturers introduce filters of this quality into their converter chips? The evidence all indicates that it will be a lot less expensive for end-users if the manufacturers of converter chips upgrade the filtering software in their chip sets instead of directing us to this mad,

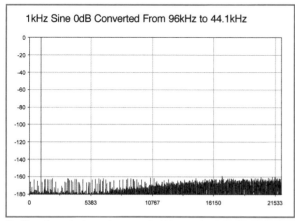

1kHz Sine 0dB Converted From 96kHz to 44.1kHz

The distortion and noise performance of a Weiss sample frequency converter.

expensive sample rate and format war. Objective experiments must be performed using state-of-the-art digital filters to determine what is the lowest practical sample rate which can be used without audible compromise.

It's A Matter of Time!

Let's be logical: since the human ear cannot hear above (nominally) 20 kHz, then any artifacts we are hearing must be in the audible band. It is well-known that low-Q parametric and shelving filters sound better than high Q; it's not a stretch to conclude this is also true for low-pass filters. Audio researcher Jim Johnston,[*] who knows as much about the time-domain response of the ear as anyone, has shown that steep low-pass filters create pre-echos which the ear interprets as a loss of transient response, obscuring the sharpness or clarity of the sound.

The pre-echo length is the inverse of the transition bandwith, so a sharp filter with a 500 Hz transition would create a 2 ms. pre-echo. Steep filtering and its attendant transient degradation is probably a reason why 44.1 kHz SR sounds less clear than 96K. Likewise, the increased clarity and purity of 1-bit recordings is probably due to their use of gentle filters rather than some mumbo-jumbo about the "magic" of 1-bit. Jim has experimentally calculated that the minimum sample rate which would support a Nyquist filter gentle enough to elude the ear would be 50 kHz.[3] I suggest that manufacturers and engineers must test as soon as possible the audibility of gentle low-pass filters, at the more common sample rate of 96 kHz. It would

---

[*] In correspondence. JJ is the inventor of the science of perceptual coding, which led to coding developments such as MP3, Atrac, etc.

be trivial to build a 96 kHz SR A/D/A system with the gentlest possible filter that's flat at 20 kHz and removes aliasing at 48 kHz, but no current chip manufacturer has done so. This system can be compared against the analog source, and against the competing DSD recording system. If the gentle-filtered PCM wins or sounds as good, it would be the triumph of psychoacoustic research over empirical design. Still, if it can be shown that good-sounding DSD at the consumer end is cheaper to implement than good-sounding gentle-filtered PCM reproduction, it is cheaper for us to record and process with gentle-filtered PCM and finally convert to DSD for the consumer (this is how most 1-bit DACs operate anyway).

I firmly believe that some minimal sample rate (perhaps 96 kHz) will be all that is necessary if PCM-converters are redesigned with psychoacoustically-correct filters (hopefully inexpensively). For the benefit of the myriads of consumers and professionals, we need to make a cost-analysis of the whole picture instead of racing towards bankruptcy.

The Advantages of Remastering 16/44.1 Recordings at Higher Rates

Researchers such as J. Andrew Moorer of Sonic Solutions, and Mike Story of dCS have demonstrated theoretical improvements from working at a higher sampling rate. Moorer pointed out that post-production processing, such as filtering, equalization, and compression, will result in less distortion in the audible band, as the errors are spread over twice the bandwidth—and half of that

bandwidth is above 20 kHz.[4] Measurements discussed in Chapter 16 confirmed these conclusions. In addition, as we've seen above, if after processing the destination is DVD-A or SACD, then the master can be left at the higher sample rate and wordlength, avoiding another generation of sound-veiling 16-bit dither and yet another sharp filter at the end of the process. Thus, consumers should not scoff at DVDs which have been digitally remastered from original 16-bit/44.1K sources. They will be getting real, audiophile-quality sonic value in their remasters.

---

1   Other engineers who do not fully understand the nature of PCM argue that the higher sampling rate sounds better because it would seem to create a more accurate 20 kHz sine wave, as there are more "dots to connect" to describe the wave. But this is erroneous; while there are more "dots," in reality only 2 samples are necessary to describe an undistorted 20 kHz sine wave; the low-pass filtering smooths out the waveform and eliminates all the glitches.

2   DSD, also known as 1-bit or **Direct Stream Digital**, a trademark of Sony and Philips is the format of the SACD and employs a form of Delta-Sigma modulation. Delta-Sigma modulation is the very dense native coding format of the first stage of modern-day oversampling converters, about 2.8 Megabits per second, as opposed to 44.1 kHz/16-bit PCM, **Pulse Code Modulation**, which runs at about 1.4 Megabits per second. When you study the block diagram of a record-reproduce chain, the significant difference between using DSD format and PCM is that PCM requires a steep Nyquist filter at half the sampling rate (about 20 kHz with 44.1 kHz SR).

3   This is based on the length of the shortest *organic* filter in the human ear, and Jim Johnston notes that the 50 kHz number nicely matches the original work with antialiasing filters done by Tom Stockham for the Soundstream project.

4   Julian Dunn (in correspondence) clarifies: A 3 dB reduction in distortion results because the error products are spread amongst twice the bandwidth. This is true for **uncorrelated** quantization errors which fall evenly throughout the frequency range from dc to fs/2. And does not work for distortion products which will correlate with the signal. Jim Johnston (in correspondence) indicates that processing at higher rates is **required** for any non-linear processing, such as compression. These non-linear processes produce new frequency components, some at higher frequencies. A high enough sampling rate avoids aliasing of these new frequency components (see Cranesong and Weiss FFTs in Chapter 16).

CHAPTER 19

Jitter–Separating the Myths from the Mysteries

I. Introduction

One of the least-understood, and hardest-to-explain phenomena in digital audio is **jitter**. To truly understand the influence of jitter on your digital recordings, you will have to reject years of analog experience. In a classic Marx Brothers movie, Groucho's girlfriend catches him embracing another beautiful woman. In defense, Groucho quips, "Are you going to believe me, or your own eyes?" Let me apply this to audio and ask, "Are you going to believe the facts, or your own ears?" For in this topsy-turvy digital audio world, sometimes you have to abandon the evidence of your senses and learn a totally new sense, but one that is fortunately based on well-established physical principles.

In 1980, because most sound systems, A/D and D/A converters and processors had such low resolution, jitter errors were far down on the priority list. Nonlinearity, noise modulation, truncation, improper dithering, aliasing and other errors created audible problems that tended to swamp the effects of jitter. But today, where audio performance frequently reaches 20-bit level[*] and sometimes exceeds it, jitter has reared its ugly head. The symptoms of jitter mimic the symptoms of other converter problems—blurred, unfocused, harsh sound, reduced image stability, loss of depth, ambience, stereo image, soundstage, and space—though usually in a subtle way, and it can take time for even a critical ear to learn to identify them.

What causes these problems? Is our digital audio actually being affected by jitter in our clocks?

---

[*] Rarely does typical equipment exceed 20-bit performance, as we shall soon see.

The simple answer is: Sometimes yes, mostly no! Should we believe our ears? It'll take a whole chapter to sort this one out.

II. What is Jitter?

Digital audio is based upon the concept of sampling at regular intervals. To keep those intervals constant requires a consistent clock. If the frequency of the clock varies during A/D conversion, then because the waveform will be at the wrong amplitude at the wrong place when the digital audio is played back, the audio will be permanently distorted.

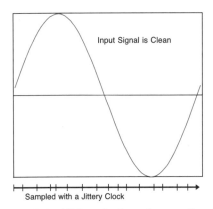
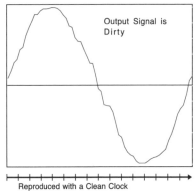

Input Signal is Clean

Output Signal is Dirty

Sampled with a Jittery Clock

Reproduced with a Clean Clock

Jitter During A/D Conversion Creates Permanent Distortion

That's why it is critical to have a consistent clock during A/D conversion. Similarly, an inconsistent clock will yield distortion during D/A conversion. We call this inconsistency *jitter*. One period of a 44.1 kHz clock is 22.7 μs.[*] Amazingly, variations in the duration of that period as short as 10 picoseconds may cause audible artifacts, depending on the quality of the reproduction system and your own hearing acuity. As sample rate increases and

wordlength lengthens, jitter must be proportionally lower to maintain sound quality, because jitter affects the absolute noise floor. Jitter produces sidebands (additional frequencies, or tones) that mask inner detail in a recording.

We can measure jitter in two places: 1) **interface jitter**, the jitter present in the interconnections between equipment, or 2) **sampling jitter**, the jitter in the clock which drives the converter. And we can measure the effects of jitter on converters, using special analog and digital test signals. If a converter has excellent internal **jitter rejection**, then high interface jitter may not result in sampling jitter. In other words, you can have a jittery interface or cable, and it won't matter a bit to a well-designed converter. In this chapter, we are mostly concerned with sampling jitter, because, as we shall see, interface jitter is rarely important unless it causes a breakdown in communication between devices.

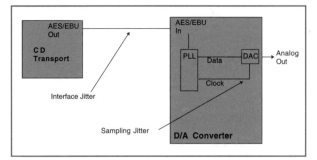

In the figure above, you can see that it is up to the PLL[†] inside the DAC to create the sampling

---

[*] One microsecond (μs) is one millionth of a second. One picosecond is one millionth of one millionth of one second, or 10^{-12} second.

[†] PLL is a *Phase Locked Loop*. Its operation is explained later in the chapter.

clock. If it is a superb (very rare) PLL, then none of the artifacts of incoming interface jitter will be transmitted to the sampling clock.

III. Jitter, When it Matters, When it Doesn't

If leaping to conclusions about jitter were an Olympic event, sound engineers would win the gold medal. An entire audiophile subculture has developed around digital cables and jitter reduction units in an attempt to achieve better reproduction, which has led engineers to change cables everywhere they hear that such a replacement makes a difference, or to experiment with "stable" external clocks, each of which produces a different sound.* I don't blame them for trying, but in general, cables and wordclock generators are only bandaids for jitter problems which must ultimately be solved **within** the converters. No cable can remove the inherent jitter problems in the AES/EBU-SPDIF interface, because the imbedded clock interacts with the data stream. Thus, external jitter reduction units will always be limited in their effectiveness because jitter may be increased at the output interface between the jitter reducer and the D/A.

Since engineers hear improvements with the better cables[19] (and jitter reduction units) feeding their D/As, they conclude these same cables will improve their digital audio processors. But this is (largely) a misconception.† Remember: **audio processors process data, not clock**. If they hear a difference, it is because a cleaner clock is passed to the D/A converter; but there is no difference in the data being processed. Believe the facts, not your own ears! The listening problem is ephemeral, and also has an immediate solution—get a better DAC!

How to Lie With Measurements

Clock jitter can produce insidious audio artifacts in converters. Most manufacturer's specifications hide these artifacts because we have not yet established a measurement standard for the effects of jitter on converters. For example, some recent A/D (and a few D/A) converters now report exceptional >120 dB signal-to-noise ratios, theoretically equivalent to >20-bit performance, but is this true in practice? These figures are obtained by the traditional method of calculating signal-to-noise ratios: first measuring a full-scale signal, then removing the signal and measuring the residual analog noise. But this does not take into account additional noise (or distortion) when the signal is present. As far as I'm concerned, *traditional audio signal-to-noise ratio measurements have (almost) no relationship to the sound of a converter when it is receiving signal.* It is this which accounts for some of the previously-unexplained sonic differences between converters. Most *signal-to-noise ratio* measurements quoted in manuals are therefore irrelevant, and most people have never heard true 20-bit performance, let alone 24.

> *"Traditional audio signal-to-noise ratio measurements have (almost) no relationship to the sound of a converter when it is receiving signal."*

---

* leading to the "Wordclock Du Jour" effect, as we shall see. And an erroneous audiophile magazine DAC review marvelling at a DAC "revealing" cable differences!

† Shortly we'll describe the infinitesimal number of exceptions.

Digital Print-Through. Ideally, the converter's PLL should completely reject incoming jitter with its clock-smoothing circuit, but if the PLL has inadequate *jitter attenuation*, it will pass some of the interface jitter to the critical conversion clock. The most egregious-sounding type of uneliminated jitter is **signal-dependent jitter**, caused by the designs of external interfaces such as AES/EBU and SPDIF. Although signal-dependent jitter is analogous to analog tape flutter, it is very much like analog tape print-through because it is signal dependent and adds a blurred quality to the sound. Around 1975, analog tape manufacturer BASF demonstrated that an analog tape with lower print-through can sound cleaner and quieter than a tape with lower hiss level and higher print-through.[1]

Similarly, a converter which successfully rejects jitter can sound much cleaner than another with a lower absolute noise floor. Talk about lying with statistics! Jitter can produce signal-dependent effects (which yield distortion from intermodulation between the sample rate and the audio signal), random effects (which translates to a higher random noise floor which can also be signal-dependent), and discrete frequency effects (such as other clocks in the box producing random tones and inter-modulation between the other clocks and the main sampling clock). Some of these effects are more benign to the ear than others, which is why it is so difficult to put a single meaningful number on jitter.

> { *"Most digital processors are completely immune to jitter"* }

Storage Media

There is no jitter on a storage medium—only the data is stored, not the clock. Likewise, there is no clock on a compact disc. A new clock is generated on playback, and thus jitter comes into play only when data is clocked out of the medium. Bits are usually stored in a very irregular fashion; on hard discs, the data may be out of order, non-contiguous, and widely spread. Data stored on CD (in EFM format) must be unscrambled and decoded during playback, and DAT data is stored in separated blocks, but none of these storage formats can be called *jitter*, since time is not involved until the data is played back. So, if you're looking for the causes of playback jitter, you have to study the complete mechanism.

During playback, the amount of clock jitter on the output device is determined by the quality of the servo, buffering, and clocking circuitry that drives the data. Manufacturers differ widely in their abilities to keep outgoing clocks under control and clock stability is simply not important to the original computer-based technology that we have now adapted to digital audio. In fact, the standard computer hard disc interfaces (e.g., SCSI, IDE) are *asynchronous* (non-clocked), they have a completely irregular output. The equivalent jitter of a SCSI interface is enormous, for at one moment, there may be no data; at another moment, it's streaming at many times real time. When such non-clocked interfaces are used, it is the duty of following circuitry to make the data conform to a steady clock.

Digital Mixing and Processing
Jitter does not affect the data...

...when you are performing an all-digital mix in most digital consoles. After the initial analog-to-digital conversion (we hope with a low-jitter clock) the data can pass from processor to processor, from medium to medium regardless of clock jitter—just as long as the interface jitter is low enough to allow an error-free transfer. Similarly, clock jitter has no effect on the performance of most outboard digital equalizers, limiters, or compressors, which are nearly all *state machines*. A state machine is defined as any type of processor which produces identical output for the same input data, and which does not look at data timing or speed, but only at the state or recent history of the data. **In other words, most digital processors are completely immune to jitter.** With a state machine, you could make the clock completely irregular, or even slow it down to 1 sample per second, and eventually, the processor would output all the correct data words. When these words were played back at the right speed and with a clean clock, all would be well.

All current professional oversampling processors—such as equalizers and compressors—are state machines. They use synchronous converters to double the internal sample rate, and since synchronous converters are state machines, the same rule applies.[2] Any state machine can be implemented offline in a computer and without a clock, where *real time* and *jitter* have no meaning.

With digital pitch processors such as Autotune™ the explanation is a bit confusing, but these are not affected by jitter. Pitch processors are not state machines; due to their randomizing algorithms many of these repitching processors produce a different output from the same piece of music each time they are run. But they do look at each sample coming in, one at a time, regardless of the regularity of the clock feeding the box. As you know, these repitchers can run offline in a DAW, at any speed, without a clock.[3]

{ *"Don't confuse the messenger with the message"* —Andy Moorer }

Jitter affects the monitoring

Jitter usually becomes meaningful in a digital mix only during monitoring, when the data is clocked out of a D/A converter. This is where everyone gets hopelessly confused, like the girlfriend who caught Groucho Marx during his hijinks (he probably was guilty anyway). Let me emphasize: high jitter during the monitoring **seems** to affect the overall sound quality, but it really only affects that individual listening experience, and has no effect on the data. **Don't confuse the messenger with the message.[4] The message (the data) remains intact; so if it sounds funny, blame the messenger (the clock inside the monitor DAC).** This is what I call "ephemeral jitter." If you improve your connections and the sound gets better, this does not mean that the digital equalizers are suddenly performing better—it only means that a cleaner clock is getting to the D/A converter.

Jitter affects the data during a digital mix only...

- when you leave the digital realm to use outboard analog processors, hence superior converters and clocking must be used for outboard equipment feeds.
- Some digital consoles contain asynchronous sample-rate converters (ASRC). These types of SRCs use variable filters based on a continuously-running estimate of the incoming sample rate, and thus are sensitive to clock jitter. An asynchronous SRC is not a state machine and will produce a different output each time it is run. You should question the quality of any ASRC, and try to deliver to it the highest quality clock. This is a serious issue, especially in low-cost consoles, where clocks are often compromised for economy, and especially since the console is typically driven by an external (word) clock, which puts the burden of low-jitter on a cheap PLL inside the console. I am not a fan of consoles that contain ASRCs, unless they can be completely bypassed when not needed. Modern-day ASRC chips contain sophisticated jitter-reduction algorithms and have relatively low distortion, so console performance is slightly degraded. The audible effect is a slight veiling or diminishing of stereo image stability, to my ears, about 90% of the original sound quality. Can we accept 90%? I'll leave it to you to decide.

Analog Mixing

Clearly, jitter matters anytime a conversion takes place. Thus, when mixing with an analog console and digital multitrack, jitter is extremely critical. In contrast to the advice given by manufac- turers of word-clock distribution devices I recommend that mix engineers try running the multitrack or D/A converters on **internal** clock; it may sound better. The manufacturers of outboard clocking boxes are trying to sell you equipment which in all cases is a bandaid and not a cure—so investigate, inspect the measurements and test before you buy. In an ideal world, the converter should handle any reasonable clock feed or cable interface without affecting the sound, and there are now a handful of converters that meet that requirement. Authoritative measurements and good subjective tests are hard to come by, so cherish the magazine article or book that provides good information on the jitter performance of your favorite converter. Be aware that it is a lot easier to design a stable crystal clock than a PLL, which has to perform double-duty as an oscillator and reject incoming jitter.[5] Thus, any reasonably-designed converter or multitrack recorder can perform better on internal clock, and in a superior converter, the performance on external clock can only do as well as internal, but not better. If a converter does better on external, this should be seen as a criticism of the quality of the internal clock.

However, outboard word clocks are useful for syncing non-conversion processes and in a perfect world should be used to drive **anything but** converters! I know this goes against the common "wisdom" but it does not contradict the basic principles of digital audio design. Later in this chapter, we present some measurements to help guide you, measurements you can duplicate with readily-available equipment and test signals. It's amazing how few manufacturers take advantage of these simple measurement techniques, or perhaps they're too embarrassed to publish the data.

Clock Stability Requirements for Converters

An ordinary crystal oscillator is sufficient for a computer that processes data, but audio converters require an extraordinarily stable master oscillator. To get 20-bit performance at 44.1 kHz SR requires oscillator stability (jitter) at or below 25 picoseconds peak to peak.[6] One nanosecond (1000 picoseconds) in the time domain equates to 1 GHz, which is why a critical converter's circuitry must be shielded and isolated from even the tiniest RFI or clock leakage that can enter via power supply, grounds, or emissions. Now it should be obvious why good-sounding converters are rare and expensive.

The Effects of Internal Sync vs. Various External Sync Methods on Converter Performance

There are two ways to clock a converter:
a) via **Internal Sync**, where a stable crystal clock located inside the converter directly drives the circuitry. In an excellent design, a crystal clock located very close to the sampling clock pin of the converter chip will yield the best audio performance.
b) **External Sync**, which usually requires a phase-locked loop (PLL), a critical and cantankerous circuit, the fundamental culprit of jitter-induced converter artifacts. The PLL has to filter jitter caused by poor source clocks and by interference along the cable which brings in the clock. Thus, the common use of unbalanced wordclock cables can produce ground loops in the clock signal itself.

Examples of External Sync:
i) **AES/EBU sync**, which is prone to *signal-related jitter*, as first illustrated by Chris Dunn and Malcolm Hawksford in their seminal AES Journal paper.[*] Thus AES/EBU "black" will produce a cleaner clock than AES/EBU with signal, with a typical PLL. But a "smart" PLL will not produce *signal-related jitter*, also known as *program-modulated jitter* or *data-dependent jitter*.

ii) **Wordclock sync**, which can yield extremely low jitter, because the PLL required is simpler. Despite this, only a handful of the converters I've tested have inaudible degradation due to jitter under wordclock, and even fewer under AES/EBU! This means that you may have to re-evaluate your current converter choice if you want to obtain audiophile performance when locking to video.

iii) **Superclock sync**, which may or may not require a PLL, depending on the frequency of the superclock and design of the converter receiving it. There is no such thing as a free lunch, and manufacturers must still pay attention to jitter issues with superclock.

iv) **Other Interfaces.** The computer industry is continually reinventing the wheel, and the audio industry is about to adopt the latest wheel, a very jittery computer interface commonly known as Firewire or MLAN. For lower jitter, a supplementary wordclock or internal sync cable will be required. Let Firewire carry the data, but not the clock. I predict sound-quality will initially go downhill when Firewire takes over, until manufacturers pay better attention to jitter issues.

---

[*]   Dunn, Chris & Hawksford, Malcolm. Is The AES/EBU/SPDIF digital audio interface flawed? *Journal of the AES* preprint 3360 October 1992.

IV. How to Get the Best Performance from Converters

A/D-Jitter Permanently Affects the Recording

In 1988, all available A/D converters left me cold, so I built the world's first working implementation of Bob Adam's DBX oversampling technology, later purchased and refined by Ultra Analog. A few engineers latched onto this technology, and in my opinion, the quality of custom-built Ultra Analog converters was unbeaten for almost 10 years, when finally, a few high-end professional A/Ds arrived that sounded as good, and eventually, better. I always operate well-designed A/Ds on internal sync for best performance, unless doing video, when they must be locked externally.[7] The A/D should be the master clock in any system when recording, and the D/A the master when playing back. Remember, jitter in an A/D translates to distortion which can never be removed.

D/A-Low Jitter Important for the Listening

Until recently, professional D/A converters also left me cold, and for over 10 years, I resorted to using customized consumer (audiophile) units that exhibited, to my ears, superior depth, space and tonality. However, while professional units were slowly advancing in terms of jitter-immunity, most consumer and audiophile units - which were never meant to reject the high jitter levels encountered in a complex digital recording studio - were not. So I had to suffer from inconsistent sound depending on the source feeding the D/A converter. Only recently have a few professional DACs appeared with both good-sounding analog circuitry and virtual immunity to incoming jitter.

In the year 2000, I installed a new converter into our mastering suite whose key to low jitter performance is having **all** converters operate from a common bus master clock, so there is no longer the question of switching clock when recording or playing back. The source of the bus clock can be an internal oscillator, AES/EBU, or wordclock.[8] There must be only one master clock in a system at any time. Every playback device (e.g., DAT, CD) must either slave to that clock, or must become the master. This raises a fundamental question of technique. How do you put the master clock where it belongs (inside the converters), and still be able to play back DATs and CDs? The solution is to use professional-quality transports that have external wordclock connections.

I have tried all clocking possibilities with this new converter, which is highly immune to jitter on all its interfaces. Yet I heard and measured a slight improvement with each enhancement in clocking, with internal clock performing better than WC and much better than AES/EBU (as theory would predict). When I installed a CD transport that would slave externally, it was such a very pleasant surprise to hear CDs sounding better than ever that I took a pleasureful day off to enjoy to some of my favorite music before going back to work! Jitter measurements seem to confirm these results, and lead to the conclusion that jitter artifacts must be near the noise floor to become inaudible.

So, what does it take to make a superior converter that produces only **inaudible** effects from jitter? The answer is time, research, and critical design implementation. The engineers who produced this superior converter spent one **man-year** on the phase locked loop alone, and a further year on the converter details. Successful converter manufacturers must master the techniques of PC board layout, grounding, internal clock distribution, and immaculate separation of digital and analog signals. Things are looking up. But caveat emptor.

Do we need to worry about cables, which produce sonic differences with jitter-susceptible converters? When I was using a jitter-susceptible converter, I spent a long effort cleaning up cable runs, using proper-impedance cable, avoiding ground loops, etc., and this resulted in improved monitoring. But really, you can mismatch impedances (e.g. 110 ohm to 75 ohm) with no concerns that jitter will affect the data. However, at high sample rates, impedance mismatches are more likely to cause poor signal transmission (and incidentally, high interface jitter), resulting in glitches or dropouts, so it's wise to get your cabling act together. Balanced digital connections can also reduce RF radiation into sensitive analog stages, and improve the performance of jitter-sensitive converters.

The Internet and Jitter

As studios begin to collaborate through the Internet, jitter issues will be even more challenging, since DSL and T1 lines are notoriously jittery. Perhaps it may be possible to use a master clock based on a GBS satellite clock, provided that a GBS-derived clock can drive a converter with the required low jitter. Or perhaps the solution will be to install an elastic buffer where the Internet sources enter the building.

V. Stop Leaping to Conclusions: Real World Examples

Let's apply some of the principles we've discussed. The names have been changed to protect the misinformed!

Example A: Digital Copying and Jitter Reduction.

Engineer Betty would like to do some Digital Copying (cloning), from CD to DAT. First she notices that her CD recorder sounds better than her DAT machine. The reason is that the internal clocks of typical DAT machines are not as clean as those in CD players (perhaps because they have more motors to interfere with the electronics).[9] But mostly she's concerned about the sound differences she hears; her DAT machine sounds **better on playback than on record**! She tries inserting a "jitter reduction unit" before the DAT, hoping to make better dubs, but this only creates more puzzles—now it sounds better during dubbing than when it is played back! What is going on here?

---

* *E-E* is a term commonly used in video meaning "Electronics To Electronics," when a machine is in record monitor mode as opposed to playback.

† Or you may choose to conclude from this example that DAT dubbing via AES/EBU may only be susceptible to jitter in the most subtle or imperceptible way, which I do not believe to be the case.

The Miracle of the Blessed DAT Resurrection

Always wanting to see how far equipment can be pushed, I decided to demonstrate the veracity of the laws of physics at an AES Convention around 1992. I built a special clock oscillator whose jitter could be altered, and connected it to a DAT machine, thus simulating conditions such as mismatched cable impedances or extreme problems with clocking. The digital out of this machine was then connected to another DAT machine to make a dub. While monitoring the record machine in E-E,* I increased the jitter of the source machine until the record machine exhibited serious distortion on its analog output; it sounded like a vocoder in overload. Anyone listening to this machine's output would conclude it was broken and that it was making a defective recording. *But believe the facts, not your own ears*, because on playback, there was no trace of the distortion; the playback sounded very clean. Thus demonstrating that digital dubs are not susceptible to jitter.†

Digital copies really are perfect (as long as the playback deck is in good condition and not interpolating digital errors). Illustrated below, the DAT machine drives its DAC from two choices of clock; during record it depends on the phase locked loop to generate a clock from incoming clock, and during playback it uses its internal oscillator. The reason the DAT machine sounds better on playback is that its internal clock is probably more stable than its PLL. **However, the message isn't changing, only the messenger delivering it**. And when Betty inserts the jitter reduction unit, it's no surprise that record mode now sounds better than playback—since DAT machines typically are built to a price, a $2000 jitter reduction unit helps the PLL produce a cleaner clock than the machine's own 25 cent oscillator![10]

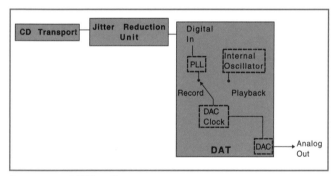

A consumer-model DAT machine switches its master clock source between record and playback mode.

But Betty's jitter reduction unit does not improve the copy in any way (though you won't hear this story from the manufacturers of jitter-reduction units).[*] She can prove that there is no problem with the DAT by listening under identical jitter conditions. For example, she can copy the DAT back to the CDR and play the two CDR tracks back to back. What conclusion must she draw about the DAT tape if the two CDR tracks sound identical?

---

* The author has a collection of surplus high-end jitter-reduction units in his garage, available at bargain prices.

Example B: Copying via SDIF-2 versus AES/EBU

Engineer Don has concluded that DASH recorders make cleaner digital copies through the SDIF-2 interface than through AES/EBU, because he knows that SDIF-2 is a "cleaner" interface. His experience has been that the SDIF-2 interface makes a DAC sound better via its separate, clean word clock, while AES/EBU embeds a (jittery) clock in the data stream. And since the DASH tape recorder sounds better to Don, he concludes that the DASH tape copy is better than the DAT tape copy. But it is equally feasible that it's the DASH machine itself that "sounds better," not the tape. Both the DAT and DASH tape make equivalent masters, except the DASH tape uses more robust error correction and will probably last longer.

Don can prove his own conclusion to be false by taking the "questionable" DAT copy and playing it on a DAT machine equipped with the SDIF-2 interface, preferably slaving the DAT to wordclock. He'll probably find the DAT copy now sounds as good as the DASH. Regardless, Don should also invest in one of the new jitter-immune DACs, which can make the SDIF-2 interface unnecessary.

Example C: Clock Accuracy?

Ray was told that an accurate crystal wordclock fed to all of his gear would make it sound better. The operative word here is not **accuracy** but rather **stability**. For jitter removal, **stability counts more than absolute accuracy**. A crystal may produce 44,100 Hz **on the average**, but a jittery crystal oscillator deviates above and below that average. In a totally digital production studio, even if the master crystal is several Hertz off, and even if that causes an audible pitch error, the end result will sound correct when reproduced with a correct crystal. If I'm in a hurry, I can speed up my clock to 48 kHz, or even faster if the equipment supports it, and still make a valid dub at high speed. This illustrates the fact that jitter cannot influence the accuracy of a dub: we can speed up the source to a frequency 10,000 times greater than the frequency deviation due to jitter and still make a perfect data copy! Dubbing is done on a sample by sample basis; the job of the clock is simply to deliver succeeding samples into the queue.[11]

Example D: Mixing down via internal or external sync?

A recent magazine article purported to evaluate the "sound" of wordclocks. But wordclocks have no "sound;" what counts is the ability of the converter to reject jitter on the incoming wordclock and pass a clean clock on for conversion. Engineer Fred says his multitrack sounds much better with a new wordclock generator than with the old one. I don't doubt it, but Fred should investigate putting his multitrack on internal clock, which is a lot easier to design well than a PLL. Note that if Fred is performing an analog mixdown, he can run his mixdown A/D on its own (independent) internal sync, and get the best of both worlds.

Example E: Load-in jitters?

Engineer Jeff thinks that digital load-ins made through his DAW's S/PIF input sound better than those made through its jittery Toslink optical input. But he's mistaken, the sonic difference is ephemeral. It will only be present during the loadin,

and the DAW's playback will actually sound better than the loadin! And Jeff will only notice this if he uses an inferior DAC which is susceptible to differences in clocking. Rest assured that interface jitter or clocking differences have no effect on the integrity of a digital load-in from a digital source.

In summary, we should not blame clocks for problems that should be fixed in the converter. And we should stop working on minimizing jitter in the digital processing chain, instead concentrate on ways to reduce the jitter at the sampling clock inside the converters.

VI. Concern for the rest of the world...

Since the whole world is probably listening to music on inferior D/A converters, it's very important that the CDs (and DVDs, or SACDs) we cut for them have the best possible sound. As I said, there is no jitter on a storage medium, but there is some (controversial) evidence that CDs cut at high speeds sound inferior to CDs cut at low speeds, and that CDs cut with a jittery clock sound worse than those cut with a clean clock.[12] We theorize that certain mechanical parameters of the disc are altered by the cutting speed, making it more difficult for the CD player's servo mechanism, passing the varying servo load to the CD player's power supply and thus affecting the stability of the master clock. It only takes a few picoseconds to make an audible difference. Regardless of the theoretical reasons why this might be happening, it's important to note that the CD difference is an ephemeral and correctable phenomenon, clearly related to some difficulty of the CD player **only**

during playback, and that the data itself is unchanged. The differences are no longer audible when played over a jitter-immune DAC. Because there is no permanent distortion in a D-D dub (as would be the case in A/D conversions) the output of the CD player can be reclocked to make the apparent audible differences inaudible. Time and again I have observed that when the clocking has been fixed, formerly audible differences disappear. However, until everyone else has perfect D/As, it's important for the CD production plants to heed the audible evidence, and cut glass masters at 1x speed and find other ways to make the best-sounding CDs.

By the way, for those listeners with inferior DACs (the majority), I always find that I can restore the sound quality of an "inferior" CD by copying it back to a workstation and then outputting on a good SCSI writer at 1x speed. In this case the dub does sound better than the original! It is technically impossible for previous jitter to be passed through an asynchronous interface such as SCSI to the final sampling clock.

VII. Things That Go Bump In The Night
Framing and Timing Errors
Wordclock to AES timing error

Although jitter is often made the scapegoat for a motley of problems in digital audio the fact is that 99% of the time, glitches, clicks, dropouts, noises and lockup problems, are caused by **framing problems**, not by jitter at all. Framing problems are caused by timing differences in critical signals and cannot be solved without equipment software or hardware modifications. At left is an oscilloscope photo, at the top of which is the start of the AES preamble (which defines the beginning of the AES data word), and on the bottom, the point where wordclock changes from high to low.

To complicate matters, there is no standard that defines which wordclock transition (low to high or high to low) the AES preamble should line up with. This is a timing difference of 180 degrees, or approximately 11 µS at 44.1 kHz, which is enough to drive workstations, processors and consoles batty, producing glitches, or no signal at all. Fortunately, my workstation has a menu choice that allows us to choose the wordclock phase, making it more compatible with products of various manufacturers.

AES to AES framing error

Digital audio is a small industry, still experiencing growing pains. Many current digital consoles and DAWs are oversensitive to timing problems, which I must stress are unrelated to jitter. And since some digital audio processors produce an AES output that is out of timing with their AES input, intolerant consoles and workstations have trouble locking to them (illustrated at right). Once I was forced to insert a simple reverb unit via analog, because the digital console would not lock to it on a digital send/return path. The fault was caused by the console's intolerance to AES framing errors,

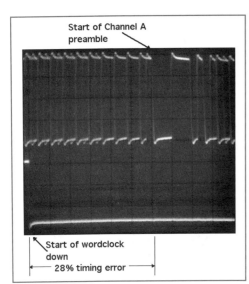

Start of Channel A preamble

Start of wordclock down

28% timing error

This oscilloscope photo compares the timing of the start of the Channel A AES preamble against the start of wordclock at the output of a digital processor. This timing offset of 28% of the length of the AES frame is 3 points greater than the permissible tolerance in standard AES11 and would cause locking trouble to intolerant consoles or DAWs or other receivers.

aggravated by the reverb unit's output being slightly out of framing (timing), as seen in the following figure. You can probably prove it's a framing problem without measurement equipment: in this situation, set the digital processor to run on its internal clock, and lock the console to the external processor on its reverb return. If the console will lock and pass audio from the external processor, then the previous problem was a framing problem.

Locking the console to wordclock would probably not help and may even worsen the situation since the timing difference between the AES sources would remain. Framing errors are cumulative in a chain of processors if they are

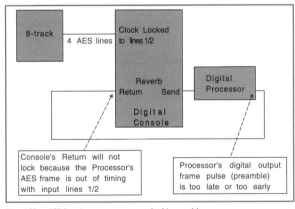

How AES to AES framing error can cause locking problems

chained via AES/EBU (or S/PDIF). If the framing error of each box is in the same direction, then the total error could be enough to cause locking problems in sensitive consoles and DAWs. You may be able to stabilize the system by locking the last processor in line to external sync (wordclock or AES). If the last processor in line is framing-tolerant on its AES input, then locking it to external sync will force its output to a known framing and hopefully to within the tolerance of the DAW. It's also possible to build an outboard box that will fix this sort of framing problem, but really, the burden is on the manufacturers to produce consoles and processors that are within the AES standard tolerances.[13] Again: Caveat emptor.[14]

Off-Center Clocks

Another problem mentioned earlier is loss of lock caused by an off-frequency master crystal and a sensitive PLL. Some digital inputs have a very low tolerance to incorrect center frequency (which also makes them uncomfortable when varispeeding). If you have locking problems not due to framing errors, confirm that the source frequency (e.g., 44.1 kHz) is correct, and if not, have the master crystal oscillator trimmed.

VIII. How It Works

Simple in Theory...

Most engineers don't need the heavy technical details of how equipment works, but there are usually a couple of nagging questions, like... *What is a reclocking circuit? Why do we need a high-frequency clock?*

Reclocking Circuit. The data inside typical audio processors travels from chip to chip *serially*, that is, bit by bit. A clock pulse moves this data along. This clock bus is distributed to all the critical chips inside the box. As we've seen, it doesn't matter if this clock is jittery, proper data still makes it to the next chip in line. But sometimes data needs

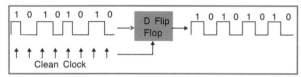

A Simple Reclocking Circuit

to be *reclocked*, for instance when feeding a D/A converter. Pictured here is a simple reclocking circuit; on the left side is an incoming data word that's been clocked by a jittery clock; the data value is (conveniently) 10101010. This word passes, one bit at a time, into a logic circuit called a *D-type Flip flop*, which is being fed a clean clock. Almost magically, the data neatly marches out of the flip flop, and in theory, all the jitter is gone and the data is ready to feed the DAC. Notice how the clean clock's pulses permit the flip flop to properly "sample" each data value, but only if the clock pulse lands within the acceptance time of each incoming bit. In this illustration, the fourth (and eighth) data bit is in danger of being missed if it arrives a moment later, in which case the clean clock would land on the previous bit and the wrong data would be output. Fortunately, typical audio sources have much less jitter than in this illustrative example. Otherwise the system would break down and we would get glitches, clicks or hash instead of clean audio, and then the output data is really being changed![15]

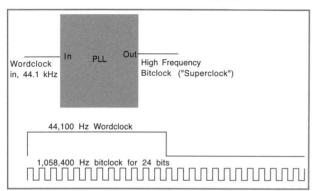

A PLL is needed to generate the higher frequency clock required to move the individual bits from place to place.

Why PLL? The figure at bottom left illustrates why a phase-locked loop (PLL) is needed. If we are passing 24-bit audio bit by bit, then we need a high-frequency clock pulse that is 24 times the frequency of wordclock. Wordclock enters the device, and has to be multiplied up to the higher frequency to drive those bits around, known as the *bitclock*. It's easy to divide down without creating jitter, but very difficult to multiply up, and it's the job of the sophisticated circuitry of the PLL to create the higher frequency while reducing incoming jitter.[16] A PLL is a sort of electrical flywheel; it tries to find a center, holding reasonably steady while still following the average frequency of the incoming source.

...Complicated In Practice

What makes these circuits so difficult to design well is that at high frequencies, leakage from the jittery portion of the circuit can travel through back paths to contaminate the clean portion of the circuit. These paths include power supply and ground. Couple that with outside interference and ground loops, and you have an analog designer's nightmare. 10 picoseconds error can make the difference between an 18 or 20-bit noise floor. Some manufacturers use a dual-PLL, where the first is an analog circuit, and the second a voltage-controlled crystal oscillator (VCXO), in an attempt to get the jitter down to that of a quartz crystal. Unfortunately, designs using VCXOs cannot varispeed because of their narrow frequency tolerance. It is difficult, yet possible to design a jitter-immune PLL that's as good as a crystal, has wide frequency tolerance and quick lockup. No matter how many PLLs it says on the label, the quality of a designer's work should be tested objectively.*

---

* I once owned a Swiss watch that said "17 jewels" on the label but only 5 of them rattled when I shook the watch case. What does the label really mean?

IX. Jitter Measurements

Here are some jitter measurements made on D/A converters. Before you buy an expensive console or conversion system, you can take measurements like these yourself, using readily-available test equipment. You'll be shocked at the variance in performance from one model to another. It's unfortunate that magazine reviewers and editors like to see single-number performance (e.g., *this converter has an intrinsic jitter of 40 ps*), which means little technically and nothing psychoacoustically. What we need to see are detailed graphs of the deterioration of a converter's performance when it is fed a jittery signal, and this is the least that we should expect from a magazine reviewer.

A/D and D/A converters can be tested for the effects of jitter using a very high frequency sine-wave test signal, but the test signal must be very pure and frequency-stable, probably crystal or digitally-generated. For these DAC tests, I used the J-Test signal invented by Julian Dunn, an independent consultant best-known for his work on the Prism brand of converters.[17] The 24-bit J-test signal was not available, so the 16-bit version was used; we'll have to ignore some artifacts that are part of the source signal. Here are a few guides: The lower the noise floor, the less jitter. We have not fully learned which jitter spikes are psychoacoustically important, but, as I have said before, my listening tests show that jitter must be very low (close to the system noise) to be inaudible. Also, since test equipment varies, your J-Test results will be different from mine, but relative rankings will likely remain.

In the the color plates section, **color figure C19-01** shows, in red, the noise floor of my UltraAnalog A/D (which I used to sample the outputs of various D/As under test), and in blue, the artifacts of the 16-bit J-Test signal, which are at −132 to −135 dBFS.[18] This means if we appear to measure jitter in the device under test below −132, it may simply be due to artifacts of the test signal. I think it's more important to look at how the jitter artifacts affect the DAC's own noise floor and at what particular frequencies, than to calculate the actual jitter value in picoseconds.

Color figure C19-02 in the Color Plate section shows a considerable measured difference in jitter performance when an inexpensive consumer D/A is fed from two different sources. A cheap consumer CD player yields the highest output jitter, with the output of Sonic Solutions even less. If this were a linear display instead of semi-log, it would be more obvious that jitter usually produces paired artifacts around the center frequency, usually at equal deviation about the center. Compare the consumer D/A's performance to that of the excellent "jitter-immune" TC Electronic System 6000 D/A. When fed from either of two sources, the TC's jitter is effectively identical and just about as low as its quiescent noise floor!

Color figure C19–03, in the Color Plates section, shows that sync mode hardly affects the TC's jitter performance, with extraordinary measurements in internal sync and slight differences when locked via AES/EBU. When slaved via AES/EBU it produces very slightly more jitter (only the two discrete frequency blue lines circa −117 closest to the center frequency). When on internal sync (red trace), its jitter is nearly as low as the UltraAnalog's noise floor, and realize that most of the *grass* is the 16-bit J-test signal itself. I can hear a slight degradation in sonic clarity, a smeared image and brightness when the TC is slaved to AES/EBU, which implies that the black-colored spikes at approximately -117 dBFS may be audibly significant.

The Weiss is the first DAC I have measured with no apparent trace of discrete frequency jitter in its output when locked via AES/EBU (**Figure C19-04** in the Color Plates section). Instead, its noise-floor rises with the test signal and the jitter "skirts" appear to widen; all incoming jitter has been converted to random noise. Or is the sonic improvement due to euphonic coloration (higher noise floor masking discrete jitter components, since we can no longer see the floor of the test signal itself)? This also brings up concerns about potential converter noise modulation with signal, which may mask low level signals or reverberation. However, low-amplitude, random noise is the most benign signature one could wish for and the DAC sounds great. I did not test the DAC on internal sync.

In Conclusion: When it comes to jitter, there's a lot more to know than what meets the ear! Until our audio systems have advanced to the point where **all data-identical sources sound identical**, then we cannot make valid judgments about sound character.

1   For those of you who were born after the era of analog tape, *print-through* is a phenomenon where one layer of magnetic tape magnetically imparts some of

its signal on the adjoining layer. After years of storage, it is possible to hear two or three repeating echos in the tail decay of a song recorded on analog tape (which can be repaired by the "adding tails" technique explained in Chapter 7). But even when print-through does not provide a distinct echo, it is always there to some extent, affecting the clarity of the sound. A "low-print" tape has less print-through than a high print.

2 Technically speaking, any processor which adds random dither is not a state machine. All good SRCs and equalizers employ internal dither to linearize the process. The randomizing effect of dither means that each output pass will produce slightly different data at each instant. However, on the average, the output stream is really the same, and if we could subtract the random dither from the output signal, each pass would be identical. You can prove this by running two passes through the same equalizer, lining them up, and subtracting one from the other. You would be left with no residual of the signal, only random noise, thus proving the processor is a state machine. By the way, if you tried this with an ASRC, you would hear a small amount of residual signal in the noise floor, proving that an ASRC is not a state machine. Many digital processors create dither using a *pseudo-random sequence*, which predictably repeats after a period of time, so they are perfect state machines; if when comparing two successive passes you can find the moments where the two dither signals exactly line up!

3 When transferring from file to file, there is no clock at all and the process just deals with one sample after another. I sometimes explain jitter with a bowling ball analogy. Throw a series of bowling balls, some white and some black, down the alley. Although their timing is irregular, when they land back on the stand, the white and black are in the same data order. Digital processors look at the samples (bowling balls), not at the time they arrive, so the output data is identical, even if the timing is irregular.

4 As illustrated in a demonstration of extreme jitter that I performed at the AES Convention, when the audio was so distorted by a jittery clock that it was unrecognizable, but the data remained intact. See the accompanying sidebar. I'd like to thank Andy Moorer for coining the *message/messenger dichotomy* in understanding jitter.

5 The external wordclock replaces the signal from the quartz crystal with a PLL. In the vast majority of converters manufactured today, "the jitter caused by wordclock is typically 15 times higher than when using a quartz based clock," according to the manual for the RME model ADI-8-DD format converter. For mid-priced converters where little attention was paid to the internal clock design, I've seen some surprising situations where internal clock is not as good as external, but it is much cheaper and easier to design a good quartz clock than a powerful PLL. External clock can never perform better than a reasonably-designed internal clock, because of all the jitter-producing obstacles involved in clock extraction and regeneration (the job of a PLL). In fact, I go so far as to say that any digital console or DAW interface that does not perform equally or better on internal clock is a defective design that is definitely not living up to its full sonic potential. Insist on full-disclosure and FFT-based jitter measurements before buying.

6 According to a simplified formula from Don Moses. Enclosure Detuning for 20-Bit Performance, *Journal of the AES* preprint 3440 October 1992.

> The following expression utilizes Carlson's similar triangle analysis method and is useful for the case where: (1) the jitter deviation is small compared to the sampling interval, (2) distortion is measured at the zero-crossing of a sine wave, (3) the peak-to-peak amplitude is normalized to 1-V, and (4) the maximum slope is approximated as 2 X the information bandwidth:
> Resolution (in dB) = 20 log (time deviation x 2 x information bandwidth)
> For example, 25 ps of jitter, 20 kHz information bandwidth, yields:

20 log (25 ps x 2 x 20 kHz) = -120 dB, which provides 20-bit resolution.

In other words, if you double the sample rate to 88.2 kHz (the information bandwidth becomes ~40 kHz), the same amount of jitter reduces signal to noise ratio by 6 dB. For 20-bit performance, at 88.2 kHz, if you consider the information bandwidth goes to 40 kHz, you would need to halve the jitter to less than 12 picoseconds. And for each 6 dB improvement or 1-bit increase in wordlength, you must halve the jitter yet again. Even if you limit the information bandwidth to 20 kHz, in order to get excellent performance with long wordlength, it boggles the mind the degree of care required to lessen external EMI/RFI, bypass power supply problems, to say nothing about the stability of the PLL required! Making it clear the myth of the >20 bit converter, which may have >20 bit quiescent noise, but how does it perform with real-world signals?

7 Not exactly *must*. In a self-contained audio for video post studio, it is possible to make the A/D the master clock for everything, by using an AES/EBU to video sync converter, thus forcing the video clock to slave to the audio instead of the other way around. Presto: Low jitter while mastering for video, and a product-design opportunity for fussy audio mastering engineers who must work with video.

8 I'm sure you're curious as to the brand. It's the TC System 6000. Other current converter manufacturers who claim to have produced "jitter-immune" units include Prism, dB Technologies, Benchmark, and Weiss. The lockup time in the Weiss, Prism and TC units is virtually instantaneous, pointing out that jitter-free design does not require a long lockup. In fact, the mathematics proves that any buffer longer than about a sample is superfluous, since jitter is a small fraction of a sample period.

9 Bob Harley of *Stereophile* measured output jitter as low as 10 to 100 picoseconds with some of the best audiophile CD transports, but as high as 1500 picoseconds (1.5 ns) with a DAT machine. Different methods of measuring jitter yield different results, but the approximate relative values will remain.

10 In the case of a cheap crystal oscillator, the external clock is cleaner than the internal. The cleaner the incoming clock, the less work the PLL has to do. But the output jitter of a PLL can never be better than its nascent or intrinsic jitter, and is typically worse. The output jitter of a PLL is a combination of three things: its intrinsic jitter, incoming jitter, and the PLL's jitter attenuation.

11 However, a crystal which is off the standard center frequency can cause locking problems, since many low-jitter PLL designs have a narrow lock range. But if the system components lock, then an off-standard crystal won't affect digital dubs at all. Some PLLs have a narrow and wide setting to deal with sources that are a bit off the standard. Switching to wide increases frequency tolerance, but also increases the PLL's jitter. Don't be concerned, as long as the PLL is not driving a converter.

12 Reports from the musical artists themselves led engineers at Sony Corporation to work on improving the jitter in their CD cutting systems. Without outside influence, some major artists had been reporting that their CD pressings did not sound as good as the reference CDRs they had received. We theorize that irregular pit spacing or inadequate pit depth on the CDs themselves is affecting the player's servo mechanism. The servo mechanism and sample clock share a common power supply, so with poor power supply bypass in the player, simple power or ground leakage may affect the stability of the clock. It doesn't take much leakage to change a few picoseconds. Critical listeners making CDRs have heard superior sound with SCSI-based CD Recorders than with standalone CD recorders. In standalone recorders, the master clock driving the laser is slaved to the AES/EBU input, while computer-based recorders use a FIFO buffer and a crystal clock to drive the laser.

We theorize that the reason we have not noted such differences with DAT machines is that even if the incoming differences are passed on to the medium, they are swamped by the much-higher jitter in a typical DAT machine than in a typical CD player. What's a few picoseconds out of thousands?

13 Julian Dunn clarifies: "This could be considered to be a synchronisation issue and these are covered in AES11. These define the permitted output alignment error (+/-5% of a frame period) and the tolerance to input timing offset (+/-25% of a frame period) before the delay becomes uncertain.

The specifications for the interface itself (AES3, IEC60958) do not allow a receiver's ability to decode data to depend on the relative alignment of clocks - as long as the dynamic variation is within the jitter tolerance spec. (about +/-4% of a frame period at low jitter frequencies)."

14 If you're spending $30,000 and upward on a digital console, request the manufacturer to sign an agreement that the digital inputs and wordclock framing tolerances must meet or exceed the AES11 synchronization specs or the manufacturer will correct the problem at no charge. This amounts to a sad wakeup call to the manufacturers, but consumers should be entitled to interface real-world equipment to their consoles.

15 Those dyslexics in the audience will appreciate that I am taking slight liberty with this discussion for ease of understanding. Since the left hand end of the bitstream is the last to go into the flip flop, the "fourth bit" counted from left to right is actually the fifth bit to go in! This leads to the requirement that software has to decide whether to make the left or right end of the bitstream be the most significant bit. Intel and Motorola have been fighting over this for decades, so if you don't follow this part, you're not the only one!

16 Many bitclocks are 32x the wordclock, or greater, to allow for a longer internal wordlength. A typical PLL may generate a superclock which is 128, 256 or even 384 times the wordclock frequency, and is then divided down using a simple divider.

17 The J-Test is a special signal designed to aggravate a D/A converter's jitter. It contains a fundamental signal at 1/4 the sample rate, which is 11.025 kHz at 44.1 kHz SR and a low frequency component added to deliberately add data-jitter on the AES input. The test is particularly designed to pick out the interaction to the sample clock from the data on the AES/SPDIF interface that is used to derive the sample clock. When AES/EBU is not involved, it would be more practical to use a simple clean high frequency tone.

Julian: "There are four 24 bit numbers in a sequence that is 192 samples long that repeats.

0xC00000, 0xC00000, 0x400000, 0x400000 (x 24 i.e. 96 samples)
0xBFFFFF, 0xBFFFFF, 0x3FFFFF, 0x3FFFFF (x 24)

Hexadecimal		Binary
C00000	=	1100 0000 0000 0000 0000 0000
400000	=	0100 0000 0000 0000 0000 0000
BFFFFF	=	1011 1111 1111 1111 1111 1111
3FFFFF	=	0011 1111 1111 1111 1111 1111"

The 16-bit version of the J-Test signal is currently available on a CD from Audio Precision and on another test CD from Checkpoint Audio in the Netherlands.

Further information can be found at Julian's website www.nanophon.com

18 The measured amplitude of the noise depends on the number of points (bins) in the FFT, the window which is used, and the A/D converter in the measurement equipment, which is why each reviewer's results will be different. These measurements were taken with a 32K point FFT with an averaging time of about 2-4 seconds, and a Hanning window.

19 There is only one right "kind" of digital cable, one whose impedance is a correct match for the circuit (e.g. 75 or 110 ohms). Some audiophile manufacturers have made so-called *digital* cables which are improper for the circuit, but since they affect the sound of a typical, consumer-grade D/A converter in some unpredictable way (usually adding jitter, not reducing it), consumers have been known to play with such cables to tune their systems. It's a losing battle, because the cable-induced jitter reduces resolution and colors the sound.

CHAPTER 20

Tips And Tricks

I. Introduction

This little chapter reveals some previously-untold secrets of how to maintain and run a digital audio studio, including dealing with the vagaries of timecode that just won't stay stable, how to make clean AES/EBU connections, advice on hard disk formatting, and more.

II. Timecode and Wordclock in a Digital System

Drifting drifting drifting

An engineer attempted to synchronize an analog tape deck, sequencer and digital tape deck by slaving everything to the timecode coming from the analog deck. The sync seemed to work fine for a little while, but after a couple of minutes, he noticed that the analog deck was drifting out of sync with the rest of the system. The reason for the drift was that there must (and can) be only one master in any system, and in this case there was already a master clock in the digital system—the digital audio clock. When a computer (or interface) receives timecode, it takes a *stamp* or *trigger* from the first valid timecode it sees. From that point on, the interface ignores incoming timecode; it runs its own timecode, locked to the digital audio clock. The two sources would drift apart if the source of the incoming timecode is not locked to wordclock. In this instance, timecode

> { *"There must be only one master in any system"* }

from the analog tape deck is independent, based on the speed of the tape deck.

One method to synchronize an analog deck with a digital system is to slave the analog deck; in order to avoid introducing wow and flutter, a special type of *flywheel* synchronizer speeds or slows the analog deck, holding it within an acceptable margin. The other method is with a digital clock generator/timecode regenerator specially designed to lock to analog-style timecode, that locks to the analog deck and slowly adapts the master clock to the rate from the analog deck. The latter method is likely to cause higher jitter; it is much better (as in the first method) to have the A/D on internal sync.

When locking two digitally-based systems together via timecode, drifting will result if one or the other is set to the wrong timecode or wordclock rate. As we mentioned, usually the sequencer triggers to the first burst of timecode and then runs on wordclock. To prevent drifting, make sure the sequencer is set to receive the same wordclock and timecode as the DAW is transmitting. It would be nice if all sequencers did this automatically, but this only happens in a perfect world.

Pull-ups

Things are far simpler without video. At 44.1 kHz SR and 30 (25) timecode fps, there are exactly 1470 (1764) samples per frame, so just divide the audio rate by an exact **integer** to arrive at the timecode rate. But when NTSC video is involved, the timecode rate is slower, 29.97 fps, which yields a non-integer number of samples per frame. Normally the wordclock is slaved to the video, so we require a sophisticated wordclock generator which takes in video and produces wordclock with the right ratio, called a *pull-up*, by approximately 0.1%. If not, then the two systems will drift apart and audio-visual sync (e.g., lip sync) will be lost.

Wordclock Voltages

The problem with standards is there are so many of them! With Johnny-come-lately digital, no voltage standard was developed for wordclock, and this lack of standard has produced a chaotic situation. Many of the earliest wordclock generators were based on video sync (blackburst) generators, which produce 4 volts peak to peak into a 75 ohm load (abbreviated *4 v p-p*). This is a fairly expensive circuit, so soon wordclock generators appeared based on the video standard of only 1 v. Other manufacturers settled on a TTL-level standard, which if terminated is 2.5 volts, and unterminated could be 4-5 volts! Chances are if a device will not lock to incoming wordclock, either the receiver is insensitive or the generator is not putting out enough voltage. At this point, the only way out of this mess is to insist on wordclock generators that produce 4 volts and wordclock receivers that can accept anything between 1 and 5 volts. Impedance-matching is not that critical on low-frequency lines such as wordclock, so if the cable run is short, you may be able to make the circuit work by removing the load termination, or in extreme cases, lowering the value of the source resistors in the generator below 75 ohms. If this doesn't work, then you need a new wordclock generator. An oscilloscope can verify the amplitude of the wordclock. Caveat emptor.

House Video and DARs Sync

Surprisingly, in the year 2002, video and digital audio interfacing is still in a primitive state.[1] This is because neither wordclock nor video contain markers as to the beginning of the digital audio frame or channels. If you use house video directly you will produce a word clock of the correct frequency but not the correct phase.[*] It is highly likely that two video recorders containing digital audio will not be phase locked. This will cause unpredictable phase shift because there is no phase reference in video sync, so think twice if you are forced to lock multiple audio devices via video sync. Wordclock has fewer such problems but it still does not define channel beginnings, and this lack can result in channel reversals, for example. The only dependable sync reference for multiple devices is AES-11, also known as DARS (digital audio reference signal) which is equivalent to AES/EBU with muted (black) audio; it maintains channel beginning markers (blocks) and channel-to-channel relationships. If a video house reference must be used, I suggest hooking it to a single distribution amplifier that then derives AES-11 sync. At least from that point on all subsequent devices will be properly referenced **to each other**. And in multichannel work, do not split channel processing to multiple devices which only accept wordclock sync, which can cause latency differences (many samples of error) or indeterminate sync, a variance of 1 or more samples between channels. There should be no interchannel sync problem if devices carry all channels and are used in a chain, each one locked to the previous. But see Chapter 19 for jitter and framing considerations when daisy-chaining. I caution against indiscriminate use of new interfaces such as USB and Firewire for multichannel audio until we understand the latency issues therein. When in doubt, insert coherent test signals and test for phase shift between channels.

III. Debugging AES and S/PDIF Digital Interfaces

When the AES/EBU and S/PDIF[2] interfaces were created, the use of standard audio connectors and cabling seemed like a godsend, but people were tempted to use regular audio cables, which were never intended to carry the high frequencies of digital audio (about 6 MHz bit rate for 48 kHz SR). So eventually we ended up with special RF-rated cables attached to our old-fashioned XLR connectors. This identity problem will likely go away as we move from 2-channel to multichannel, which generally will require specialized connectors.[†] The easiest way to debug interface problems is to divide the issues into two parts: hardware and software. The hardware includes the cables, connectors, and signal levels, and the software is the bitstream and how it is interpreted.

> { *"The first step in fixing interface problems is to separate the issues into two parts: **hardware** and **software**."* }

For example, getting consumer DAT machines to record from a digital source used to be a headache

---

[*] Thanks to Julian Dunn (in correspondence) for this good advice.

[†] The multichannel MADI standard interface uses BNC connectors, which are already fully-compatible with a video installation.

(DAT machines are already officially obsolete as of this book publication[*]). Here's the typical behavior of an apparently malfunctioning machine: Set the machine to digital input; without a digital source, the record-ready button will not light. Connect a digital source to the DAT, and the record-ready will operate as well as the sample rate indicator; however, the machine will not go into record. Some machines show incoming level on their meters, others' meters appear to be dead. If we divide the problem into two parts, it will be clear that the problem is not in hardware but in software, for the machine's record-ready light tells us the machine is locked, it's also indicating the incoming sample rate and sometimes even the meter is functioning. The lock indicator is the line of demarcation between hardware and software problems, as in this figure at left.

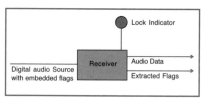

The AES/EBU or S/PDIF-format consists of digital audio data mixed with status flags. The Receiver locks to the incoming signal and separates it into its component parts for use by the DAW, DAT machine, etc.

If there is no lock, or the lock indicator is intermittent, or the output audio cuts in and out, then look to the left side of this diagram for hardware problems (voltage level or cabling problems). Otherwise, the problem is in software, either with the flags being transmitted or the way in which the device interprets the flags (more on flags in a moment).

Fixing Interface Hardware Problems

To the outside world, a digital interface either appears to work or not; we never have much idea how well it's working unless either it stops or we stop to measure it. I'd love to see signal-quality indicators in a receiver; even a green-yellow-red lock indicator would be nice. Currently, the only way to assess a hardware interface is to measure its objective performance by looking at the width of an eye pattern on an oscilloscope. Always use matched-impedance cabling, especially for long runs (either 75 ohm low-loss coax for the unbalanced interface, or 110 ohm cable for the balanced). The balanced signal should measure between 2 and 7 volts p-p, while the unbalanced signal should not be below 0.5 v p-p, into a terminated load (all S/PDIF and AES/EBU inputs are terminating).

With the balanced interface, shields are actually unnecessary, as can be illustrated by the success of Belden's Mediatwist™, consisting of four bonded-twisted pairs for up to 8 channels, and performing as well as the highest-grade coax. In fact, standard Cat 5 twisted pair Ethernet cable makes a very good AES/EBU cable, second only in quality to Mediatwist. The biggest problem with the hardware of the unbalanced consumer interface is the low voltage (0.5 v p-p), which can easily degrade with lossy coaxial cables or long cable lengths. These problems could have all been avoided if only the S/PDIF interface protocol had specified 1 volt like the **AES 3-ID** standard, which uses a BNC connector, popular with video houses.

Improving the stability of the unbalanced interface. The stability of the unbalanced interface can be considerably improved by upgrading to special **low-loss** 75 ohm cable, and/or by raising the output voltage from 0.5 volts to 2.5 volts, easily done by replacing the voltage divider at the transmitter with a single 75 ohm resistor, as in this next figure:

---

[*]   Replaced by hard disc, CDR and DVD-R

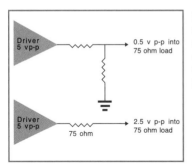

The low voltage of the coaxial digital interface can be overcome by replacing the transmitter's 2-resistor pad with a single 75 ohm.

This modification to the transmission side works so well because it raises the noise margin of the receiving circuit at no significant cost or interference with other circuits. *Warning: modifying circuits usually voids the warranty.* Although the AES standard is between 2 and 7 volts, note that commonly the same audio receiver chip is used for both AES/EBU and S/PDIF decoding and it can accept from as low as 200 mv p-p to as high as 7 volts, so higher voltages are usually not a problem with S/PDIF but extreme low source voltages reduce noise margin and may introduce dropouts or glitches. Input transformers are almost always used for both the balanced and unbalanced interface, so the major difference between AES and S/PDIF at the input is a change of connector and termination resistor between 75 and 110 ohms.

Impedance Mismatches. A mismatched impedance (as well as circuit imbalance) will result from putting an RCA connector on one end of an XLR cable without changing the source or load resistors. However, at short cable lengths and lower sample rates, impedance mismatches and voltage variations are far less of a problem than is commonly thought. As long as the signal gets through with adequate voltage and few reflections (probably not a problem with a short cable), the receiver chip will decode it regardless of the impedance, albeit at the cost of jitter and possibly reduced noise margin. And noise-margin in the digital circuit does not affect sound quality unless the digital signal is so low the receiver drops out and loses sync. There are several proper-impedance methods of connecting a digital balanced source to unbalanced load or vice-versa, descriptions of which are beyond the scope of this book.[3]

Cable Lengths. The higher the sample rate, the shorter the tolerable cable length, because of the possibility of interfering reflections from the impedances and connectors at each end of the cable. The AES3 standard specifies usable lengths up to 100 Meters at 48 kHz, which is possible with careful termination and high-bandwidth, matched-impedance cable. However, 1/4 wavelength is the critical length where reflections can become their worst, so impedance and termination errors will be aggravated with cables that are close to about 20 meters, or 66 feet (48 kHz SR),[*] or 33 feet at (96 kHz SR). The critical length issue is one probable reason why standard-length mike cables make bad digital audio interfaces. Neither the XLR nor the RCA connector was designed with exacting impedance specifications, so avoid passive hardware patchbays, splices, and multiple intermediate connectors which will tend to exacerbate impedance problems.

Fortunately, cables do not have to be cut to the same length, since the AES11 standard permits a framing tolerance of 25%, and 25% of a 192kHz

---

[*] The 48 kHz AES/EBU interface has a prime bit frequency of 3.072 MHz, extending to about 6 MHz.

frame is 1.3us, the production of which would require a cable length difference of over 200 meters. In addition, should cable lengths or equipment delays exceed the 25% error, the only signal degradation would be to insert a delay of a sample to the signal (or more samples for much larger mismatches).[*]

Optical Cables and length. Obviously, electrical impedance is not a consideration when the interface is optical. The main concerns are bit integrity and jitter. And as we explained in Chapter 19, as long as the bit integrity is maintained, then jitter is only a consideration when delivering signal to a D/A converter or sync signal to an A/D. When using jitter-susceptible DACs try to avoid Toslink optical connections because their low bandwidth (3 MHz for the Sharp brand, up to 6 MHz for the Toshiba) exacerbates interface jitter. But the bit integrity is perfectly acceptable on a plastic Toslink interface as long as the lengths remain under 5 meters (some receiver models support up to 10 meters) beyond which there is unacceptable signal loss.

If you have to run long optical cable, a perfectly legitimate test for cleanliness of an optical interface is *margin distance before dropout*. While looking at the lock indicator on an AES receiver, or, simply listening to the audio, disconnect the cable from the input and slowly pull it outwards. The amount of distance you can pull the connector before losing lock is an indicator of the margin of sensitivity of the receiver and the strength of the signal at the

receiving end of the cable. If you cannot pull the cable out at least 1/8th inch, preferably 1/4" or more, then there is probably too much loss in your cable length, or the transmitter is weak, or the receiver insensitive. It is possible to get more output from a Toshiba transmitter by changing some resistors, which can then give up to 60 meter transmission, but then a short cable, which has less loss, will overload the receiver. Glass fiber has much less loss than plastic, and can transmit for thousands of feet; it also has superior bandwidth and therefore causes fewer interface jitter problems, jitter as low as any good copper connection. Glass fiber connections can have even lower interface jitter than unbalanced copper connections because they eliminate ground loops and EMI sensitivity. I've seen some manufacturers adapt Toslink connectors to glass fiber, but if you want dependable long-length optical transmission, the best solution is to change receivers and transmitters to glass-type, which are electrically compatible once you have converted from optical.

Fixing Interface Software Problems

If difficulties still remain after eliminating hardware problems, then software issues are obviously the cause, and, sadly, these are much less straightforward to pin down and eradicate. The DAT machine cited above probably failed to record because the data stream was copy-protected, or the machine was expecting a professional channel-status bit when the consumer bit was presented, or because a sample-rate flag was misrepresented. The same software considerations apply both to copper connections or optical, as it is possible to feed the

---

[*] Thanks to Julian Dunn (in correspondence) for clarifying the framing information.

consumer or professional bitstream down an optical cable, or multichannel protocols such as the multichannel MADI or Sony's DSD (multichannel protocols are beyond the scope of this book).

The **flags** are the road signs of the bitstream, officially known as **channel-status bits**. Over the years, the standard has been abused, evolved, multiply-interpreted, mutilated, or just plain forgotten about, like the "detour ahead" sign which some worker never put away after repairing the road. This may sound like heresy, but I think the current implementation of the standard is so poor that many times it would be best for all receivers and recorders to ignore the flags and ask for human help. For example, one common problem is a flag saying the sample rate is "unindicated," which stops some DATs from recording. Ironically a receiver can't read a flag unless it's already locked to the sample rate, so it must know what the rate is without the flag! Therefore, it's illogical for a machine to reject a digital audio signal because the sample rate is not indicated. And with the advent of dual-AES connections for double sample rates, each channel is at half the final rate, so the flag may be wrong anyway. The human being should be the traffic cop making the final decision, not the machine; thus the smartest DAWs make only certain assumptions about the bitstream, otherwise letting the user make adjustments from menus and checkboxes. In the case of the recalcitrant DAT machine, it may be necessary to insert a channel-status-bit analyzer and/or modifier, changing flags until the machine begins to record.[4]

The Critical Flags

- The status of a bit (flag) called the **PRO** bit distinguishes the consumer bitstream from the pro. However, the pro bitstream can run on consumer connectors and vice-versa, and it's done all the time. This includes the Toslink optical interface, XLR, RCA and BNC, any of which can be used to carry consumer or pro information (by de facto but not official standards). So, never assume that the bitstream matches the connector unless you have investigated the equipment manuals or menus, or measured the contents of the bitstream with a tester. Fortunately, the audio itself is in a common place in both PRO and CONSUMER bit streams, and with some care, the two bit streams can be somewhat interchanged.

- Although the interface can send full-bandwidth 2-channel PCM data, it also has been used to transmit coded (data-compressed) multi-channel data such as Dolby Digital and DTS. The **Normal-versus-Data** bit is used to define coded multichannel data, which cannot be read by an ordinary DAW or D/A. As a precaution, these decoders will produce no audio unless the PRO bit is set (even on an RCA connector). The main danger is that an ordinary D/A or digital recorder may ignore the data bit and send full-level noise over the loudspeakers. For this reason I usually turn the monitor gain down whenever beginning to monitor an unknown bitstream!

- The three **emphasis** bits in the professional stream partially overlap the **copy-prohibit** bit and the single emphasis bit in the consumer stream. Fortunately, professional DAWs ignore the copy prohibit bit and most recordings now are made without emphasis.[5] In general, the copy-prohibit bit and SCMS bits are only used when recording to consumer-grade CD and DAT Recorders. Future digital interfaces will encompass far more rigid copy-protection schemes, which will probably introduce further complications for audio professionals.

- The consumer bitstream can transmit **program IDs** (for automatic tracking in DATs and CD Recorders) but the pro bitstream has no such provision, except that some Sony machines will interpret program IDs on the pro interface.
- The consumer bitstream was originally designed to carry 20 audio bits, with the remaining 4 **auxiliary bits** available to carry a low-resolution auxiliary channel (e.g., talkback). But the consumer bitstream can utilize all 24 bits to carry up to 24-bit audio, and this has become de facto regardless of how the flags are set. Although the standards committees spent much time carefully revising the standard so the consumer bitstream could flag the use of those 4 bits, currently most transmitters and receivers ignore those flags, and most current receivers default to assume 24-bit audio. It's up to the user to take appropriate action; a bitscope (see chapters 2 and 16) may help sort out the issues. One D/A converter (Prism) and one console manufacturer (Yamaha) rigidly follow the standard and automatically truncate bits beyond 20 on the consumer connector; pro connectors must be used on those devices if you want to use all 24 audio bits.
- The esoteric consumer flags that govern **category code** practically affect two classes of recording

equipment: standalone 16-bit CDR and DAT recorders. If the source machine's category code is set to CD, the recorders interpret user bits as track IDs, and if this category code is wrong, the recorders may write undesired track IDs. One DAW (SADiE) is capable of sending DAT and CD track IDs, and thus it must be set to consumer status and CD category code and there must be no bitstream modifiers in the line between SADiE and the recorder in order for automatic tracking to work.

Two-wire 96K and 192K

Originally, the highest sample rate that could be carried on an AES/EBU interface was 48 kHz (slightly higher with varispeed). In order to double the sample rate with recorders that can only handle 48 kHz, a system was invented that places half the samples on one cable and half on another, each cable running at half the final rate. One cable carries all the left channel samples and the other right. If you plug one of these cables into a standard stereo DAC you will hear a mono signal that sounds a little strange since the timing between the two ears is incorrect. The main concern when using the two-wire method is to write impeccable documentation, since there is no standard flag to indicate the dual-wire method is being used, nor which channel is which. At the time of this writing, there is no official single-wire interface for 176 and 192 kHz SR, so at least 2 cables are needed, and often 4 for stereo; good luck to anyone who scrambles those cables!

Given all these violations and exceptions, it's amazing the AES/EBU standard works at all!

IV. How To Get Good Audio Extraction

I rarely recommend brands outright, but in this case I'll make an exception: Plextor. Plextor CD-ROM readers have been specifically designed for excellent audio.[6] Audio extraction from CD is not as easy as the computer industry has implied; it is not the same as reading data from CD ROM, which can be done at high speed. Most drives fail at this chore, and default to speeds which cause dropouts or glitches in the audio. In contrast, the Plextor drives have available a special protocol which will read and reread any portion of a disc, slowing down when needed, until they get a good read. The audio program or operating system must be designed to work with the Plextor, which needs proper handshaking in order to speed up and slow down. As of this writing, only certain programs on the PC provide this functionality, ironically, not on the Macintosh.

V. Compilation CDs/CD-On-Demand

Producing compilation CDs is a problem for the quality-conscious engineer. In an ideal world, the same mastering engineer who produced the original discs should produce the compilation, which helps ensure a unity of sound. In an ideal world, compilation CDs should be made from original or early-generation sources, not by copying from final masters or pressed CDs. For if a track's level (or EQ) needs to be adjusted, then the sound quality will deteriorate when going from 16-bit to 16-bit, especially when using a highly-processed 16-bit master as the new source. A final master represents the end of the line of a processing chain, including limiting, which cannot be reversed, only the level can be lowered (yet sound deteriorates because of additional DSP calculations and the accumulation of 16-bit dithers).

But in the real world, record companies usually don't want to pay to redo that which they've already amortized, plus, it's extremely difficult and expensive to acquire the source masters from many different places. Nevertheless, even in the real world, we can still make some decisions to maximize quality of the compilation CD. We try to produce bit-for-bit copies (clones) of as many cuts as possible, the ones which work together level- and sound-wise. For the same reason, if one of the cuts is out of line and much louder, we try to convince the record company not to take the least common denominator approach; that it is better to take the level of the loud cut down, which avoids using degrading processing on the majority of cuts.

The phrase **good-sounding CD-on-demand** is an oxymoron. As soon as users are given the ability to create their own CDs from previously-mastered product, change levels and then (hopefully) redither to 16-bit, the sound-quality will suffer. There is no shortcut nor substitute for a good mastering engineer working from early-generation (unmastered) sources.

VI. CD Text

Mastering engineer Jim Rusby, an expert on CD text, explains the process:

> **CD Text** is a facility that provides titles, authors, and even lyrics on the display of specially-equipped CD players. Be aware that most of the replicators (pressing plants) are ready for CD text, but many of the CD brokers are not, even if they say they are. There are two schemes for CD Text. One (and the most common) places the text in the lead-in area. The other extends it into the program zone; this scheme is also used for special applications like Karaoke.
>
> Sony has been freely distributing CD Text software at their Austria DADC web site. The mastering engineer organizes the text using the software and a set of binaries are generated that reside on a floppy, which is then sent along with the disc master to the broker or replicator. Fields are available for such things as ISRC, album name, etc. This scheme is used by many (other than Sony) - the Doug Carson system supports it, for example.
>
> There are two general sources of problems with the process:
>
> 1. Expectations. CD Text is only guaranteed to work on CD Text-enabled CD Players. Performance of CD Text on computers (e.g. Windows Media Player) will vary depending upon the drive, software, and phase of the moon. Some clients confuse CD Text with CDDB databases—these are servers that your computer logs into and gets info about the disc in your unit.
>
> 2. Product ID. In recent years a number of CD burners began supporting CD Text. Consequently, the client types in the information, burns a disc and sends it in for replication. The pressed discs come back with no text on them. Be sure to tell your broker or replicator that this is a CD Text title. Don't just send the disc in expecting all will be well. Most replicators have product codes. They will assign a piece number to your product that may indicate what's on the disc — and this code may tell the cutting system what information needs to be passed along. Many times there is bogus character information in the text fields of non-text titles. The replicator doesn't want to pass along information that isn't valid, so if they think it is just straight audio they will not activate the text feature. Some replicators may require that text be submitted using the floppy method.[*]

---

[*] Jim Rusby, on the Mastering Webboard.

VII. Why do many mastering engineers use unbalanced connections between analog gear?

My philosophy is: All other things being equal, unbalanced is better, which boils down to a *less is more* philosophy.

Here are the caveats: In a small room, where all the power is coming from a central source, and all the analog gear is plugged into that power and no analog audio enters or leaves the room, and you have your signal-to-noise and headroom issues all straightened out, then unbalanced is almost always better-sounding than balanced. Most balanced gear is created out of unbalanced internal connections by adding additional stages of amplification, which often creates a loss of transparency; however, it's important to study the schematics and determine if this is the case. In those cases, I may remove the extra stages, also being aware of the internal gain structure, headroom, and driving capacity of the internal parts, which are going to be exposed to the outside world.

Exceptions: a) Equipment whose balanced stages are so-well-designed that it is impossible to design the same piece of gear with fewer stages unbalanced than the balanced version. b) Equipment which uses balanced topology throughout, with impeccably-designed internal components in a mirror-image configuration. But I'm not so sure it sounds better because it's balanced or just because it's better!

VIII. Analog tape simulation in the mixing process?

I am concerned about recommending the use of analog tape simulators in the mixing process unless you have world-class monitoring which can tell you unequivocally when (if) you've gone "too far." There's nothing worse than the sound of oversaturated analog tape; turn the drive knob on the simulator one step too far and the sound will turn from "good" distortion to "bad." Once any damage has been done, it cannot be undone without a remix and it's a lot easier to do alternate mixes at the time of the first one! Furthermore, I've found that after good mastering, a little bit of analog tape simulation is enough; so using such a device in the mixing chain can be a problem, because it's not possible to anticipate its interaction with the mastering processors. As usual I recommend that mix engineers send two versions of a mix to the mastering house, one with and one without processing. This applies to any processor(s) on the mix bus, unless the processor is so adjusted that removing it would seriously alter the intent of the mix.

Speaking of Flux

For those trying to get *that sound* with analog tape, personally, I have found that analog tapes sound too *saturated, undefined, and muddy* at +9, for 9 out of 10 projects in my experience. To be more explicit, reduce the level till 0 VU = + 6 dB over 200 nW/M (known colloquially as *+6)*, which is the same as 0 VU = +4 dB over 250 nW/M, is the practical limit for GP9, the hottest tape made by Ampex. It is better, in my opinion, to run at +6 or lower and use the extra as headroom, especially when using VU meters. (See Appendix 5)

IX. ISRC Codes and UPC/EAN

The **UPC/EAN** code is also called Mode 2 data and is a barcode that contains information about the product. Most times the Mode 2 data is added at the plant, but DAWs include a space for that data to be added by the mastering engineer. The **International Standard Recording Code**, defined by the RIAA is a unique code for each track on the CD. This allows use of automated logging systems to be used at radio stations to track copyright ownership/royalties. The system is very popular in Europe and slowly gaining acceptance in the U.S. The record label provides the codes to be entered for each track.

ISRC contains exactly 12 digits; only the digits without any dashes should be entered in the DAW. In the ISRC code: ES-BO1-01-10503, the first two digits are the country code (in this case, ES for España), the next three digits are the code for the original issuing record label, which owns the rights. The next two digits are the year the song was recorded, and the last five are recording codes designated to the version of the song itself. That is, Elton John's version of *Your Song* will have a different ISRC code from any cover of the same song.

X. What's special about the PMCD?

The term **PMCD** was invented by Sonic Solutions as a method of allowing glass masters to cut directly from CDRs. However, as of this date I doubt there are any plants which continue to cut glass masters using the PMCD method, since Doug Carson systems introduced a different system which allows glass masters to be cut from any standard pressing or CDR. So there's nothing special anymore about PMCD and most mastering engineers who may write "PMCD" on the label are probably creating standard orange-book CDRs.

XI. The writable DVD confusion

These are relatively new media and there is much confusion over their capabilities and distinctions, and the "standards" are in a state of flux. I advise clients to send mix files on CD-ROM even though DVD could save a few discs, because CD ROMs are still the most compatible with typical readers.

DVD-RAM is a rewritable medium, claiming 100,000 re-write cycles, but most existing DVD players cannot read DVD-RAM discs. DVD-RW can be read on more players, and the technology is limited to 1000 re-write cycles. Look for a player labeled *RW-compatible*. The recorders for each format require specific blanks, which are not interchangeable. DVD-R can be played on most set top DVD players, yet have difficulty with older computer drives. DVD-R can only be written once, not a problem since the costs of blanks have become affordable.

XII. Mastering for Vinyl or Cassette

The full considerations required for vinyl and cassette mastering require more space than is available in this book. These days, most mastering engineers do not have a cutting lathe, and should let the experts do the final processing for vinyl, which usually includes narrowing the separation at the bass end to protect the groove excursion, and some high-frequency limiting to protect the cutterhead. The LP cutting engineer will also determine the level of the record; there is nothing a mastering engineer making a DAT or CDR can do about the absolute level of the final vinyl. When making masters for vinyl, the one thing to be concerned about is **duration,** especially when there is a lot of bass on the record. A ten-minute side is usually no problem when there is heavy bass. It's technically possible to put a half an hour on an LP side, but almost inevitably with loss of level, stereo separation and/or bass.

The cassette replication house may not have a skilled mastering engineer, so the original digital engineer should make a special premaster for cassette, following the processing and level guidelines in Chapter 15. Try to make Side A the longer side, otherwise in the car there will be an irritating pause in the music at the end of side A.

XIII. Low Level vs. High Level Hard Disc Formatting

Most operating systems and disc utilities provide an option to **format** a hard disc, but the engineer should be aware that there are two different degrees of formatting: low level and high level. High level formatting is the most common type. **High level formatting** installs the operating file system and a new directory, e.g., Mac HFS, or FAT 32, and is the most reliable way to initialize (remove and erase) the directories on a disc. It should take only a couple of minutes to high-level format a hard disc of any size. Note that high level formatting does not erase the whole disc; your old files are probably still there and a clever thief can find traces of them even though the old directory is gone, as long as the old files have not been written over.

Low-level formatting completely erases a hard disc, and thus may take from several minutes to several hours depending on the size of the disc. Low level formatting reinitializes the sectors and compensates for physical changes in the disc as it ages, and it also maps out bad sectors that have errors. It's a good way to check out any suspect hard disc, and a good thing to do to rejuvenate a drive that's a couple of years old and in apparently good shape. Read the error reports afterwards to see how many sectors or blocks were mapped out, for anything more than, say, 5 to 10, indicates the disc is on its way to IBM heaven.

XIV. Digital Monitor Controls vs. Analog

As we learned in Chapter 16, some analog systems perform better than digital, and vice versa. Digital monitor level controls used to sound quite poor, but a few well-designed systems sound as transparent as their analog counterparts, provided that we use low amounts of attenuation. A well-designed high resolution digital level control correlated with RP 200 gains (see Chapter 14) will not require much attenuation to produce a proper loudness. Prior to purchase, test the proposed system's distortion using an FFT and also by careful listening comparisons to an analog-based system, at equal loudness. The same goes for D/A converters with built-in digital monitor level controls; some sound extremely transparent, and others quite grainy due to quantization distortion.

1 The SMPTE is developing a universal standard, to replace video reference. This proposed standard defines a format for transmitting a universal date and time reference, called the Absolute Time Reference (ATR) for the purposes of distributing synchronization information and for the distribution of time. This is under the jurisdiction of the SMPTE group as well as the AES working group SC-02-05.

2 S/PDIF stands for *Sony Philips Digital Interface*, which grew up into the IEC60958 standard, which supercedes IEC 958. Officially, type 1 is consumer with the consumer bitstream (protocol) on unbalanced RCA or Toslink optical connectors. Type 2 is professional, with the professional bitstream over XLR balanced connectors. There is also the AES-3ID standard, which transmits the professional bitstream over a 75-ohm BNC connector at 1 volt p-p. However, as this Chapter points out, the devil is in the details.

3 An internet search for IEC 958 yielded this resourceful URL: http://www.epanorama.net/documents/audio/spdif.html, which includes some balancing and unbalancing circuits. However, most of the time, I recommend using an official RS-422 receiver/transmitter chip as the common-mode rejection will be superior.

4 Digital Domain manufactures a simple channel-status bit modifier/analyser known as the FCN-1. More sophisticated analysers can be obtained from Audio Precision, Neutrik, Prism, Audio Digital Technology (ADT) and others.

5 **Emphasis**, also known as **preemphasis**, is an equalization curve. If emphasis is off, then the recording and playback are both made flat. If emphasis is on, then the recording has a specified high-frequency boost and the playback a corresponding high-frequency cut, intended to improve signal to noise ratio. However, since the SNR of flat 16-bit is more than adequate, and since headroom is reduced when emphasis is used, the practice has been pretty much abandoned. Furthermore, the flag which pertains to this has been abused by the pro-consumer conflict and has fallen into disfavor. If you suspect a recording has been made with emphasis, it is advisable to re-equalize it (roll off the highs).

6 We all owe a great debt to mastering engineer Glenn Meadows for having worked with Plextor to produce drives which meet the needs of audio engineers, and informing the audio community of their performance.

"WE'LL FIX IT IN THE SHRINKWRAP."

—FRANK ZAPPA

"EVERY DAY
IN EVERY WAY
I'M GETTING
BETTER AND BETTER"

—BOB'S MOM

CHAPTER 21

Education, Education, Education

What Have We Learned?

As we reach the end of this book, it has become clear that the craft of **Mastering** requires tremendous attention to detail, technical and musical knowledge, plus the ability to get along well with a wide range of people from artists to record company executives. In other words, *able to leap over tall buildings in a single bound*. But since Superman is not available, humans have to substitute, and all we can do is try to reach an ideal. Nobody's perfect—we make mistakes all the time, the trick is to get to correct them before the product goes out. That's what a system of **quality control** is about, reducing the level of mistakes until they're *below the radar*. All we can do is try to measure ourselves against the tough words my Mother taught me: "Every day in every way I'm getting better and better."

Another area we've stressed is that **good-quality mastering** requires a dedicated room with refined acoustics and accurate reproduction. But with good audio equipment and a talented engineer, a typical project studio can produce a good-sounding master, although with noisy fans, low-resolution monitors, interfering console and rack surfaces, the work involves a time-consuming, trial-and-error process. Check the material in as many alternate environments as possible. Project studios wishing to do mastering ought to construct a dedicated room for that purpose and hire an engineer inclined to the skills of mastering. However, if for economic reasons you must master your own mix,[*] and in a less-than-ideal environment, then use as much of this book's advice as possible. Also, master with the aid and

---

[*] Perhaps misguided, since cutting corners at the last stage before producing a record may prove very costly in the long run.

perspective of an experienced producer present, or an objective professional whose ears you trust. Another person's opinion will ensure that you aren't so close to the material that you're missing something essential, especially if you are the artist. For example, if you know the lyrics by heart, then you are probably the wrong person to judge the vocal level! This is why mastering engineers avoid mastering their own mixes; I try to go to another engineer to master work that I have mixed—to get their valuable perspective. Mastering your own album is like marrying your first cousin. You never know how the children will turn out, or maybe you do![*]

> Without collaboration the music is not being given its full potential. There is a reason that you have the talent, the engineer and the producer because each one can worry about their own *thing* and they can collaborate on the final outcome. When music is done in a *virtual vacuum* it does not sound as good.[†]

The Cure for the Ear

Our critical listening audience is diminishing, because the average hearing acuity of the modern listener has been getting worse, decade by decade! Living and working in the city causes a threshold shift in our hearing sensitivity, and exposure to high-level, distorted music in clubs can cause permanent hearing damage. The ear contains tiny, delicate parts which can take only so much battering

before they give up. Club owners should be required to pass out ear plugs to customers walking in the door, because alcohol dulls all our senses and we don't notice that our ears are being bombarded. It's the physical equivalent of sticking thousands of needles in our arms and legs all night, but ignoring the pain! Our job as audio professionals is to educate our audience to these very real dangers.

When clients are going to be driving or flying a long distance to the mastering session, I advise them to wear ear plugs, or any ear protection—cotton or tissue is better than nothing. This greatly reduces the fatigue of traveling. I suggest they travel the night before and get a good night's sleep locally before the session, which reduces their threshold shift, and improves their perception during the mastering.

The Cure For Our Art: Think Long-Term

We need to educate record companies that a singles-oriented approach is self-defeating. It looks good for the quarter's bottom line, but leaves no equity for the future. Instead, they should cultivate artists who have staying power, and long-lasting value.

The Cure For Stress: Dynamic Range

Not every recording benefits from having dynamic range, but I feel that recent trends in pop music recording have taken the fatigue of *slamming it against the wall all the time* to an extreme. So I'd like to briefly discuss the phenomenon and ways in which we can educate people to see just what they have been missing. These days, audio and visual media are perceived as advertising, continually trying to get our attention. This bombardment is

---

[*] One mastering engineer likens mastering your own mix to giving yourself your own haircut!

[†] Tom Bethel, from the Mastering Engineer's Webboard.

very stressful, and because of that, we tune it out, turning it into audio wallpaper, just noise to us. While not an advertisement, a club where records are spun is singles-oriented, and in that context, relentless, rhythmic sound may work; the dance exercise is also stress-relieving and very exciting—though I don't know how single people can meet if they can't hear each other over the music! But beyond the singles and party environments, a musical record album is not an advertisement for itself, it's (hopefully) a work of art. Fortunately there's a large crossover where music which is suitable for dancing also makes an enjoyable *sit-down* listening experience. But what enriches the *sit-down* experience is a well-programmed album with artfully-used dynamic range, fast and slow numbers, loud and soft pieces, which exercises our senses and may relieve stress more than relentless banging for an hour.

The problem is that dynamic range in pop music has become an increasingly rare phenomenon, due to the fruitless volume wars and pressure from A&R to make a record that can get through the noise-to-signal ratio of restaurants, record-store kiosks, car-play, etc. Years ago, music in cars was heard only from the already-compressed radio, and at home we listened to *record albums*. But today's public listens to CDs in the noisy car, and at home does more casual and background listening than before, so the number of critical listeners is a smaller part of the total audience. Some of the public has gotten lazy and expects their CD changer to perform like a radio, keeping a constant loudness with each CD. This makes some anxious record

producers ask for compression to the extent where sonic quality is damaged. Ironically, new sound palettes (such as **shred**) have been discovered out of the distorted processes we use to make things louder, but let's hope not all music has to go this way! The answer is **EDUCATION**…

We need to educate producers that fatiguing, hypercompressed CDs will not be auditioned more than once—the record loses critical word-of-mouth advertising. Teach them that a decent amount of dynamic range helps make an album more enjoyable, lively, even clearer in most cases, and that sound quality suffers as the average level goes up. Teach them that hypercompression is incompatible with radio and lossy (MP3) encoding (see Chapter 10, Appendix 1). Ironically, the recordings which sound **loudest** and most impacting on the radio are usually the ones which have the **lowest** absolute CD loudness.

We need to educate the public that it is normal to adjust the volume control from CD to CD, and to turn it up and down in a noisy car. Teach them to use the compressor button if they're annoyed by riding levels. Unfortunately, the disappearance of the audio cassette has removed our opportunity to release in two formats, one with reduced dynamic range; all the new formats have tremendous dynamic range capability. So now we have to turn to solutions in the consumer equipment, including metadata (Chapter 15), if it ever catches on.

We need to educate car audio equipment manufacturers that recordings are and should be variable in their levels, so a compressor should be an essential part of every noisy car's system. Some sophisticated cars have automatic level controls tied with the speed and ambient level, which is a tremendous engineering advance. When most cars have this equipment, there will be less producer demand to overcompress material.

We need to educate home audio equipment manufacturers that all CD and DVD changers should have compressor buttons. Call it the *party button*. As we move into media that accept metadata, such as DVD and DVD-A, manufacturers should include ergonomic controls that look at dialnorm levels (see Chapter 15), permitting casual listeners to switch media without riding the volume control.

We need to educate new mastering engineers by teaching them to study the great-sounding pop recordings of yesteryear. Many of today's hypercompressed recordings sound worse than 60's and 70's analog recordings and have much less dynamic range. Yet the older pop recordings play well everywhere, again illustrating that hypercompression is unnecessary.

The Cure for Hypercompression: How Loud Should I Make It?

Not all producers and engineers will master the concepts of the K-System (Chapter 15), but I ask mastering engineers and producers to please consider *How Loud Should I Make It?* An acceptable answer could be: *Turn it up until it sounds bad and then back it off by several dB.*

Actually, the best way to "win" the loudness race is to be far from first place. Be prepared to be at least 3 dB lower than "the winner" if you want your record to even sound acceptable! And considerably lower if you are looking for an *open, clear, dimensional* sound.

I have never lost a job by suggesting to a producer that I have already mastered a record as hot as it should be; most producers appreciate the advice of an experienced professional. If he prefers differently, then of course I turn it up, for the customer is always right. But mastering engineers should gain the producer's confidence; it's often useful to demonstrate the sonic deterioration if a recording is turned up any further. Then on the next record you do together, he will (hopefully) accept your word that the record is mastered as hot as it should be. This bit of education and effort is one sure way that we can combat the sound-ruining loudness war.

> { *"Be prepared to be at least 3 dB lower than the current "winner" if you want your record to even sound acceptable!"* }

CHAPTER 22
At Last

At last we come to the end. But this is not the end, it is a new beginning, brought to life in this poem on the next two pages. My hope is that we will all work together to make this vision come to pass.

A Wonderful Musical World

I see:
talented artists making the hits,
30-piece big bands thrilling the kids;
A&R folks nourishing art,
and dynamic leaders lighting the spark.

I see:
popular music that dares to be original,
records which are highly prized;
artists who care to be experimental,
and sound that is not synthesized.

I see:
song lyrics that come direct from the heart,
with superb sonic quality, that's the art;
lively music which leaps off the shelves,
and musicians interacting, that's ourselves!

I see:
a world which recognizes craft and training
in audio itself which is not disdaining;
where overall excellence is what we seek
and art comes from long-worked technique.

I hear:
a *vibraphone* that's not from a patch,
playing *ensemble* with baritone sax;
varieties of classical and jazz,
exciting us all with their pizzazz.

I hear:
live music everywhere
and only background sounds compressed;
tonally aware people
make us truly blessed.

I hear:
recordings that inspire,
new music that I admire;
radio that's innovative,
forms of music so creative.

I hear:
full-length albums in my own abode
and only catch singles when on the road;
There's music too in all public schools
nurturing nature's future jewels.

We will all work toward this change:
let's put artists back in charge;
a cordial composition arranged
by creative minds at large.

Bob Katz, January 2002

> "
> Never doubt that a
> small group of thoughtful,
> committed citizens can
> change the world.
> Indeed, it is the only thing
> that ever has.
> "
>
> —Margaret Mead

About the Margaret Mead quotation on the previous page, which signals the last chapter of this book: The Institute for Intercultural Studies (www.mead2001.org) has been unable to locate exactly when and where this famous admonition was first cited. They believe it probably came into circulation through a newspaper report of something said spontaneously and informally. They know that it was firmly rooted in Mead's professional work and that it reflected a conviction that she expressed often, in different contexts and phrasings.

[Appendix 1]

Radio Ready: The Truth

I. Introduction

Radio, like all technology, is constantly changing. Digital radio will eventually change the way that our records sound, and we now have to contend with low-fidelity Internet radio. But for the immediate future, most of our recordings will be reproduced on standard analog FM radio. Have you ever wondered what happens to your recording when it is played on the radio? Ever wondered how to get the most out of radio play? I am pleased to introduce the guest authors who have largely written this section — Bob Orban and Frank Foti.* Both of them are considered to be the world's authority on radio processing. Bob is the engineer and designer of the Optimod line of audio processors, while Frank, who has an extensive radio engineering background, is the creator and lead designer of the Omnia product line. Together, their products are used by nearly every radio station around the world.

In 2000, participants in the Mastering Webboard engaged in a friendly collaboration to find out what range of levels we are using. Engineer Tardon Feathered of San Francisco put a rock and roll mix on his FTP site,

which was then downloaded by a great number of mastering engineers, mastered, and uploaded back to Tardon, who then made a two-CD collection called "What Is Hot?" The absolute loudness of the cuts on this compilation ranges from extremely hot and highly distorted (monitor turned down to about −14 ref. RP 200) to very light (monitor position about −5), a loudness difference of 9 dB!

After "What is Hot?" came out, the Webboard participants felt that it would be important to demonstrate what happens to these cuts when passed through radio processing. Enter Bob Orban, who volunteered to process the music with typical radio station presets. Tardon then produced a compilation CD comparing the songs before and after radio processing.

This next figure, courtesy of Tardon, is a comparison of several sample mastered cuts before and after Orban processing.

At top, five mastered cuts of the same music, with increasing loudness and visual density. At bottom, the same cuts passed through the Orban radio processor.[1]

* Robert Orban, Orban Inc. (A CRL Company). Frank Foti, Omnia Audio

Notice that regardless of the original level, after radio processing every source cut ends up with similar apparent density: soft passages are raised radically, and loud passages slammed to a maximum limit. I've auditioned this revealing comparison CD: Every track ends up at the same loudness, proving beyond a shadow of a doubt that there is no advantage to extreme compression in mastering when a cut ends up on the radio. I also observed that the radio processing severely distorted just about all the originals, except for the softest track, which came in at about a K-14. The rightmost and most squashed source track was unlistenable after it was processed through the Orban. The radio processing also somewhat randomizes the stereo image and lowers the high end, but listening revealed that adding severe highs in mastering only aggravated the distortion; it did not help the clarity of the final product. Let's hear what Bob Orban and Frank Foti have to say about *what's inside the box...*

II. What Happens to My Recording When it's Played on the Radio? by Robert Orban and Frank Foti[*]

Few people in the record industry really know how a radio station processes their material before it hits the FM airwaves. This article's purpose is to remove the many myths and misconceptions surrounding this arcane art.

Every radio station uses a transmission audio processor in front of its transmitter. The processor's most important function is to control the peak modulation of the transmitter to the legal requirements of the regulatory body in each station's nation. However, very few stations use a simple peak limiter for this function. Instead, they use more complex audio chains. These can accurately constrain peak modulation while significantly decreasing the peak-to-average ratio of the audio. This makes the station sound louder within the allowable peak modulation.

Garbage In—Garbage Out

Manufacturers have tuned broadcast processors to process the clean, dynamic program material that the recording industry has typically released throughout its history. (The only significant exception that comes to mind is 45-rpm singles, which often were overtly distorted.) Because these processors have to process speech, commercials, and oldies in addition to current material, they can't be tuned exclusively for "hypercompressed," distorted CDs. Indeed, experience has shown that there's no way to tune them successfully for material which has arrived so degraded.

For 20 years, broadcast processor designers have known that achieving highest loudness consistent with maximum punch and cleanliness requires extremely clean source material. For more than 20 years, Orban has published application notes to help broadcast engineers clean up their signal paths. These notes emphasize that any clipping in the path before the processor will cause subtle degradation that the processor will often exaggerate severely. The notes promote adequate headroom and low distortion amplification to

---

[*] Edited and adapted from a 2001 AES presentation.

prevent clipping even when an operator drives the meters into the red.

About 1997, we started to notice CDs arriving at radio stations that had been pre-distorted in production or mastering to increase their loudness. For the first time, we started seeing frequently recurring flat topping caused by brute-force clipping in the production process. Broadcast processors react to pre-distorted CDs exactly the same way as they have reacted to accidentally clipped material for more than 20 years—they exaggerate the distortion. Because of phase rotation, the source clipping never increases on-air loudness—it just adds grunge.

The authors understand the reasoning behind the CD loudness wars. Just as radio stations wish to offer the loudest signal on the dial, it is evident that recording artists, producers, and even some record labels want to have a loud product that stands out against its competition in a CD changer or a music store's listening station.

In radio broadcasting this competition has existed since about 1975,[2] when radio stations used simple clipping to get louder, and this technique has now migrated to the music industry. The figure at right shows a section of a severely clipped waveform from a contemporary CD.

The area marked between the two pointers highlights the clipped portion. This is one of the roots of the problem as described in this paper; the other is excessive digital limiting that does not necessarily cause flat-topping, but still removes transient punch and impact from the sound.

The problem today is that we now have sophisticated and powerful audio processing for the broadcast transmission system and this processing does not coexist well with a signal that has already been severely clipped. Unfortunately, with current pop CDs, the example shown above is more the norm than the exception.

The attack and release characteristics of broadcast multiband compression were tuned to sound natural with source material having short-term peak-to-average ratios typical of vinyl or pre-1990 CDs. Excessive digital limiting of the source material radically reduces this short-term peak-to-average ratio and presents the broadcast processor with a new, synthetic type of source that the broadcast processor handles less gracefully and naturally than it handles older material. Instead of being punchy, the on-air sound produced from these hypercompressed sources is small and flat, without the dynamic contours that give music its dramatic impact. The on-air sound resembles musical wallpaper and makes the listener want to turn down the volume control to background levels.

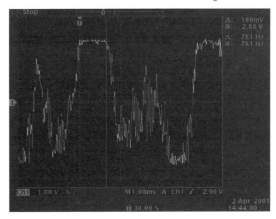

A severely-clipped waveform from a contemporary CD.

There is a myth that broadcast processing will affect hypercompressed material less than it will more naturally produced material. This is true in only one aspect—if there is no long-term dynamic range coming in, then the broadcast processor's AGC* will not further reduce it. However, the broadcast processor will still operate on the short-term envelopes of hypercompressed material and will further reduce the peak-to-average ratio, degrading the sound even more.

Hypercompressed material does not sound louder on the air. It sounds more distorted, making the radio sound broken in extreme cases. It sounds small, busy, and flat. It does not feel good to the listener when turned up, so he or she hears it as background music. Hypercompression, when combined with "major-market" levels of broadcast processing, sucks the drama and life from music. In more extreme cases, it sounds overtly distorted and is likely to cause tune-outs by adults, particularly women.

A Typical Processing Chain—What Really Goes On When Your Recording is Broadcast

A typical chain consists of the following elements, in the order that they appear in the chain:

Phase rotator. The phase rotator is a chain of allpass filters (typically four poles, all at 200Hz) whose group delay is very non-constant as a function of frequency. Many voice waveforms (particularly male voices) exhibit as much as 6dB asymmetry. The phase rotator makes voice waveforms more symmetrical and can sometimes reduce the peak-

to-average ratio of voice by 3-4dB. Because this processing is linear (it adds no new frequencies to the spectrum, so it doesn't sound raspy or fuzzy) it's the closest thing to a "free lunch" that one gets in the world of transmission processing.

There are a few prices to pay. In the good old days when source material wasn't grossly clipped, the main price was a very subtle reduction in transparency and definition in music. This was widely accepted as a valid trade-off to achieve greatly reduced speech distortion, because the phase rotator's effects on music are unlikely to be heard on typical consumer radios, like car radios, boomboxes, "Walkman"-style portables, and table radios.

However, with the rise of the clipped CD, things have changed. The phase rotator radically changes the shape of its input waveform without changing its frequency balance: If you measured the frequency response of the phase rotator, it would measure "flat" unless you also measured phase response, in which case you would say that the "magnitude response" was flat and the phase response was highly non-linear with frequency. The practical effect of this non-linear phase response is that flat tops in the original signal can end up *anywhere* in the waveform after processing. It's common to see them go right through a zero crossing. They end up looking like little smooth sections of the waveform where all the detail is missing—a bit like a scar from a severe burn. This is an apt metaphor for their audible effect, because they no longer help reduce the peak-to-average ratio of the waveform. Instead, their only effect is to add unnecessary grungy distortion.

---

* Automatic Gain Control. A type of compression that brings up low-level passages. See Chapter 11.

There has been a myth in the recording world that broadcast processing will modify these clipped, over-compressed CDs less than it will modify clean, dynamic CDs. Thanks in part to phase rotation, this contention is absolutely false. In particular, any clipping in the source material causes nothing but added distortion without increasing on-air loudness at all.

AGC. The next stage is usually an average-responding AGC. By recording studio standards, this AGC is required to operate over a very wide dynamic range—typically in the range of 25dB. Its function is to compensate for operator errors (in live production environments) and for varying average levels (in automated environments). Average levels vary mainly because the peak to average ratio of CDs themselves has varied so much from about 1990 on. Therefore, normalizing hard disk recordings (to use all available headroom) has the undesirable side effect of causing gross variations in average levels. Indeed, 1:1 transfers (which are also common) will also exhibit this variation, which can be as large as 15dB!*

The price to be paid is simple: the AGC will eliminate long-term dynamics in your recording. Virtually all radio station program directors want their stations to stay loud always, eliminating the risk that someone tuning the radio to their station will either miss the station completely or will think that it's weak and can't be received satisfactorily. Radio people often call this effect "dropping off the dial."

AGCs can be either single-band or multiband. If they are multiband, it's rare to use more than

two bands because AGCs operate slowly, so "spectral gain intermodulation" (such as bass' pumping the midrange) is not as big a potential problem as it is for later compression stages, which operate more quickly.

AGCs are always gated in competent processors. This means that their gain essentially freezes if the input drops below a preset threshold, preventing noise suck-up despite the large amount of gain reduction.

Stereo Enhancement. Not all processors implement stereo enhancement, and those that do may implement it somewhere other than after the AGC. (In fact, stand-alone stereo enhancers are often placed in the program line in front of the transmission processor.)

The common purpose of stereo enhancement is to make the signal stand out dramatically when the car radio listener punches the tuning button. It's a technique to make the sound bigger and more dramatic. Overdone, it can remix the recording. Assuming that stereo reverb, with considerable L–R energy, was used in the original mix, stereo enhancement, for example, can change the amount of reverb applied to a center-channel vocalist. The moral? When mixing for broadcast, err on the "dry" side, because some stations' processors will bring the reverb more to the foreground.[3]

Because each manufacturer uses a different technique for stereo enhancement, it's impossible to generalize about it. The only universal constraints are the need for strict mono compatibility (because FM radio is frequently received in mono, even on

---

* No wonder CD changers are a predicament. See Chapter 15 [BK].

"stereo" radios, due to signal-quality-triggered mono blend circuitry), and the requirement that the stereo difference signal (L–R) not be enhanced excessively. Excessive enhancement always increases multipath distortion (because the part of the FM stereo signal that carries the L–R information is more vulnerable to multipath). Excessive enhancement will also reduce the loudness of the transmission (because of the "interleaving" properties of the FM stereo composite waveform, which we won't further discuss).

These constraints mean that recording-studio-style stereo enhancement is often incompatible with FM broadcast, particularly if it significantly increases average L–R levels. In the days of vinyl, a similar constraint existed because of the need to prevent the cutter head from lifting off the lacquer, but with CDs, this constraint no longer exists. Nevertheless, any mix intended for airplay will yield the lowest distortion and highest loudness at the receiver if its L–R/L+R ratio is low. Ironically, mono is loudest and cleanest!

Equalization. Equalization may be as simple as a fixed-frequency bass boost, or as complex as a multi-stage parametric equalizer. EQ has two purposes in a broadcast processor. The first is to establish a signature for a given radio station that brands the station by creating a "house sound." The second purpose is to compensate for the frequency contouring caused by the subsequent multiband dynamics processing and high frequency limiting. These may create an overall spectral coloration that can be corrected or augmented by carefully chosen fixed EQ before the multiband dynamics stages.

Multiband Compression and Limiting. Depending on the manufacturer, this may occur in one or two stages. If it occurs in two stages, the multiband compressor and limiter can have different crossovers and even different numbers of bands. If it occurs in one stage, the compressor and limiter functions can "talk" to each other, optimizing their interaction. Both design approaches can yield good sound and each has its own set of tradeoffs.

Usually using anywhere between four and six bands, the multiband compressor/limiter reduces dynamic range and increases audio density to achieve competitive loudness and dial impact. It's common for each band to be gated at low levels to prevent noise rush-up, and manufacturers often have proprietary algorithms for doing this while minimizing the audible side effects of the gating.

An advanced processor may have dozens of setup controls to tune just the multiband compressor/limiter. Drive and output gain controls for the various compressors, attack and release time controls, thresholds, and sometimes crossover frequencies are adjustable, depending on the processor design. Each of these controls has its own effect on the sound, and an operator needs extensive experience if he or she is to tune a broadcast multiband compressor so that it sounds good on a wide variety of program material without constant readjustment. Unlike mastering in the record industry, in broadcast there's no mastering engineer available to optimize the processing for each new source!

Pre-Emphasis and HF Limiting. FM radio is pre-emphasized at 50 microseconds or 75 microseconds, depending on the country in which the transmission occurs. Pre-emphasis is a 6dB/octave high frequency boost that's 3 dB up at 2.1 kHz (75µs) or 3.2 kHz (50µs). With 75µs pre-emphasis, 15 kHz is up 17dB!

Depending on the processor's manufacturer, pre-emphasis may be applied before or after the multiband compressor/limiter. The important thing for mixers and mastering engineers to understand is that putting lots of energy above 5 kHz creates significant problems for any broadcast processor because the pre-emphasis will greatly increase this energy. To prevent loudness loss, the processor applies high frequency limiting to these boosted high frequencies. HF limiting may cause the sound to become dull, distorted, or both, in various combinations. One of the most important differences between competing processors is how effectively a given processor performs HF limiting to minimize audible side effects. In state-of-the-art processors, HF limiting is usually performed partially by HF gain reduction and partially by distortion-cancelled clipping.

Clipping. In most processors, the clipping stage is the primary means of peak limiting. It's crucial to broadcast processor performance. Because of the FM pre-emphasis, simple clipping doesn't work well at all. It produces difference-frequency IM distortion, which the de-emphasis in the radio then exaggerates. (The de-emphasis is flat below 2-3 kHz, but rolls off at 6dB/octave thereafter, effectively exaggerating energy below 2-3 kHz.) The result is particularly offensive on cymbals and sibilance ("essses" become "efffs").

In the late seventies, one of the authors of this article (R.O.) invented distortion-cancelled clipping. This manipulates the distortion spectrum added by the clipper's action. In FM, it typically removes the clipper-induced distortion below 2 kHz (the flat part of the receiver's frequency response). This typically adds about 1dB to the peak level emerging from the clipper, but, in exchange, allows the clipper to be driven much harder than would otherwise be possible.

Provided that it doesn't introduce audibly offensive distortion, distortion-cancelled clipping is a very effective means of peak limiting because it affects only the peaks that actually exceed the clipping threshold and not surrounding material. Accordingly, clipping does not cause pumping, which gain reduction can do, particularly when gain reduction operates on pre-emphasized material. Clipping also causes minimal HF loss by comparison to HF limiting that uses gain reduction. For these reasons, most FM broadcast processors use the maximum practical amount of clipping that's consistent with acceptably low audible distortion.

Real-world clipping systems can get very complicated because of the requirement to strictly band-limit the clipped signal to less than 19 kHz despite the harmonics that clipping adds to the signal. (Bandlimiting prevents aliasing between the stereo main and subchannel, protects subcarriers located above 55 kHz in the FM stereo composite baseband, and protects the stereo pilot tone at 19 kHz). Linearly filtering the clipped signal to remove energy above 15 kHz causes large overshoots (up to 6dB in worst case) because of a combination of spectral truncation and time dispersion in the filter. Even a phase-linear lowpass filter (practical only in DSP realizations) causes up to 2dB overshoot. Therefore, state-of-the-art processors use complex overshoot compensation schemes to reduce peaks without significantly adding out-of-band spectrum.

Some chains also apply composite clipping or limiting to the output of the stereo encoder, which encodes the left and right channels into the multiplex signal that drives the transmitter. It's actually the peak level of this signal that government broadcasting authorities regulate. Composite clipping or limiting has long been a controversial technique, but the latest generation of composite clippers or limiters has greatly reduced interference problems characteristic of earlier technology.

Conclusions

Broadcast processing is complex and sophisticated, and was tuned for the recordings produced using practices typical of the recording industry during almost all of its history. In this historical context, hypercompression is a short-term anomaly and does not coexist well with the "competitive" processing that most pop-music radio stations use. We therefore recommend that record companies provide broadcasters with radio mixes. These can have all of the equalization, slow compression, and other effects that producers and mastering ngineers use artistically to achieve a desired "sound." **What these radio mixes should not have is fast digital limiting and clipping. Leave the short-term envelopes unsquashed. Let the broadcast processor do its work. The result will be just as loud on-air as hypercompressed material, but will have far more punch, clarity, and life.**

A second recommendation to the record industry is to employ studio or mastering processing that provides the desired sonic effect, but without the undesired extreme distortion from clipping. The alternative to brute-force clipping is digital look-ahead limiting, which is already widely available to the recording industry from a number of different manufacturers (including the authors' companies). This processing creates lower modulation distortion and avoids blatant flat-topping of waveforms, so is substantially more compatible with broadcast processing. Nevertheless, even digital limiting can have a deleterious effect on sound quality by reducing the peak-to-average ratio of the signal to the point that the broadcast processor responds to it in an unnatural way, so it should be used conservatively. Ultimately, the only way to tell how one's production processing will interact with a broadcast processor is to actually apply the processed signal to a real-world broadcast processor and to listen to its output, preferably through a typical consumer radio.

---

1 These tracks were ordered according to increasing loudness using the Waves PAZ meter. However, the apparent waveform density implies sample #2 is louder than #3. Neither measurement method is perfect.

2 Bob Ludwig (in correspondence) mentions that competition in radio broadcasting was already happening in the late 1960's, noting WABC "color radio" added EMT plate to everything to increase average density.

3 BK: On the other hand, the other radio processing, especially the compression, reduces the sense of depth, plus, typical reception areas tend to lose separation so, improving the stereo image in mastering may not be such a bad thing.

[Appendix 2]

The Tower of Babel
Audio File Formats

Platforms, Extensions and Resource Forks

Macintosh files are divided into two parts, the **data fork** (which is the main part and which is transferable to a PC), and **resource fork.** Most Macintosh programs look for the file type in the **resource fork**, unique to Macintosh computers. The resource fork is the Macintosh way of telling programs who created a file, its file type, and additional information proprietary to the particular file type; it is analogous to the three letter **extension** on the PC (e.g. .aif, .wav). These were invented to allow users to double-click on a file and automatically open a program, an advance over the DOS command line. I don't know whether the Windows or Mac approach is better, because both can cause serious headaches when things go wrong. Resource forks cannot be transmitted over the Internet (except with Mac-specific compression utilities), and can only be transferred between platforms in a limited manner. So on the Mac, if the resource fork is empty (e.g., if the file came from a PC) or has an error in it, then a simple four-letter variable may be all that's keeping the audio from playing. More advanced programs, such as **Barbabatch** and **Soundhack** on the Mac, ignore the resource fork and look inside the data fork of the file for the

header, which contains far more information, including the file type, wordlength, and sample rate. If a Mac program restricted to reading the resource fork does not recognize the file type, try using a file-typing program. Replace the incorrect value with the letters AIFF, WAV (sometimes WAVE), or BWF. But turn down your monitor gain before playing in case you chose the wrong one!

When transferring files between platforms, the **WAV, AIFF** and **BWF** file types (described below) are the most universal, because they do not depend on resource forks for anything except file type, and the file type is also duplicated within the Header (in the data fork) if the resource fork is missing. We often receive files on Macintosh-formatted CD or DVD-ROMs, and these may be read on a PC using a simple system addition such as **MacOpener.** MacOpener reads the resource fork on the CD-ROM and uses a table (user-configurable) that automatically supplies an appropriate file extension; you can tell the process is working because Windows will supply the icon for that file type. I do not know of a way to mount a Mac-format hard disc on Windows and read the resource fork. However, it is a blessing that the SADiE (through ver. 4) proprietary SCSI bus can **read and write** to all common audio formats as well as Mac (including resource fork) and PC-formatted hard discs. In fact, SADiE can freely intermix file formats and wordlengths within its EDL, also a blessing. Sonic Solutions has historically been a closed platform, but Sonic Solutions HD 1.7 can read AIFFs, and 16-bit (not 24) WAVs; the only format it can write is AIFF. This necessitates frequent use of a universal conversion application such as **Barbabatch** on the Mac to exchange files between Sonic and the rest of the world. Barbabatch also performs excellent sample rate and wordlength conversion as well as batch renaming and splitting regions within files if desired, and acceptable dithering.

SADiE identifies the file type by the file extension on PC-formatted discs, and the resource fork on Mac-formatted discs. If a file somehow arrives on a PC with no extension,* try adding the extension, but turn down your monitor gain before playing! When in doubt as to the type, try adding the extension **.raw** or **.pcm** and tell a program which can read raw files (such as **Wavelab**) the suspected wordlength and sample rate and attempt to play the raw file. From there it may be transferred via AES/EBU into SADiE, for example. But, again, watch out for full scale white noise if you guessed wrong! Conversely, if you add an extension to a Mac file while it is on the Mac (or accidentily use a . character in any Mac file name), when it eventually gets to SADiE it may end up with an extra extension to its name, or SADiE will get confused as to the file type, or the PQ list may say *Love Me Do.aif* instead of just *Love Me Do*. The lesson is not to add extensions to Mac-formatted discs and let the smart utilities do their thing.

* In Windows, turn on the option which lets you view the extensions.

File Formats-non Lossy

There are four popular audio file formats in current use: **AIFF**, **WAVE**, **BWF** and **Sound Designer II** (SD2).

AIFF

Audio Interchange File Format supports standard bit resolutions in multiples of 8, up to 32 bits fixed point, although most AIFFs are 16-24 bits. While most professional PC programs can read and write AIFFs, this format was created for use on Macintosh computers. A **mono** or **split** AIFF contains one channel, as opposed to **interleaved** AIFFs, which can contain multiple channels. We prefer to receive interleaved files wherever possible, because it is easier to group them and prevent interchannel time-slippage. There is reportedly a floating-point AIFF file type, but as of this writing, the high-end mastering programs interchanging data insist on **fixed-point notation**. Sample rates up to 192 kHz and beyond are supported. On the PC, the standard extension is **.aif**. Data is stored in *chunks*, and manufacturers can write proprietary chunks. Byte order is **big-Endian (msb)** first, which is the Motorola standard, as opposed to **little-Endian (lsb)**, the Intel standard. If a program misreads the wrong end, the result will be nearly full-scale white noise, a not uncommon result when exchanging files between platforms. Reversing the ends wastes one instruction cycle, so manufacturers are often a bit fussy about which file format they prefer. There is no official provision for time-stamping except in a proprietary manufacturer's chunk.

A variation of AIFF is called **AIFC** (short for AIFF-C), which employs optional lossy data reduction (coding) and can use floating point notation. I have not seen AIFC supported by a high-end mastering program, but I have seen the AIFC file type accidentally applied to a plain AIFF by Mac programs such as Quicktime.

WAVE

The **WAVE** file format, developed by Microsoft, is probably the most popular audio format, using a standard extension of **.wav**. It supports a variety of bit resolutions (both fixed and floating point), sample rates, and channels of audio. As with AIFF, Wave files can be split or interleaved. There is provision for time-stamping in one of the standard SMPTE timecode formats, supported by some PC and Mac programs, but the BWF (see below) is more reliable in that respect. I recommend saving files as fixed-point (integer) WAVEs, as they are the most compatible between platforms. Understandably, many programs have difficulty with the several esoteric varieties of floating-point WAVEs. Byte order is **little-Endian (lsb)** first, most appropriate for Intel-based processors.

As in the AIFF, data in WAVEs is stored in **chunks**, which can also be manufacturer-specific. The format has grown in a somewhat disorganized manner, and now supports many variant and sometimes unstandardized types of chunks. But the high-end programs seem to be successful ignoring the chunks that don't make sense to them! WAVE data may optionally be **coded** (psychoacoustic wordlength reduction, sometimes confusingly called *data compression*), though mastering engineers expect that all files sent for mastering are **linear PCM** (i.e. *uncoded, high-resolution*).

BWF

The Broadcast Wave format is based on the Microsoft WAVE file format and continues to use the WAV file name extension. The EBU has added a "broadcast wave extension" chunk to the basic wave format, which contains the minimum information that is considered necessary for all broadcast applications, such as unique source identifiers, origination station data, etc. The EBU has legislated this format to be a standard of interchange, so most high-end mastering programs will be required to support it, and its built-in timestamp will be welcome. Files may be linear or (lossy) coded via MPEG-1 or −2. As of this writing, there is no provision for linear multichannel, so BWF multichannel (greater than 2) files must be lossy-coded. Of course, you can send multiple mono linear-signal-format BWFs or stereo pairs.

Sound Designer II

SDII format was invented by Digidesign for use on the Mac. SD II (or SD2) files are **landmines** on the PC, particularly because of their reliance on resource forks, where file type, sample rate, wordlength and time-stamp information are kept. SD2s can either be multiple-channel mono, or dual-channel interleaved stereo. Sample rates up to 48 kHz are officially supported by Digidesign, although Mark of the Unicorn (MOTU) uses SDII files exclusively, up to 96 kHz, in the Macintosh program **Digital Performer 3.0.** Performer can import and export WAV and AIFF but unfortunately cannot use those file types within an EDL. SD2 can be written and read from within **Pro Tools**, but only up to 48 kHz as far as I know. SADiE 4.2 has a bug which does not permit reading interleaved SD2s, and since PC-formatted backup tapes cannot store Mac resource forks, there is no way to archive an SD2 session from within SADiE except by bouncing first to a new format. So, routinely, I convert all SD2 files to AIFF or WAV using **Barbabatch** on the Mac, and move the removable disc over to SADiE for mastering. Reportedly, SD2 has been officially obsoleted by Digidesign, but its memory lingers on!

Length Limitations

A major problem with both WAVE and AIFF file formats is that the chunk sizes (including the overall chunk describing the whole file) use 32-bit integers, holding the size in bytes. For a quad 24-bit file at 96 kHz SR, the longest possible duration is some 3728 seconds, so you get only just over an hour. Go all the way to 5.1 surround at 192Khz, 24bit, and the limit descends to some 20 minutes.* Short of inventing a new file format that can support longer length files, the solution is to use split files if interleaved format proves too long for the length to be correctly specified.

Metafile Formats

Metafile formats are designed to interchange all the information needed to reconstruct a project. Unfortunately, some manufacturers are reluctant to adopt another's format, so this valuable effort has not made enough progress.

AES-31

The **AES-31** file Interchange standard was developed by the SC-06-01 AES standards committee jointly with several manufacturers. The goal is to interchange basic projects, timestamp and crossfade information as well as audio files. There has been some success but as of this writing the format is not supported by Digidesign.

OMF

The **Open Media Format** was produced by Digidesign to interchange Pro Tools Session and audio data with other workstations. At this writing, it is in a primitive state. The last time I tried to import OMF data into Digital Performer I got a fatal error.

Lossy File Formats

MP3 and ATRAC (used on the Minidisc) are **lossy file formats**, that is, some audio information has been sacrificed in the effort to save space and increase transmission speed. Once sound has been encoded into MP3, sound quality can never be restored, which is why it's a *lossy format!* Since these have become widespread and mislabeled *CD Quality* we sometimes get them as original sources! This violates the source-quality rule. Whenever possible, we ask to have these replaced with higher-quality, earlier-generation sources, or the sound quality will obviously suffer, especially after mastering processing.

---

* Richard Dobson, as reported in the Surround Sound maillist.

[Appendix 3]

Preparing Tapes and Files for Mastering

One major theme in this book has been the mastering engineer's comprehensive attention to sequencing, spacing (aka *assembly*), leveling, clean-up and processing. The better-prepared your tape or file, the better we all will look. Make the best mix you can, then let the mastering engineer do the rest of the magic, including the "heads, tails, fade-ins and fadeouts," for if something is cut off or faded prematurely, it will be lost. Don't be tempted to fade even if there is a noise, because we have some tricks that can create real-sounding endings on tunes that everyone thought had to be faded, as described in Chapter 7. You can also include a "fade example," which we can use if this proves to be the best choice. Given freedom, the mastering engineer can produce a seamless, flowing record album from the "loose parts" sent by the mix engineer. Leaving the tunes loose also permits the mastering house the most flexibility to change the order of the album (if necessary), or produce segues in the most artistic fashion.

In the last century, the most common formats we received for music mastering were linear, e.g., analog and digital tape and standalone CDR (which is linear for writing, but random-access for reading). But now the most popular formats are completely random access (file-based). Here's how to make a mastering engineer happy when submitting finished mixes on the medium of your choice.

Communication

Mastering is a collaborative process, even if you cannot attend the mastering session; the mastering engineer's job is to realize your desires and if possible to go beyond your wildest dreams! Give the mastering engineer a call to discuss your music and what you think it needs. Get the mastering engineer involved early in the mixing process; if you work nearby, bring over a sample to hear on the high-resolution, wide-range mastering monitors. Ask yourself: Does it sound like music? Does it live and breathe? Do the climaxes sound somewhat like climaxes? Do the choruses have a bit more energy than the verses (the usual natural case)? Is the bass drum to bass ratio right or do you have doubts? Is the sound as spacious and deep as you want it to be? Have you checked the material on several alternative systems? When it comes time for the mastering, don't hesitate to provide or suggest a CD of similar music that appeals to you, yet leave your mind open to the creativity of the mastering engineer. After the mastering session is over, you can listen to a reference (CD) on your own playback system and if desired, suggest revisions or improvements.

Logs

The logs that accompany mixes are very important. Thorough logging is essential: it keeps a project from being delayed as we don't have to chase down the catalog number or other essential information on the mastering day. Some engineers forget that **a CD ROM has no order.**[1] So all logs should indicate the full title of each song, the corresponding abbreviated file name on disc, and the order the song is to appear on the final medium, plus your comments about fades, noises or anything that concerns you. Please see the example log in Appendix 4.

Stems, Splits and Alternate Mixes (e.g. Vocal Up/Down). By all means provide alternate mixes or synchronized stems if possible. See Chapter 13.

Linear Media (DAT, Analog tape, Stand-alone CDR)

Don't bother to reorder DAT tapes or CDRs by copying, because the copy process may introduce more trouble than the time saved at the mastering house (if any). Leave the tunes out of order, leave the outtakes and alternate mixes (which may prove useful), and mark all *keeper* takes. Don't bother to space the tunes on linear media other than leaving enough time to cue and to use leaders or program IDs to identify the cuts.

When mixing to disc or digital tape, never make just one. Always record two at once digitally (make data-identical mixes labeled "A" and "B"), and hold onto that safety—never send the only copy in the mail. Record one or two minutes at the head of the tape with test tones or simply blank audio and begin the first tune after that with program ID #1. Start IDs do not have to be exactly placed, but they guide us to loading the proper tune. Remember that digital tapes need time to lock up—start recording on the mix tape, and for safety's sake wait a full 10 seconds before running the multitrack (you can use the lockup time to lay down a verbal slate*). When writing to standalone CDRs, which lock up instantly, a second's pause before the downbeat should be enough, but leave those critical breaths and noises in (see **handles**, below)!

Tape to Tape Dubbing procedures. Always monitor the output of the recorder while copying. If you must pause a tape-based recorder during the dubbing process, make sure to roll in record for at least 10 seconds before the tune begins, to prevent record glitches. This means that DAT tapes dubbed from other DATs can **never** have the short spacing we like on an album. Learn how absolute time is used on DATs, and maintain continuous ABS throughout the various mixing sessions by using **end search** before beginning the next session or after any playback.

Level Check

As described in Chapter 5, mix with conservative levels, which is not a problem with 24-bit media. Print the mix with levels well under the top and no OVERs! I recommend - 3 dBFS maximum. Roger Nichols reminds mix engineers using DAWs to visit each plug-in, reset the clip indicator and check the mix. If there's a clip, then redo the mix to avoid internal clipping, which can cause pops and snaps that usually aren't heard until mastering.[5]

---

* We still appreciate having verbal *slates* when dealing with non-file-based media. A *slate* is a verbal identification of the title or take number.

Preserve Data Integrity

In general, send the earliest generation, unprocessed material to the mastering house—avoid copying or going to second-generation in a DAW. If you must edit, keep everything at unity gain if at all possible (do not normalize), even if the material is peaking low, as explained in Chapter 5. The same goes for temptation to equalize, compress, limit or otherwise process a mix after it has been made. If you must, please send both versions to the mastering house, because we may be able to better the process with our tools, or combine it with other processes and reduce cumulative distortion.

Maximum CD Program Length

Every plant specifies a maximum acceptable length, and some charge more for CDs over approximately 77 minutes. The final CD Master tape, including songs, spaces between songs, and reverberant decay at the ends of songs, must not exceed the limit, which at one popular plant is 79:38. The mastering engineer can determine the exact time after the master is assembled. DVD program lengths vary because of the data coding and must be determined at the time of authoring.

Labeling tapes or discs. Which is the Master?

Don't forget to put a name and phone number on the source media in case it gets separated from the documentation! **A DAT is not a CD master**, and neither is a mix CDR submitted for mastering. The sources for an album are **NOT** the master; the album (production) **MASTER** is the final, PQ'd, equalized, edited, assembled, and prepared tape or disc that

needs no further audio work, and is ready for replication.[2] Please label the source media: **Submaster** or **Work Tape**, or **Mix**, or **Final Mix**, or **Session Tape**, or **Edited-Mix**, or **Compiled-mix**, or **Equalized Mix**, to name several possibilities. This will avoid confusion in the future when looking through the tape library for the one and only *real* (production) master.[*]

> { *"The Source tapes/files for an album are NOT the Production MASTER."* }

Analog tape Preparation

Begin and end the reel with some "bumper," followed by leader. If possible, put leader between songs (except for live concerts and recordings edited with room tone). Tape should be slow wound, tails out. Label each reel as recommended in Appendix 5. Indicate tape speed, record level for 0 VU in nw/M, record EQ (NAB or IEC), track configuration, whether it is mono, stereo or multichannel. Indicate if noise reduction is used and include the noise reduction alignment tone. Include alignment tones 30 seconds (or longer) each, minimum 1kHz, 10kHz, 15kHz, and 100Hz plus (highly recommended) 45Hz and 5kHz at 0VU without noise reduction. Also highly recommended is a tone sweep (glide) from 20 Hz through 500 Hz. Needless to say, the tones must be recorded by the same tape recorder that recorded the music, and ideally, recorded through the same console and cables that were used to make the mix. Many mastering engineers prefer having the tones at the tail of a reel or on a separate reel.

---

[*] NARAS has produced Master Recording Delivery Recommendations. Serious recordists must study *http://grammy.aol.com/recommendations.pdf*

Many historic analog tapes do not include proper tones and sometimes it is not possible to put tones on new masters. If it was not possible to lay down tones on the session, then we will use sophisticated methods to guarantee azimuth and equalization accuracy.

Give Handles

For live concerts and many other forms of music, it's useful to include *handles*, that is *raw footage* on either side of the intended music. This can include out takes, unfaded applause, breaths, coughs, noises, speech between tracks, etc. Also include your production notes and desires, such as "please leave that ugly laugh in between songs 2 and 3, I think it's funny." Handles are especially needed when a track might have to be noise reduced, for the noise sample we need can sometimes only be found just before the downbeat.

What Sample Rate?

Until circa 2000, I recommended that mix engineers try to work at 44.1 kHz if possible, considering the abysmal state of sample rate converters. This is no longer the best advice; my current recommendations are for mixers to work at the highest practical sample rate and longest available wordlength. However, if you are mixing digitally, **do not sample rate convert yourselves**, but remain at the same sample rate as the multitrack. If you are mixing with an analog console, there is a marginal advantage to using a higher sample rate for the mixdown recorder than the multitrack. For example, even if mixing analog with a multitrack at 48 kHz, you will get better results with a mixdown recorder at 96 kHz.

{ *"CD-ROM Preparation is a nest of land mines waiting to explode."* }

Random Access Media: Preparing Files

CD-ROM preparation requires attention to detail. It's a nest of land mines to navigate which should not be taken lightly, and experience is the best teacher. A poorly-prepared CD ROM can waste a tremendous amount of time at the mastering house. Make sure the mastering house will accept the file types you want to send. I recommend you work around Murphy's law by cutting a test disc and sending some files ahead of time that we can check out. Here are some critical do's:

· **Leave blank sound at the head of the file**, in other words, start the first music at least 1 second into the file, **not at zero time.** (This is to prevent glitches that often occur at the file start).

· For stereo and multichannel, **Interleaved files** are preferred, AIFF, BWF, or WAV. SDII is also acceptable (see Appendix 2). Ask to avoid costly conversion time. **No MP3's, please!** Start and end with high-resolution, linear-format sound files.

· Try to do a single project at **one sample rate**. It involves considerable extra work to deal with multiple sample rates in a project and often involves a compromise as we must rate-convert some files to get a common rate for the project. But if for some reason your project includes different rates, carefully mark (log) the rate of the files for our information.

• Give each file a meaningful name related to the song title, like **Love Me Do**, not some meaningless serial number.

• Choose a high-quality name-brand CDR blank. To my experience, Taiyo Yuden, the oldest CDR manufacturer, continues to make the most compatible and reliable CDRs.

• For lowest error rate, obtain 74-minute blanks from a professional supplier. Avoid the error-prone 80s, which eliminates going into Costco on a Friday night to search for blanks!

• For lowest error rate, cut at 2X to 4X speed, no faster.

• Write a **Fixed** disc, i.e. a **closed session**. To verify the disc has been fixed, pop it into a PC or Mac CD reader (not a writer) and make sure it can read the file names.

• **DO NOT USE PAPER LABELS!** Stick-on paper labels may look impressive, but in my experience they appear to increase error, perhaps by altering the rotational speed of the disc, and are especially problematic at high disc spin speeds, multichannels, high sample rates and wordlengths. Paper labels can also become partly or completely unglued and tear off in the CD reader, which is not a pretty sight! Also, do not label the disc with a ballpoint pen, but with a soft marker, on the protected (overcoated) part of the top surface.[3] While I personally believe that the coating on professionally over-coated CDRs is sufficient protection from scratches and organic solvents (as in an aromatic Sharpie-brand marker), the most conservative mastering engineers recommend using water-based markers for labeling. Perhaps someone will do a long-term study measuring errors on CDRs with a coated-marked surface.

• Write to **fixed-point 24-bit files** (also known as *Integer Format*). It's unlikely that the mastering house can read any other format; e.g., **do not use** 32-bit floating point for files. This situation may improve in upcoming years and we are beginning to have success reading Samplitude-format, one of the several incompatible 32-bit float file formats.

• Use any standard sample rate up to 96 kHz. Verify the mastering house can use files with a higher rate before cutting.

• **File names** should not include hyphens (-), use an underline instead. Do not use the / or \ character. For best multi-platform compatibility, stay away from spaces and use alphas, numerics and underlines only.[4] SADiE v. 4.2 has a strong aversion to accent marks and non-English characters, keeping it from generating waveforms, archiving and other essentials, something which we hope they will change. Macs are far more forgiving in this regard than PCs.

• **We love receiving files that include the intended track number in their name**. One trick for naming files is to include the intended track number at the beginning (using two digits), which makes it much easier to assemble them in the intended order. For example: **01 I Need Somebody, 09 I Got Rhythm, 10 She's So Fine.**

• **Avoid periods (dots) in Mac file names on Mac discs** because they might be transferred to PC and be confused with extensions; use one and only one dot on PC discs in front of the 3-letter extension.

· Verify the mastering house can read DVD-ROMs before choosing that medium.

Split Files

Interleaved files are less subject to accidents since all the channels are guaranteed to start at the same point. For multichannel, include a note indicating the channel order used, e.g., L, R, C, LFE, SL, SR or L, R, SL, SR, C, LFE. If you must send split files, use a standard nomenclature to distinguish the channels, e.g. **Do It_L, Do It_R, Do It_SL, Do It_SR, Do It_C, Do It_LFE.** Letter abbreviations are preferable to ambiguous channel numbers.

When You Get Your Master Back

If the CD master is sent back to you instead of directly to the CD plant, **don't handle it or play it.** Play the ref, not the master![5]

---

[1] There is no *track order on* a non-linear, file-based medium. Often, clients ask me, "put the master in the order it's on the CD ROM," but they forget the only *order on* the CD ROM is the alphabetical directory of files.

[2] Andre Subotin on the Mastering Webboard reminds us that there may be several true Masters, each of which we must clearly label, e.g. **Production Master for Cassette; Master for foreign countries;** etc.

[3] Thanks to Clete Baker and Mike McMillan on the Mastering Webboard for clarifications on these points.

[4] Thanks to Clete Baker on the Mastering webboard for reminding me of this essential!

[5] Thanks to Roger Nichols for the nudge to put these recommendations in the Appendix.

Logs and Labels for
Tapes, Discs and Boxes

Labeling Those Tapes

I don't dare put an unlabeled DAT or CDR down on my mastering desk, for it will immediately be lost in a crowd! Please do put the following minimal information on every piece of source media, in case it gets separated from the box:

- Artist
- Album Title [or working title]
- Contact Name, phone number
- Tape or reel number
- Date [important to help separate out revisions]

Labeling Those Boxes

The box label contains much more information than what's written on the reel or disc itself.

Analog Tape Boxes: An example label

Some studios have preprinted labels with checkboxes for each option.

Mix tape, Unedited, songs head leadered [or other descriptive]

Artist: _____

Album Title: _____

Record Label: _____

Reel number: _____ of _____

Catalog Number: _____

Studio, Address, Contact Phone #: _____

Engineer: _____

Assistant: _____

Producer: _____

Date: _____

Format, EQ, Speed, Level: *[e.g. 1/2" 2-track AES stereo, no noise reduction, 30 IPS, 0 VU = 320 nW/M, or 0 VU = 250 nW/M + 2 dB]*

Test Tones @ Head _____ @ Tail _____ consisting of _____ Hz at 0 VU

Name of Song or track
Length
Comments *[e.g. "vocal up" or "needs fadeout" or "leave countoff at the beginning"*

Name of next song, etc.

Further comments can be written in a letter that accompanies all the media.

Discs: Example Label

There is not enough room on a CDR or DVD-R surface to write everything we want to know. Some studios have prescreened discs with checkboxes. At minimum, the top surface of the disc itself should include:

Mixes, Unedited *[or submaster or other descriptive]*

Artist: _____

Album title: _____

Record Label: _____

Disc and File Format: *[e.g. ISO-9660 or HFS, or Masterlink, Stereo AIFF Files, 48 kHz/24 Bit]*

Disc # _____ of _____ Date: _____
[date is very critical]

Plus, if possible:

Contact name and Phone #:_____

Catalog Number: _____

Since there is not enough room to list all information on the disc itself, be sure to include the remaining information on the box, jewel box, and/or printed log (pictured opposite page) which accompanies the media. If possible, the log can be duplicated in a READ_ME.doc file which resides on the disc, so it will never be lost.

Discs, Jewel Box or Paper cover label

Instead of using up several jewel boxes, some studios cleverly put CDRs inside a taped and folded printout of the disc's directory, which covers all the names of the tunes inside the disc. When shipping, put these paper-covered discs in a foam-lined hard-box to prevent scratching or breakage. As described in Appendix 3, the title names can also include their eventual sequence order, if this is known at the time of disc creation.

Printed Log/letter

Accompany the discs or tapes with a printed log/letter to the mastering engineer. This is where you can also include all your comments and thoughts on the eventual mastering. You can put this in the form of a letter, which includes your story and feelings about the album and its sound. Some comments (especially the need for a fade!) may be superfluous but put down anything you are concerned about.

Don't forget to include:

Artist: _____

Album title: _____

Record Label: _____

Disc, File Format, Sample Rate, Wordlength: *[e.g. ISO-9660 or HFS, or Masterlink, Stereo AIFF Files, 48 kHz/24 Bit]*

Contact name and Phone #: _____

Contact Address: _____

Catalog Number: _____

Title/File Name	CD track Order	Length (approx.)	ABS time/DAT or CD Program ID (not relevant if this is a disc of files)	ISRC	Comments [e.g. by engineer, producer or artist]
I Wanna Make You Happy/ 05_makehappy.wav	5	4:02		ES6080132805	This song needs a fadeout. Try starting circa 3:45 and be out by 4:00 from the downbeat so as not to hear the snickering! Please include the sticks at the beginning.
Love Me Do/ 02_lovemedo.wav	2	2:55		ES6080132802	This is an obvious tribute to the Beatles. The more Beatle-like you can make the mastering, the happier I will be!
Why Me?/ 04_yme.wav	4	5:02		ES6080132804	This is the only ballad on the album. The artist is not happy with her intonation entering the last chorus. Is there anything you can do about this?

Decibels

Marking Analog Tapes

I once received a 1/4" tape in the mail marked "the level is +4 dBm." But dBm and dBu do not travel from house to house. dBu is a measurement of a voltage expressed in decibels and there is no voltage on an analog tape, only magnetic flux in nanowebers per meter. The 1/4" tape doesn't have any idea whether it was made with a semi-pro level of 0 VU = -10 dBu or a professional level of +4. Instead, just indicate the magnetic flux level which was used to coordinate with 0 VU. For example, label it *0 VU=400 nW/M at 1 kHz*. 400 nW/M is 6 dB over 200 nW/M, and engineers often abbreviate this on the tape box as *+6dB/200*, as you can see from this convenient chart.

Chart 1:					
Tape Fluxivity in dB and nanowebers per meter (nW/M)					
Level dB	**Reference 185**	**Reference 200**	**Reference 250**	**Reference 320**	**Reference 400**
9	521	564	705	902	1127
8	465	502	628	804	1005
7	414	448	560	716	895
6	370	400	500	640	800
5	329	356	445	569	711
4	293	317	396	507	634
3	261	283	353	452	565
2	233	252	315	403	504
1	208	224	281	359	449
0	185	200	250	320	400

Find the actual nanowebers per meter of flux for a given reference flux. For example, a tape which is 4 dB hotter than 250 nW is 396, or rounded up to about 400. This is the same fluxivity as a tape which is 6 dB hotter than 200.

Q to Bandwidth Conversions

Chart #2: dbu (reference 0.775 volts) converted to voltage	
dBu	Volts
40	77.500
35	43.581
24	12.283
20	7.750
18	6.156
16	4.890
14	3.884
12	3.085
8	1.947
6	1.546
4	1.228
3	1.095
2	0.976
1	0.870
0	0.775
-10	0.245
-20	0.078
-60	0.001

B/W	Q		Q	B/W
0.02	72.13		0.50	2.54
0.03	48.09		0.55	2.35
0.04	36.07		0.60	2.19
0.05	28.85		0.65	2.04
0.06	24.04		0.70	2.00
0.07	20.61		0.75	1.80
0.08	18.03		0.80	1.70
0.09	16.03		0.85	1.61
0.10	14.42		0.90	1.53
0.20	7.21		0.95	1.46
0.30	4.80		1.00	1.39
0.40	3.60		1.10	1.27
0.50	2.87		1.20	1.17
0.60	2.39		1.30	1.08
0.70	2.04		1.40	1.01
0.80	1.78		1.50	0.94
0.90	1.58		1.60	0.89
1.00	1.41		1.70	0.84
1.20	1.17		1.80	0.79
1.40	0.99		1.90	0.75
1.60	0.86		2.00	0.71
1.80	0.75		3.00	0.48
1.90	0.71		4.00	0.36
2.00	0.67		5.00	0.29
2.20	0.60		6.00	0.24
2.40	0.54		8.00	0.18
2.60	0.49		10.00	0.14
2.80	0.44		20.00	0.07
3.00	0.40		30.00	0.05

Use this chart for an equalizer whose controls are marked in bandwidth but when you wish to think in Q, or vice versa. Bandwidth is expressed in octaves, at the 3 dB down point. The formula to convert bandwidth to Q is Q = Square Root(2**BW) / (2**[BW-1]).

I Feel The Need For Speed

Medium	Speed MB/hour	Speed MB/min	Speed MB/sec	Speed Mb/sec	Hours to run one CD	Minutes to run one CD	Seconds to Run or copy one CD	Facts
CD Player	635.04	10.584	0.1764	1.4112	1.00	60.0	3600.0	CD total bytes in one hour — 635,040,000 About the same as a T1 link
DSL 384 kbps	173	2.88	0.048	0.384	3.68	220.5	13230.0	Assuming Internet running at maximum efficiency
10 Base T	4,500	75	1.25	10	0.14	8.5	508.0	CD speed. Bytes per minute — 10,584,000
100 Base T	45,000	750	12.5	100	0.01	0.8	50.8	CD total MB — 635.04
1000 Base T (Gigabit Ethernet)	450,000	7500	125	1000	0.0014	0.1	5.1	This is theoretical point to point with no collisions. Ethernet mileage will be much slower on a busy network. Use an Ethernet Switch instead of a Router to maximize speed and minimize collisions.
USB 1.0 slow	675	11.25	0.1875	1.5	0.94	56.4	3386.9	
USB 1.0 fast	5,400	90	1.5	12	0.12	7.1	423.4	
USB 2.0	216,000	3600	60	480	0.0029	0.2	10.6	
Firewire	180,000	3000	50	400	0.0035	0.2	12.7	This is the maximum speed of the interface. Individual drives are much slower
Seagate 18 GB Ultra SCSI 160	108,000	1800	30	240	0.01	0.4	21.2	Typical internal transfer rate of a modern LVD drive
Ultra ATA/66 10,000 RPM	147,600	2460	41	328	0.0043	0.3	15.5	
Ultra 2 SCSI 160 MB/s LVD	576,000	9600	160	1280	0.0011	0.1	4.0	This is the interface, individual drives much slower. RAID can reach this speed.

Abbreviations: MB = Megabytes, Mb=Megabits (8 bits/byte)
Mega is defined as 1 million, Kilo is 1 thousand. Some of these
figures would change slightly if kilo is defined as 1024.

All times normalized to capacity of one hour long stereo audio CD.

I Feel The Need For Capacity

Prior to 1990, I was making CD masters with linear editing using the Sony 3/4" editing systems. In 1990 I set up my first nonlinear mastering workstation, purchasing the highest capacity hard discs available, a pair of 600 MB SCSI hard discs, that cost $1500 retail, or $1.25 per MB. Fortunately, as our needs have gone up in 10 years, capacity has increased geometrically and cost has gone down. Thus it's not out of line to expect typical storage capacity to tentuple in 10 years.

Year	Type of Storage	Capacity MB	Capacity GB	Total Cost US Dollars	Cost per MB	Cost per GB	Number of 1 Hr Compact Discs	Number of 1 Hr ,6-Ch., 24-bit Surround Masters at 44.1 kHz	Number of 1 Hr, 6-Ch., 24-bit Surround Masters at 96 kHz	Number of 1 Hr, 48-Ch., 24-bit Multitracks at 96 kHz	Facts
1980	Data General	297	0.297	$35,000	$118	$118,000	0.5	N/A	N/A	N/A	Size: 2 feet x 3 feet x 3-1/2 feet high!
1990	SCSI Hard Disc	600	0.60	$750	$1.25	$1,250	0.94	0.2	0.1	0.012	CD one hour 635,040,000 bytes
2002	IDE Hard Disc	80,000	80	$137	$0.0017	$1.71	125.98	28.0	12.9	1.608	Street price
2010	Raid? Optical?	800,000	800	$16	$0.0000	$0.02	1259.76	279.9	128.6	16.075	Projected cost, as per archivebuilders.com

Abbreviations: MB = Megabytes, GB = Gigabytes (1000 MB)

Footnotes on The K-System

The VU Meter's Actual Ballistics

The **VU** meter's actual ballistics were analyzed as early as 1940. According to **A New Standard Volume Indicator and Reference Level**, Proceedings of the I.R.E., January, 1940, the mechanical VU meter used a

> copper-oxide full-wave rectifier which, combined with electrical damping, had a defined averaging response according to the formula **i =k * e to the p** equivalent to the actual performance of the instrument for normal deflections. (In the equation **i** is the instantaneous current in the instrument coil and **e** is the instantaneous potential applied to the volume indicator)....a number of the new volume indicators were found to have exponents of about 1.2. Therefore, their characteristics are intermediate between linear (**p** = 1) and square-law or root-mean-square (**p**=2) characteristic.

History and Development of the SMPTE Standard, from Errors to Knowledge

The *theatre standard*, **Proposed SMPTE Recommended Practice: Relative and Absolute Sound Pressure Levels for Motion-Picture Multichannel Sound Systems**, SMPTE Document RP 200, defines the calibration method in detail. In the 1970's the value had been quoted as *85 at 0 VU* but as the measurement methods became more sophisticated, this value proved to be in error. It has now been proved to be *85 at -18 dBFS RMS* with 0 VU remaining at -20 dBFS (sine wave). The history of this metamorphosis is interesting. A VU meter was originally used to do the calibration, and with the advent of digital audio, the VU meter was calibrated with a sine wave to -20 dBFS. However, it was forgotten that a VU meter does not average by the RMS method, which resulted in an error between the RMS electrical value of the pink noise and the sine wave level. While 1 dB is the theoretical difference, in practice I've seen as much as a 2 dB discrepancy between certain VU meters and the true RMS pink noise level.

The other problem is the measurement bandwidth: a wide bandwidth voltmeter will show attenuation of the source pink noise signal on a long distance analog cable due to capacitive losses. The solution is to define a specific measurement bandwidth (20 kHz). By the time all these errors were tracked down, it was discovered that the historical calibration was in error by 2 dB. Using pink noise at an RMS level of -20 dBFS RMS must correctly result in an SPL level of only 83 dB. In order to retain the magic **85** number, the SMPTE decided to raise the specified level of the calibrating pink noise to -18 dBFS RMS, but the result is the identical monitor gain. One channel is measured at a time, the SPL meter set to C weighting, slow, and as explained in Chapter 14, a more accurate measurement can be obtained via 1/3 octave analysis. The K-System is consistent with RP 200 only at K-20. I feel it will be simpler in the long run to calibrate to 83 dB SPL at the K-System meter's 0 dB rather than confuse future users with a non-standard +2 dB calibration point.

It is critical that the thousands of studios with legacy systems that incorporate VU meters should adjust the electrical relationship of the VU meter and digital level via a sine wave test tone, then ignore the VU meter and align the SPL with an RMS-calibrated digital pink noise source.[*]

Detailed Specifications of the K-System Meters

General: All meters have three switchable scales: K-20 with 20 dB headroom above 0 dB, K-14 with 14 dB, and K-12 with 12 dB. The K/RMS meter version (flat response) is the only required meter—to allow RMS noise measurements, system calibration, and program measurement with an averaging meter that closely resembles a *slow* VU meter. The other K-System versions measure

loudness by various known psychoacoustic methods (e.g., LEQ and Zwicker).

Scales and frequency response: A tri-color scale has green below 0 dB, amber to +4 dB, and red above that to the top of scale. The peak section of the meters always has a flat frequency response, while the averaging section varies depending on which version is loaded. For example: Regardless of the sampling rate, meter version K-20/RMS is band-limited as per SMPTE RP 200, with a flat frequency response from 20-20 kHz +/- 0.1 dB, the averaging section uses an RMS detector, and 0 dB is 20 dB below full scale. To maintain pink noise calibration compatibility with SMPTE proposal RP 200, the meter's bandpass will be 22 kHz maximum regardless of sample rate.

Other loudness-determining methods are optional. The suggested average section of Meter K-20/LEQA has a non-flat (A-weighted) frequency response, and response time with an equal-weighted time average of 3 seconds. Since loudness is generally an overall sensation, a case can be made for a monophonic loudness meter. Expert psychoacousticians designing a true loudness K-System meter must resolve that discrepancy, permit production engineers to retain the desirable individual channel meters. They will calculate the proportion of the total loudness in each channel. The average section of Meter K-20/Zwicker corresponds with Zwicker's recommendations for loudness measurement. Regardless of the frequency response or methodology of the loudness method, reference 0 dB of all meters is calibrated such that 20-20 kHz pink noise at 0 dB reads 83 dB SPL on

---

[*] Thanks to Tomlinson Holman (in correspondence) for explaining the historical source of the measurement errors, and how **85** became **83** after a long battle.

each channel, C weighted, slow. Psychoacousticians designing loudness algorithms recognize that the two measurements, SPL and loudness are *not* interchangeable and take the appropriate steps to calibrate the K-system loudness meter 0 dB so that it equates with a standard SPL meter at that one critical point with the standard pink noise signal.

Scale gradations: The scale is linear-decibel from the top of scale to at least -24 dB, with marks at 1 dB increments except the top 2 decibels have additional marks at 1/2 dB intervals. Below -24 dB, the scale is non-linear to accommodate required marks at -30, -40, -50, -60. Optional additional marks through and beyond -70. Both the peak and averaging sections are calibrated with sine wave to ride on the same numeric scale. Optional (recommended): A *10X* expanded scale mode, 0.1 dB per step, for calibration with test tone.

Peak section of the meter: The peak section represents the true, flat (1 sample) peak level, regardless of which averaging meter is used. An additional pointer above the moving peak represents the highest peak in the previous 10 seconds. Designers can add an oversampling peak movement as long as it is clearly marked and identified, especially since all our emphasis on loudness judgment is based on the averaging section and its scale. A peak hold/release button on the meter changes this pointer to an infinite high peak hold until released. The meter has a fast rise time (aka *integration time)* of one digital sample, and a slow fall time, ~3 seconds to fall 26 dB. An adjustable and resettable OVER counter is highly recommended, counting the number of contiguous samples that reach full scale.

Averaging section: An additional pointer above the moving average level represents the highest average level in the last ten seconds. An *average hold/release* button on the meter changes this pointer to an infinite *highest average* hold until released. The RMS calculation should average at least 1024 samples to avoid an oscillating RMS readout with low frequency sine waves, but keep a reasonable latency time. If it is desired to measure extreme low frequency tones with this meter, the RMS calculation can optionally be increased to include more samples, but at the expense of latency.

Ballistics: This is only relevant to the RMS meter, as the "ballistics" of the true loudness versions will be determined by the algorithm. After RMS calculation, the meter *ballistics* are calculated, with a specified integration time of 600 ms to reach 99% of final reading (this is half as fast as a VU meter). The fall time is identical to the integration time. Rise and fall times should be exponential (log).

Recommended Reading, CDs for Equipment Testing and Ear Training

Books

Burroughs, Lou (1974) *Microphones: Design and Application*, Sagamore Publishing Company, Inc. Out of Print. A classic audio work, the first book to publish the 3 to 1 rule with frequency measurements of the anomalies.

Holman, Tomlinson (2000) *5.1 Surround Sound: Up and Running*, Focal Press. Also includes guides on the problems of locating speakers near consoles.

Howard, David M. & Angus, James (2001) *Acoustics and Psychoacoustics*, Focal Press. Includes good discussion of the time/frequency relationship of filtering.

Kefauver, Alan P. (1999) *Fundamentals of Digital Audio*, A-R Editions, Madison, WI

Kirk, Ross & Hunt, Andy (1999) *Digital Sound processing for Music and Multimedia*, Focal Press, Boston.

Nisbett, Alec (2003) *The Sound Studio: audio techniques for radio, television, film and recording*, 7th Edition. Focal Press. A classic work with practical techniques which will never go out of style. I started with the 1962 edition!

Owsinski, Bobby (2000) *Mastering Engineer's Handbook*, ISBN# 0-87288-741-3. A collection of interviews with mastering engineers.

Pohlman, Ken (2000) *Principles of Digital Audio*, McGraw Hill.

Watkinson, John (1988, regularly revised) *The Art of Digital Audio*, Focal Press, ISBN 0 240 51320 7. The definitive industry *bible*. This is where you must first go for in-depth information on how digital audio works and the specifications of much of today's digital audio equipment and interfaces.

Magazines

One To One Magazine, United Business Media International Ltd, Leics, United Kingdom, **http://www.cmpinformation.com**.

Articles in Print

Blesser, A. & Locanthi, B., (1986) *The Application of Narrow-Band Dither Operating at the Nyquist Frequency in Digital Systems to Provide Improved Signal-to-Nose Ratio over Conventional Dithering*, AES 81st Convention, Preprint 2416.

Cabot, Richard C. (1989) *Measuring AES-EBU Digital Audio Interfaces*, AES 87th Convention Preprint 2819 (I-8).

Gerzon, M.A., Craven, P.G., Stuart, J.R., & Wilson, R.J. (1993) *Psychoacoustic Noise-Shaped Improvements to CD and other Linear Digital Media*, AES 94th Convention, Preprint #3501.

Lipshitz, S.P. & Vanderkooy, J. (1989) *Digital Dither: Signal Processing with Resolution Far Below the*

Least Significant Bit. AES 7th International Conference-Audio In Digital Times, Toronto, 87-96.

Muncy, Neil, Whitlock, Bill et al (1995) Collection of definitive articles on grounding, shielding, power supply, EMI, RFI. Journal of the AES Vol. 43 Number 6, a special excerpt printing.

Nielsen, Soren & Lund, Thomas (2000) *0 dBFS+ Levels in Digital Mastering*. AES 109th Convention, Preprint #5251.

Stuart, J.R. & Wilson, R.J. (1991) A Search for Efficient Dither for DSP Applications, AES 92nd Convention, Preprint #3334.

Stuart, J.R. (1993) *Noise: Methods for Estimating Detectability and Threshold*, 94th AES Convention Preprint #3477.

Compact Discs

Auralia, Complete Ear Training software for musicians, Rising Software, Australia, **http://www.rising-software.com**

Grimm, Eelco, (2001) *Checkpoint Audio Professional Audio Test Reference*, Contekst Publishers, Netherlands, ISBN 90-806111-1-5. Test and listening CD including J-Test, Bonger Test, and unique distortion and listening tests. Written in Dutch with no English translation (as of 2002).

Moulton, David, *David Moulton's Audio Lecture Series, Golden Ears audio ear-training self-study course*, KIQ Productions (Golden Ears), or **http://www.moultonlabs.com**

Various compilation and test CDs, Chesky Records, **http://www.chesky.com**.

Articles On the Internet

Dunn, Julian, *AES3 and IEC60958*, item #26 written for Audio Precision, **http://www.audioprecision.com/publications/technotes/index.htm**

Dunn, Julian, *various articles on jitter and other audio topics*, at Nanophon, **http://www.nanophon.com**.

Lavry, Dan, *various articles on sampling, oversampling, jitter, etc.*, **http://www.lavryengineering.com/** in the Product Support area.

Story, Mike, *various articles on high sampling rates, jitter, etc.* DCS, Inc. **http://www.dcsltd.co.uk/papers/**

SMPTE RP 200 proposed standard, SMPTE **http://www.smpte.org/stds/**.

TC Electronic, articles on jitter, 5.1 surround, 0 dBFS+ levels, etc. **http://www.system6000.com/system6000.asp?Section=19**

[Appendix 11]
Eric James Biography

Eric James is an Englishman, in his mid-forties, inordinately fond of chamber music and acoustic jazz, who has been a university teacher (of the history and philosophy of science and medicine) in Hong Kong for twelve years. He has decided, after giving the matter much thought, that these four facts probably have nothing very much to do with the tremendous satisfaction he derived from working, as editor, with Bob on this book. On the other hand, although Eric has been an academic for fifteen years, before he started his graduate studies (late, at Oxford, in the history and philosophy of mathematics) he spent a large part of his working life as a professional musician, and he has very recently resigned his academic tenure in order to return to the UK to develop the music recording and editing company – URM Audio – which he had been running on a part-time basis since 1998.

In 2001 Eric became a father, and he would like to thank his daughter — Jamie Martha Perry — and her mother, Sally, for putting just about everything else into its proper perspective.

[Appendix 12]
Robert A. Katz Biography

From his earliest years, Bob has been as curious as a *Katz*. He voraciously reads audio books, service manuals, product spec sheets, license plates, and bumper stickers. But his favorite reads are Science Fiction writers Spider Robinson and Frederick Pohl, which may explain Bob's punny personality. In his teens he dabbled in hypnotism and magic, but was a bit klutzy to turn that into a career. Bob is an animal lover—all dogs and cats love him back.

Coming from a family of medical doctors, musicians and composers, Bob gravitated to the B flat clarinet at the age of ten; his aunt, a viola teacher, gave Bob his first lesson in solfège and transposition. At the age of 13, he rebuilt his first tape recorder. After wiring the house for sound, he was forced by his parents to remove the microphones he had secreted throughout the house. Clearly destined for a career in audio, by high school he had begun an amateur recording career, plus studying the sciences and linguistics, practicing French and Spanish and looking for female pen pals on three continents. Perhaps out of default he was voted *most versatile* in his class. Eventually his language skills would reach the point where he can give seminars in any of three languages.

An enthusiastic young man with a passion for good sound, Bob developed a reputation as an audiophile around Hartford, Connecticut town. The local audio stores regularly invited him over, for Bob is never short of opinions. One day he was invited to audition a new pair of speakers with the designer present. After hearing a few notes, Bob ran out of the store covering his ears! Over the years, he has learned to be more diplomatic, but his opinions continue to be defined by a love for the art of audio.

In college he played in an *ad hoc* Dixieland ensemble, and the treat of his performance life was soloing *Sweet Georgia Brown* before the homecoming football crowd. Two years at Wesleyan University were followed by two more at the University of Hartford, studying Communication and Theatre, but he spent less time in the classroom and more at the college radio station, where he became recording director. A fan of the Firesign Theatre, Bob used to write and edit humorous radio ads, and he became a DJ, manning a free-form-progressive rock radio show titled *The Katz Meow*, and doing a stint on the commercial rock station.

Bob taught himself analog and digital electronics, and was influenced by a number of creative designers. In Hartford, Bob's mentor was Steve Washburn, an EE who invented a way to nearly double the power-handling of a Hartley 24" woofer and also constructed Bob's first custom-built portable audio console. Just out of college, Bob became **(1972)** Audio Supervisor of the *Connecticut Public Television Network*, producing every type of program from game shows to documentaries, music and sports, and he learned to mix all kinds of music live. When he wasn't working television, he was on location recording music groups direct to 2-track.

In **1972**, Bob wrote his first article for **dB** magazine, describing a set of mike heaters he developed to warm his AKG microphones and keep them from sputtering due to changes in humidity. This spiked a *heated* controversy as Stephen Temmer of Gotham Audio wrote a response stating that "Neumann microphones are never affected by humidity" but Bob's experience was supported by some others and in those pre-Internet times the controversy remained of modest proportions. Hooked by the writing bug, Bob is a natural-born teacher who puts himself in the mind of the learner. He has written over a hundred articles and reviews in publications such as dB, RE/P, Mix, AudioMedia, JAES, PAR, and Stereophile.

In 1977 he moved from Connecticut to New York City, and began a recording career in records, radio, TV, and film as well as building and designing recording studios and custom recording equipment. Long before the advent of the home PC, Bob taught himself several computer languages, and sold one assembly-language program used in an embedded system at a brokerage firm. During the primitive time before cell phones, the voice of *Matilda* became well known. Matilda answered Bob's phone and forwarded calls to any place Bob happened to be. Visitors to Bob's house were dismayed to discover that sultry-voiced Matilda was not flesh and blood but rather a 6502-based controller, DTMF encoders, decoders and other gear. Matilda's true identity remains a mystery today.

From 1978-79, he taught at the Institute of Audio Research, supervised the rebuild of their audio console and studios and began a friendship with IAR's founder Al Grundy, mentor and influence. Other New York era influences include Ray Rayburn and acousticians Francis Daniel and Doug Jones. **In the 80's**, one of his clients was the spoken-word label, *Caedmon Records*, where he recorded actors including Lillian Gish, Ben Kingsley, Lynn Redgrave and Christopher Plummer.

An active member of the *New York Audio Society*, Bob was the ultimate audiophile. This led to a full-page interview/article in the *Village Voice* called *Sex With The Proper Stereo*, a story about Bob's railroad apartment on East 90th with the empty refrigerator in the kitchen and mysterious monoliths in the living room.

But the refrigerator was not empty for long. In **1984**, Bob was doing sound for a motion picture in Venezuela and met multi-lingual Mary Kent, production assistant. After the filming, Bob invited Mary to come to New York for a vacation that became a permanent engagement! One day new girlfriend Mary came home and turned on the stereo system in the wrong order, blowing up the Krell amplifier and one of the Symdex woofers producing sparks and blue smoke. When Bob arrived home, he calmed her down— "Don't worry, Mary, your love for me means more than any stereo system." Bob and Mary have been together ever since (Mary jokes that she's really in love with the stereo system).

One day Bob received a call from musician David Chesky, who had read the *Voice* article and was looking for an audiophile recording engineer. In **1988** this led to a long and pleasant association with Chesky Records, which became the premiere audiophile record label. Bob specializes in minimalist miking techniques (no overdubs) for capturing jazz and other music that commonly is multimiked. His recordings are musically balanced, exciting and intimate while retaining dynamics, depth and space. **In 1989** he built the first working model of the DBX/UltraAnalog 128x oversampling A/D converter, and produced the world's first oversampled commercial recordings. Over the years, the converter was refined, until by **1996** Bob found a commercial model that performed slightly better. Bob has recorded about 150 records for Chesky, including his second Grammy-winner, and **in 1997** the world's first commercial 96 kHz/24 bit audio DVD (on DVD-Video).

This obsession with good sound has developed into Bob's passion: *Mastering with a Capital M*. Every day, he applies his specialized techniques to bring the exciting sound qualities of live music to every form recorded today. **In 1990** he founded **Digital Domain**, which masters music from pop, rock, and rap to audiophile classical. Besides mastering, Digital Domain provides complete services to independent labels and clients, graphic design and replication. Mary, who became Bob's wife, is an accomplished photographer and graphic artist, the visual half of the Digital Domain team and more than two-thirds of the charm. **In 1996**, Bob and Mary moved the company from New York to Orlando, adding numerous Florida-based artists and labels to the international clientele.

In the **90's**, Bob invented three commercial products, found in mastering rooms around the world. The first product, the *FCN-1 Format Converter*, was dubbed by Roger Nichols the *Swiss-Army knife of digital audio*. Then came the *VSP model P and S* Digital Audio Control Centers, which received a Class A rating in *Stereophile* Magazine. These devices perform jitter reduction, routing, and sample rate conversion.

Bob has delivered lectures and seminars to the Audio Engineering Society at the conventions and sections and chaired AES workshops. He has been Convention Workshops Chairman, Facilities Chairman and served as Chairman of the AES New York Section. In **1991**, Bob began the **Digido** website, the second audio URL to make the World Wide Web, an educationally-oriented site which has grown to be a premium source for audio information. Over 1000 pages around the globe have linked to www.digido.com.

Bob's first 21st century invention is patent pending. He designed and introduced an entire new category of audio processor, the **Ambience Recovery Processor**, which uses psychoacoustics to extract and enhance the existing depth, space, and definition of recordings. Z-Systems of Florida and Weiss Audio of Switzerland have licensed Bob's K-Stereo™ and K-Surround™ processes.

Bob has mastered CDs for labels including EMI, BMG, Virgin, Warner (WEA), Sony Music, Walt Disney, Boa, Arbors, Apple Jazz, Laser's Edge, and Sage Arts. He enjoys the Celtic music of Scotland, Ireland, Spain and North America, Latin and other world-music, Jazz, Folk, Bluegrass, Progressive Rock/Fusion, Classical, Alternative-Rock, and many other forms. Clients include a performance artist and poet from Iceland; several Celtic and rock groups from Spain; the popular music of India; top rock groups from Mexico and New Zealand; progressive rock and fusion artists from North America, France, Switzerland, Sweden and Portugal; Latin-Jazz, Merengue and Salsa from the U.S., Cuba, and Puerto Rico; Samba/pop from Brazil; tango and pop music from Argentina and Colombia, classical/pop from China, and a Moroccan group called *Mo' Rockin'*.

Bob mastered *Olga Viva, Viva Olga*, by the charismatic Olga Tañon, which received the **Grammy** for Best Merengue Album, **2000**. *Portraits of Cuba*, by virtuoso Paquito D'Rivera, received the **Grammy** for Best Latin Jazz Performance, **1996**. *The Words of Gandhi*, by Ben Kingsley, with music by Ravi Shankar, received the **Grammy** for Best spoken word, **1984**. In **2001** and **2002**, the Parents' Choice Foundation bestowed its highest honor twice on albums Bob mastered, giving the **Gold Award** to children's CDs, *Ants In My Pants*, and *Old Mr. Mackle Hackle*, by inventive artist Gunnar Madsen. The Fox Family's album reached #1 on the Bluegrass charts. African drummer Babatunde Olatunji's *Love Drum Talk*, **1997**, was Grammy-nominated.

Bob's recordings have received *disc of the month* in *Stereophile* and other magazines numerous times. Reviews include: "best audiophile album ever made" (McCoy Tyner: *New York Reunion* reviewed in *Stereophile*). "If you care about recorded sound as I do, you care about the engineers who get sound recorded right. Especially you appreciate a man like

Bob Katz who captures jazz as it should be caught." (Bucky Pizzarelli, *My Blue Heaven* reviewed in the San Diego Voice & Viewpoint). "Disc of the month. Performance 10, Sound 10" (David Chesky: *New York Chorinhos*, in *CD Review*). "The best modern-instrument orchestral recording I have heard, and I don't know of many that really come close." (Bob's remastering of Dvorák: Symphony 9, reviewed in *Stereophile*).

Some of the great artists Bob is privileged to have recorded and/or mastered include: Afro-Cuban All Stars, Monty Alexander, Carl Allen, Jay Anderson, Lenny Andrade, Michael Andrew, Lucecita Benitez, Berkshire String Quartet, Gordon Bok, Luis Bonfa, Boys of the Lough, Bill Bruford, Ron Carter, Cyrus Chestnut, George Coleman, Larry Coryell, Eddie Daniels, Los Dan Den, Dave Dobbyn, Paquito D'Rivera, Arturo Delmoni, Garry Dial, Dr. John, Toulouse Engelhardt, Robin Eubanks, George Faber, John Faddis, David Finck, Tommy Flanagan, Foghat, Fox Family, Johnny Frigo, Ian Gillan, Dizzy Gillespie, Whoopi Goldberg, Bill Goodwin, Arlo Guthrie, Steve Hackett, Lionel Hampton, Emmy Lou Harris, Tom Harrell, Hartford Symphony, Jimmy Heath, Vincent Herring, Conrad Herwig, Jon Hicks, Billy Higgins, Milt Hinton, Fred Hirsch, Freddie Hubbard, David Hykes Harmonic Choir, Dick Hyman, Ahmad Jamal, Antonio Carlos Jobim, Clifford Jordan, Sara K., Connie Kay, Kentucky Colonels, Lee Konitz, Peggy Lee, Chuck Loeb, Joe Lovano, Patti Lupone, Gunnar Madsen, Jimmy Madison, Taj Mahal, Sean Malone, Manhattan String Quartet, Herbie Mann, Michael Manring, Marley's Ghost, Winton Marsalis, Dave McKenna, Jackie McLean, Jim McNeely, Milladoiro, Mississippi Charles Bevels, Max Morath, Paul Motian, New England Conservatory Ragtime Ensemble, New York Renaissance Band, Gene Parsons, Gram Parsons, Danilo Perez, Itzhak Perlman, Billy Peterson, Ricky Peterson, Bucky Pizzarelli, John Pizzarelli, Chris Potter, Kenny Rankin, Mike Renzi, Rincon Ramblers, Sam Rivers, Red Rodney, Rodrigo Romani, Phil Rosenthal, Mongo Santamaria, Horace Silver, Lew Soloff, George 'Harmonica' Smith, Janos Starker, Olga Tañon, Livingston Taylor, Clark Terry, Thad Jones/Mel Lewis Big Band, Steve Turre, Stanley Turrentine, McCoy Tyner, Jay Ungar, U.S. Coast Guard Band, U.S. Marine Band, Amadito Valdez, Kenny Washington, Peter Washington, Doc Watson and Son, Clarence White, Widespread Jazz Orchestra, Robert Pete Williams, Larry Willis, and Phil Woods.

—by Mary Kent (who knows him best)

[Appendix 13]
Glossary

A

ABSOLUTE LOUDNESS: A term I use when comparing the apparent loudness of different sources without moving the monitor control.

AES/EBU: The name of a digital audio interface jointly conceived by the Audio Engineering Society and the European Broadcasting Union. See Chapter 20.

AGC: Automatic Gain Control. Compression that brings up low-level passages. See Chapter 11.

AIFF: (along with **WAVE, BWF, SD2, MP3**): A type of audio file format. See Appendix 3.

ALIASING: An alias is a beat note or difference frequency between the audio content and the sample rate, a form of intermodulation distortion. Proper filtering should eliminate aliases, but see Chapters 16 and 18. Note in an A/D converter, the higher the sample rate, the less chance of aliasing products being created against the normal audio content, but aliasing distortions could still arise from RF interference.

ASRC: Asynchronous sample rate converter. A converter from one sample rate to another which can work with a wide relationship of input to output frequency, and thus can deal with varispeeded rates. Filter coefficients are continuously variable, computed *on the fly*. See Chapter 19.

A-WEIGHTING: See **Weighting**.

C

COMPACT DISC: A 16-bit stereo 5" disc standard jointly developed by Sony and Philips in 1980. It can carry digital audio (Red Book standard) or standard computer files (Yellow Book), and other formats as well.

COMPRESSION RATIO: The ratio between input and output level of a compressor at the threshold point. See Chapter 10.

D

DAT: Digital Audio Tape Recorder. Short for RDAT, which stands for Digital Audio tape recorder with rotating heads. There was an SDAT (stationary head) standard, but this was never released.

DAW: Digital Audio Workstation. Usually a computer with dedicated hardware and software for editing and processing digital audio.

DB: Decibels. A logarithmic measure of audio level. See chapter 5.

DBFS: The level meters on digital equipment all read in dBFS, decibels below full scale. Full scale is the highest signal which can be recorded. Positive going signals with a value of 32767 or negative with a value of -32768 (at 16-bit) are at the maximum. Levels below those are translated to decibels, with 0 dBFS being full scale. For example, -10 dBFS is a level 10 dB below full scale. **0 DBFS** means "0 dB reference full scale," as on a digital meter. Full scale is 0 dB and the meter reads negatively below that.

DITHER: A process that linearizes digital audio by adding a random noise signal at the point of the circuit just before wordlength truncation. Dither is absolutely required for clean digital audio recording and processing. After dithering, the wordlength can be safely truncated or shortened, but truncation without dithering results in quantization distortion. See Chapter 4

DSD (direct stream digital) is the audio format used on the SACD (Super Audio Compact Disc), a rival format to the DVD-A. *DSD*, as opposed to multibit PCM, carries audio information using one-bit encoding. See Chapter 18.

DVD-A: *DVD* originally stood for *Digital Video Disc*, but it has now been dubbed *Digital Versatile Disc* as it can support computer, audio, and video formats. The *-A* suffix defines the multichannel audio disc standard that supports a wide range of PCM sample rates and wordlengths, and limited (still) graphics.

DVD-V: A video and audio disc standard that also supports multichannel digital audio SRs up to 48 kHz/24-bit, and 2-channel digital audio at 96 kHz SR and 192 kHz SR, but there is usually not enough room on the disc to fit high-quality video and high resolution audio at the same time. When MPEG video takes up much of the space on the disc, usually coded (data-reduced) formats such as DTS or Dolby Digital carry the multichannel audio track.

DYNAMIC RANGE: The range in decibels between the highest level which can be encoded and the lowest level which can be heard. Since this is a perceptual, or ear-based determination, it is an approximate number. In a properly dithered system, available dynamic range can be greater than its measured signal-to-noise ratio. See Chapters 4 and 5.

E

EDL: Edit decision list. Also known as Playlist. Instead of cutting the actual audio, an EDL is a list of instructions of where and how to cut and reproduce the audio when played back. Thus, many different versions or playbacks of the same audio can be reproduced from the audio files. An EDL is to audio as a Word Processor is to words.

E-E: Pronounced "E to E." Electronics to electronics. For example, when a tape recorder is put into record, its output monitors its input directly. This mode is known as E-E.

EMPHASIS: In an effort to improve the already-excellent signal-to-noise ratio of the Compact disc, CDs (as well as digital tapes) can be recorded with emphasis. If it is decided to use emphasis, the recording is made with a calibrated high frequency boost (called Emphasis), and during playback, a corresponding high frequency rolloff (called Deemphasis) is applied. Thus, in theory, signal-to-noise is improved, though in practice the loss of high frequency headroom may reduce any audible improvement. Most CDs made today do not use emphasis.

F

FIR VS. IIR: FIR stands for finite impulse response and IIR for infinite impulse response. These are types of filters which can be implemented in equalizers. All analog equalizers behave like IIR filters, in that there are no unnatural delays, just phase shift when the equalization is changed. In contrast, an FIR equalizer can only be implemented in digital circuitry, and has only been implemented in a few user-operated designs because of its cost. An FIR equalizer can be made linear phase, that is, with no change in time delay as the equalization is raised or lowered. But this is done at the price of yielding a pre-echo, a time delay before the sound occurs. Something which cannot occur in nature and so perhaps the ear may never get used to this if the echo is spaced far enough away from the original signal. See Appendix 10.

FIREWIRE: The name of a high-speed bi-directional serial interface originally developed by Apple computer, but then officially adopted as standard IEEE 1394, for use with digital audio, video, hard drives, controllers, etc. See Appendix 7.

FIXED-POINT VS. FLOATING POINT: Fixed-point is the language of the AES/EBU interface, so all devices must speak Fixed-Point on their inputs and outputs. Thus, if a processor uses floating point, it must convert to and from Fixed. The Motorola-based DSP processors use

Fixed-point math, and Texas Instruments and AT&T processors use Floating-point. Fixed-point arithmetic can only represent a dynamic range equal to the wordlength, e.g. 24-bit fixed point can only represent 144 dB of range and 48-bit (double precision) yields 288 dB. But Floating point processors can represent thousands of dB. The downside of floating point is that the noise floor changes with the precision, which can cause noise modulation.

All other things being equal, 32-bit floating point is roughly equivalent in absolute signal-to-noise ratio to 24-bit fixed, but in general, 32-bit float outperforms 24 bit fixed. This is because 24-bit fixed only has 24-bits of precision when the absolute level of the signal is 0dBFS. As the level of the signal decreases the precision decreases. For each 6dB, you lose one bit. In contrast, 32-bit float provides 24-bits of precision independently of the absolute level of the signal (until the level is extremely low or high.

Assuming equally skilled designers, 48-bit fixed point is probably better (cleaner) than 32-bit float, but 40-bit float and 64-bit float trump them all!

FRAMES: There are two commonly used "frame" standards in CD work, with different lengths: 75 CD Frames in a second, as opposed to 30 SMPTE frames per second. Modern PQ lists are usually expressed in CD Frames, but the older 1630 systems used SMPTE frames, which have less timing resolution.

G

GAIN, LOUDNESS, VOLUME AND LEVEL: Distinctive terms each with their own meaning, carefully distinguished in Chapter 14.

GLASS MASTER: Glass Mastering is the process of transferring the CD master (either on PCM-1630 tape, recordable CD, or Exabyte tape) to a physical image of the pits on a coated glass substrate. See Chapter 1.

J

ISRC: International Standard Recording Code, defined by the RIAA as a unique code for each track on the CD. See Chapter 20.

JITTER: Timing variations in the digital audio clock, producing distortions. See Chapter 19.

K

KHZ: Abbreviation for kiloHertz, meaning audio frequency in thousands of cycles per second. Commonly this usage also applies to sampling frequency. To avoid confusion, in this book, we sometimes add the letters SR to help distinguish sample rate, for example, 44.1 kHz SR from audio frequency, for example 5 kHz.

K-SYSTEM: An integrated system of metering and monitoring devised by the author (Ch.15).

K-STEREO™, K-SURROUND™: Patent-Pending (still as of 2002) processes for extracting and enhancing the already existing ambience of recordings. See Chapter 13.

M

MLP: Meridian Lossless Packing, a data-reduction technique which made it possible to fit as many as 6 high quality channels of digital audio at 96 kHz SR on a DVD-A disc.

N

NORMALIZATION: An automatic process available in most DAWs, whereby the gain of all program material is adjusted so the peak level will just arrive at 0 dBFS. There are many esthetic and technical reasons to avoid normalization. See chapter 5.

P

PLUG-IN: An extra process which can be inserted into a DAW. Some plug-ins utilize the power of an external DSP card, while others, called Native Plug-Ins, utilize the computer's CPU.

PQ CODING: The Compact disc contains a number of subcode areas, each area is named with a letter, from P to W, with information on track number, timing, and so on. See Chapter 1.

R

RED BOOK defines the standards for the audio CD as defined by Sony and Philips. No ordinary individual has a copy of the Red Book. The real Red book can only be found at authorized Compact Disc replication plants. The Blue book defines enhanced CDs with audio and ROM material. Yellow Book defines CD ROMs. Green Book defines compact disc Interactive. White book defines the Video CD. Orange Book CD-R or Recordable CDs.

RMS: Root-Mean-Square. A method of averaging levels which computes the equivalent power of the material. For all naturally-occurring music, an RMS-responding meter will read several dB below the actual peak level of the music at any moment in time.

S

SACD: See DSD.

SDIF-2: Sony Digital Interface-2. The stereo version uses 3 cables, one for each channel and one for wordclock, thus avoiding the interaction between clock and data that causes interface jitter in the competing AES/EBU or S/PDIF interfaces.

SEGUE: A crossfade between two different types of music, pronounced seg-way, from the Italian *seguire* meaning *to follow*.

SNR: The abbreviation we use in this book for **Signal to Noise Ratio**. SNR of a digital system is the decibel ratio between the highest level which can be encoded (0 dBFS) and the dither noise. Since the noise can be measured with different weightings, SNR is simply a number we can use to compare, but may have little relationship to the actual range the ear hears. Dynamic range represents more closely what the ear hears, but it's difficult to define precisely the absolute lowest levels we can hear in any particular digital system. See Chapters 4 and 5.

S/PDIF: Shorthand for **Sony-Philips Digital Interface**. Standard IEC-958 and IEC-60958 defines this interface, usually found on an RCA (coaxial) connector. See Chapter 20.

SR: The abbreviation we use in this book for **Sample Rate**, aka Sampling Frequency.

SRC: (also abbreviated SFC) Sample rate converter, or Sample Frequency Converter. See Chapters 18/19. A Synchronous SRC uses fixed filter coefficients, can only convert between certain fixed rates, e.g. 44.1, 48, 88.2 and 96 kHz, and cannot accept varispeeded sources.

STATE MACHINE: A state machine is defined as any type of processor which produces identical output for the same input data, and which does not look at data timing or speed, but only at the state or recent history of the data. Most digital processors are state machines and thus are completely immune to jitter. (See Chapter 19).

T

TRUNCATION: Reduction of wordlength by cutting off the lower bits. If dithering was not performed first, then simple wordlength truncation causes distortion.

W

WEIGHTING: When measuring noise, weighting applies a non-flat frequency response curve in an attempt to correlate better to what the ear hears. *A-weighting* is one of the most primitive curves, based on a simple model that the ear hears low frequencies and high frequencies less than mid frequencies. Other types of weighting include *CCIR* or *IEC*, also outdated by the latest psychoacoustic research. The most accurate curve is called *F-Weighting*, but even so, applying a single weighted number to the measured noise floor of an amplifier is still deceiving. A single number has little relationship to the more complex way in which the ear really works. Ultimately, the impact of noise should be interpreted based on individual time and level analysis of each critical band of the ear.

Index

Afterword

How This Book Was Written and Edited

This book was collaboratively produced by two individuals (writer and editor) located on opposite sides of the globe. Computer technology and the Internet have advanced the non-linear process of writing and editing a book—proofreader's marks and symbols have become obsolete. Instead, we have Microsoft to thank for providing two little-known features in **Word** called: **Track Changes** and **Comments**. Through these features, Eric and I were able to interact, exchange document revisions, annotate and comment the text, and view each others' changes.

I created a system for all the author's output to be odd-numbered revisions, and the editor to respond with even-numbers. Each revision was in its own document (we did not use version-tracking, which has limitations). So as each chapter progressed, it would be incrementally numbered, and it was easy to see its status and who had produced the last revision.

When it came time for fact-checking, Jim Johnston added his comments and Word correctly identified JJ as their origin. I worked on a Macintosh and Eric and Jim on a PC, but fortunately Microsoft Word transcends operating systems.

I think it most appropriate that my interior graphic designers Toni and Thuan have chosen to set this book in a typeface named *Filosofia*.

Bob Katz, Orlando 2002